FARMING AND FOOD

FARMING AND FOOD

GENERAL EDITOR

Dr John Tarrant

New York
OXFORD UNIVERSITY PRESS
1991

CONSULTANT EDITOR
Professor Peter Haggett, University of Bristol

Dr Kathy Baker, School of Oriental and African Studies, University of London
Central Africa

Dr I.R. Bowler, University of Leicester
The Low Countries

Professor G.P. Chapman, School of Oriental and African Studies, University of London
The Indian Subcontinent

Dr Hugh Clout, University College, London
France and its neighbors

Dr Nicholas Evans, Coventry Polytechnic
The British Isles

Dr Brian Ilbery, Coventry Polytechnic
The British Isles

Dr Alan Jenkins, Oxford Polytechnic
China and its neighbors

Dr Richard Joby
Australasia, Oceania and Antarctica

George Joffé
Northern Africa

Dr Alun Jones, University College, London
Central Europe

Professor W.R. Mead, University College, London
The Nordic Countries

Dr Michael Pacione, University of Strathclyde
Italy and Greece

Dr Deborah Potts, School of Oriental and African Studies, University of London
Southern Africa

Dr Jonathan Rigg, School of Oriental and African Studies, University of London
Southeast Asia

Dr Guy M. Robinson, University of Edinburgh
Canada, The United States, Central America

Dr David Seddon, University of East Anglia
Northern Africa

Richard Sexton, University of East Anglia
The Middle East

Dr John Tarrant, University of East Anglia
The World of Farming, The Soviet Union

Dr David Turnock, University of Leicester
Eastern Europe

Dr P.T.H. Unwin, Royal Holloway and Bedford College
Spain and Portugal

Dr Michael Witherick, University of Southampton
Japan and Korea

Dr John Wright, University of Birmingham
South America

AN EQUINOX BOOK

Planned and produced by
Andromeda Oxford Limited
11-15 The Vineyard, Abingdon
Oxfordshire, England OX14 3PX

Published in the United States of America by
Oxford University Press, Inc.,
200 Madison Avenue,
New York, N.Y. 10016

Oxford is a registered trademark of
Oxford University Press

Library of Congress
Cataloging-in-Publication Data

Farming and food / edited by J.R. Tarrant.
p. cm.
Includes bibliographical references and index.
ISBN 0-19-520917-6
1. Agriculture. 2. Agricultural geography. 3. Food supply.
I. Tarrant, John Rex.
S439.F37 1991
630--dc20 91-766
CIP

Volume editor Susan Kennedy
Assistant editor Victoria Egan
Designers Jerry Goldie, Rebecca Herringshaw
Cartographic manager Olive Pearson
Cartographic editor Zoë Goodwin
Picture research manager Alison Renney
Picture researcher David Pratt

Project editor Candida Hunt
Art editor Steve McCurdy

ISBN 0-19-520917-6

Printing (last digit):9 8 7 6 5 4 3 2 1

Printed in Spain by Heraclio Fournier SA, Vitoria

INTRODUCTORY PHOTOGRAPHS
Half title: *Harvesting grapes in the Yarra Valley, Australia (ANT Photo Library/Bill Bachman)*
Half title verso: *Fishermen loading a catch of sea mullet (ANT Photo Library/C. & S. Pollitt)*
Title page: *Ripe pumpkins on a farm in California, USA (Planet Earth Pictures/Ken Lucas)*
This page: *The rice harvest on Savu Island, Indonesia (Planet Earth Pictures/Rod Salm)*

Contents

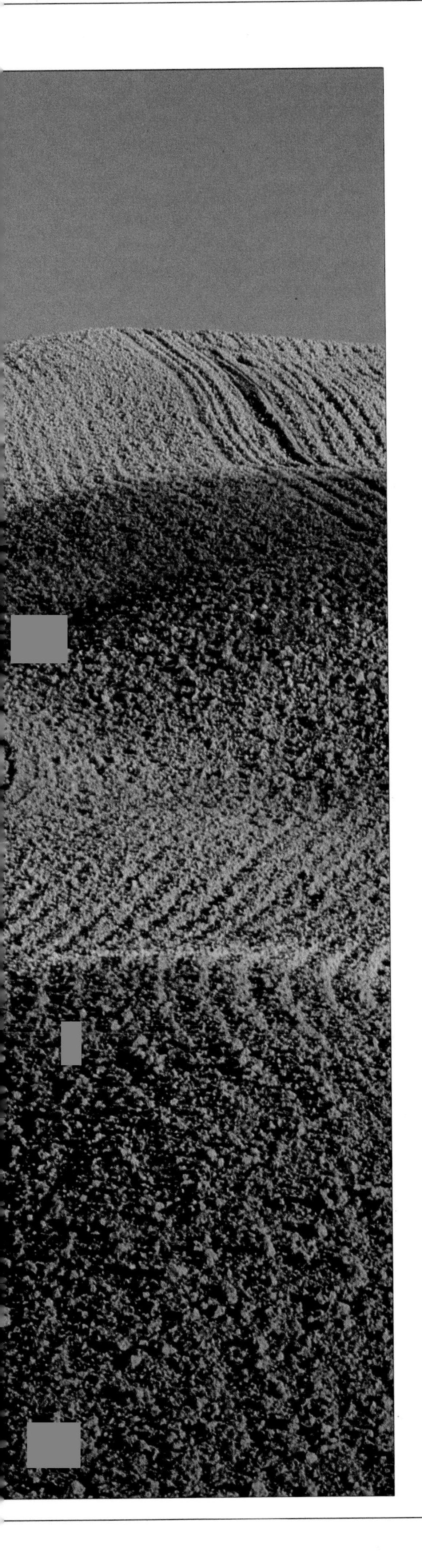

PREFACE

FEW EVENTS IN RECENT DECADES HAVE HAD MORE IMMEDIATE IMPACT than the transmission of pictures of famine from Africa. These stimulated unprecedented international efforts to raise huge sums of money for starving people, and also prompted some fundamental questions. How is it possible for our television screens to show scenes of mass starvation while well-stocked Western supermarket shelves display obvious signs of plenty? Why is the agriculture of North America and Western Europe so conspicuously successful, while the people of Eastern Europe and the Soviet Union endure endless food queues? The aim of this book is to explain some of these paradoxes, as well as to illustrate the immense variety of the crops we grow and the livestock we rear, which become the food we eat.

Farming is the single most important occupation of people in the world. About half the world's working men and women are farmers or farm workers. These 1,000 million people grow the food without which the world's population, approaching 5,000 million, cannot live. In the few remaining societies where food is hunted and collected, rather than cultivated, survival depends directly on finding enough food from day to day. With the growing of crop plants and the domestication of animals for meat came the need to store food from one year to the next, lengthening the link between food and survival. For much of the world's population, survival is still a direct result of successfully growing enough food to eat and to exchange for other necessities of life.

Just as all the world's people are touched by agriculture, so is the face of the Earth itself. Thousands of years of plowing the land for crops and of grazing it by livestock have altered much of its surface: natural vegetation, growing without the influence of agriculture, has become little more than a theoretical concept.

The development of modern transportation systems means that the economies of different countries are increasingly interdependent. Agricultural trade encourages the growth of tastes for food from all parts of the world. Knowledge of this interdependence has added to a growing responsibility to produce enough food to feed the world's population – but in such a way as not to destroy the Earth's resources of soil and water. It also increases awareness of the need to do something about the inequalities that allow starvation to exist alongside abundance. By illustrating the diversity and problems of agriculture today this book contributes to our greater understanding of the importance of farming and of food production and distribution.

Dr John Tarrant
UNIVERSITY OF EAST ANGLIA

Arable land in the valley of the Orcia river in central Italy

Grain silos at Thunder Bay, Ontario, Canada *(overleaf)*

THE WORLD OF FARMING

PRINCIPLES OF AGRICULTURE

CROPS AND LIVESTOCK

THE CONSUMERS

THE CHALLENGE TO AGRICULTURE

What is Farming?

FARMING IS THE PRACTICE OF AGRICULTURE. Strictly speaking, this means tilling or cultivating the soil to produce crops, but the term may be used more widely to cover all forms of farming practice, including the raising of livestock. There are over 250 million farmers in the world. If farm workers and families are included, the number of people who are supported directly by farming rises to over 2.3 billion. The crops and livestock that they produce on their farms feeds nearly all 5 billion of the world's population.

Although agriculture accounts for the vast majority of the food produced in the world, a small and diminishing number of people are still able to support themselves on the natural produce of uncultivated soil, through gathering wild plants and hunting wild animals. Such hunter–gatherers are the rare survivors of pre-agricultural societies.

In many parts of the world fish are "farmed" as an important source of food. The raising of fish under controlled conditions may be closely integrated with other forms of farming; for example, irrigation canals may be used to rear fish, which are fed with waste agricultural products. Most of the fish consumed as food, however, are not farmed but caught in lakes, rivers and the sea, in a manner akin to hunting.

By no means all farming and agriculture is confined to food production. Crops are also grown for their fiber, such as cotton, flax and jute, for other products such as rubber, and as sources of beverages (tea and coffee), drugs, perfumes and a multitude of other things. Animals are reared for their power, and for their hides, pelts and fleeces as well as for food.

Much of the Earth's surface is unsuitable for cultivation: the climate is too arid or too cold, or the gradient of the land too steep. In addition, large areas, which could be cultivated under careful management, have not yet been extensively exploited by farmers. Many such areas are covered in forests, some of which are managed for their timber. Sometimes such forestry is closely integrated with farming (and also perhaps with fishing) to provide farmers with a diversified basis of survival. In much of the Third World tree crops are being developed as sources of fodder, fuel and food in close conjunction with farming as agroforestry.

Farming: many ways of life

Farming (by far the most significant aspect of which is food production), fishing and forestry are thus overlapping activities. The relative importance of each one varies from region to region. However, we accord particular value to the people involved in producing our food, the primary necessity of life. Farming is sometimes described as a way of life, a description not readily applied to other forms of economic activity.

Within this way of life there exists enormous diversity. A thousand or more species of crops and livestock are used in agriculture around the world. Even within the production of just one crop – wheat, for example – there is little in common between the smallscale cultivator in the Third World who uses mainly hand implements to grow perhaps 1 ha (2.5 acres), and the farmer in Western Europe who farms over 100 ha (250 acres) of the crop with the help of several machines. Even this type of farming pales into insignificance alongside the cultivation of thousands of hectares of wheat in one farm in Montana in the north of the United States. In the same way, animal

Age-old agriculture Farming is one of the oldest of human activities. For the great majority of the world's farmers, techniques and crops have hardly changed for centuries. This farmer in Pakistan is working with oxen, using a simple wooden plow to turn the soil. Nearly all the food he grows will be used to feed his family.

production by nomadic pastoralists in eastern Africa has nothing except the broad animal type in common with the intensive feedlot production of beef in the United States.

Farming, then, embraces many ways of life. It is the appreciation of this rich diversity, with its variety of crops, livestock and cultivation techniques, along with differences in the size and organization of the labor force, the nature of land tenure, and the appearance of the farmed landscape, that comes under the heading of agricultural geography.

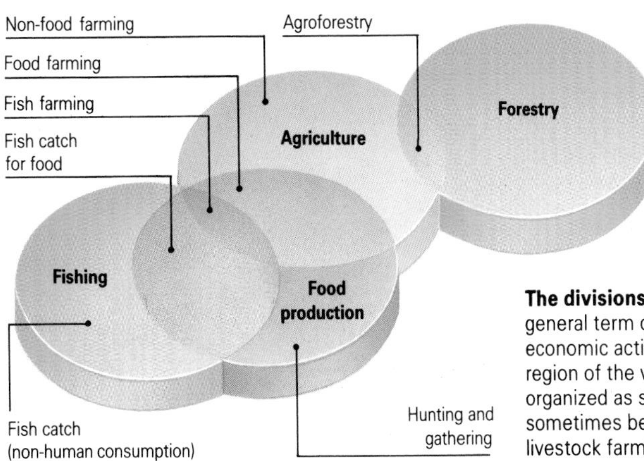

Largescale cultivation A machine drilling, or sowing, seed creates a pattern of faultless straight lines across a vast field. For this Western farmer, the emphasis is on largescale methods of production to supply grain for the market, and he relies on modern technology to help him push yields as high as possible.

The divisions of agriculture (*left*) Included under the general term of agriculture are a wide range of economic activities. They are not all significant in every region of the world. Fishing and forestry are normally organized as separate economic activities, but may sometimes be closely integrated with crop and livestock farming.

The Development of World Agriculture

THE DELIBERATE PLANTING AND HARVESTING of crops instead of gathering them probably began in Southeast Asia in about 8000 BC, though the domestication of animals had probably begun even before this. Dogs were used for hunting, and pigs, fowl, geese and ducks were also domesticated early in human history. The first methods of crop cultivation were based on vegetative reproduction, or vegeculture, whereby rhizomes (swollen underground food organs capable of producing shoots) are cut from the roots of plants and planted in holes in the soil. The sweet potato, taro, manioc and yam can all be cultivated in this way.

As well as Southeast Asia, independent centers of vegeculture arose in Central and South America, and possibly also in tropical Africa. The idea of cultivating crops spread from such centers to places less suitable for vegeculture where the climate was drier and, unlike the tropical areas, had marked seasonal patterns. Agriculture based on seed domestication developed in northern China, India, the eastern Mediterranean, northern Mexico and possibly Ethiopia. Grasses (rice, wheat and barley) were cultivated for grain, and legumes (beans and pulses) for food, oil and fiber.

A more complex agriculture developed in the Fertile Crescent of the Middle East, where grain-based agriculture was integrated with livestock production before 5000 BC. Cattle, sheep and goats were early domesticates. As ideas and expertise continued to spread, farming became regionally specialized. By about 1000 BC it had spread through southwest Asia into areas where rainfall was erratic, and methods of dryland cultivation were needed: fallow periods, when the land was left unplanted, were used to conserve soil moisture.

Further developments included the specialized water control agriculture practiced in the Nile delta and the agriculture of the Mediterranean, which combined dry farming techniques with the cultivation of olives, figs and grapes on terraces. In Southeast Asia vegeculture gave way, with the domestication of grains, to a specialized wet rice cultivation; the cultivation of dryland rice, with terracing of slopes to retain water, was integrated into vegeculture in hilly areas.

The origins of agriculture in Africa are uncertain. Grain cultivation either spread there through Egypt and the Middle East, or from independent hearths of tropical vegeculture, which expanded to a seed-based agriculture in Ethiopia. Whatever the origins, millets and sorghum were the principal grains to be cultivated, and these spread slowly southward. Parts of southern Africa remained uncultivated until European settlers began to farm there from the 17th century onward. The tropical vegeculture that was practiced in South and Central America developed into an agriculture based on maize (for grain) and on vegetables such as squashes and beans. It gradually spread northward as far as the Great Lakes in northern America.

In Europe agriculture moved rapidly northward and westward from its origins in the Danube basin from about 4000 BC, and both oats and rye were added to the inventory of domesticated cereals. The development of ironshod plows between 1000 and 300 BC enabled the cultivation of heavier clay soils to take place. Over the next millennium and a half change was slow, with the emphasis on ways of maintaining soil fertility while increasing production to feed a growing urban population: such methods included rotating crop production with periods of fallow and the heavy manuring of fields.

An early agricultural society (*above*) An Egyptian tomb painting depicts harvesters cutting wheat with sickles. From about 3000 BC water from the annual flooding of the river Nile was used to irrigate fields, allowing a sophisticated system of agriculture to develop: crops were grown to feed an urban elite.

A survival from the past (*below*) Shifting agriculture, in which forest plots are cleared to grow root crops such as manioc and sweet potatoes, is still practiced in the rainforests of South America, where vegeculture – the simplest form of crop cultivation – first evolved about 6,000 years ago.

Toward a world agriculture

In early industrial Europe, improved transportation and the rapid expansion of urban markets increased the demand for food still further. Both root crops and nitrogen-fixing clover were introduced to allow soil to recuperate between periods of cereal production. Specialization in livestock production led to the development of improved regional breeds.

Faster and cheaper sea transportation saw the expansion of European agriculture worldwide. Cheap land and the use of mechanization in Europe's overseas colonies allowed the production of grain at low cost to feed the industrial towns of Europe. As a result, the present regional pattern of world agriculture was all but in place by the end of the 19th century.

The origins of agriculture Centers of domestication developed independently in different parts of the world. The domestication of animals began earlier than that of plants. The selection of annual species, such as grain-bearing grasses, squashes, peas and beans, meant that breeding could proceed quite quickly, bringing rapid improvements in crop varieties.

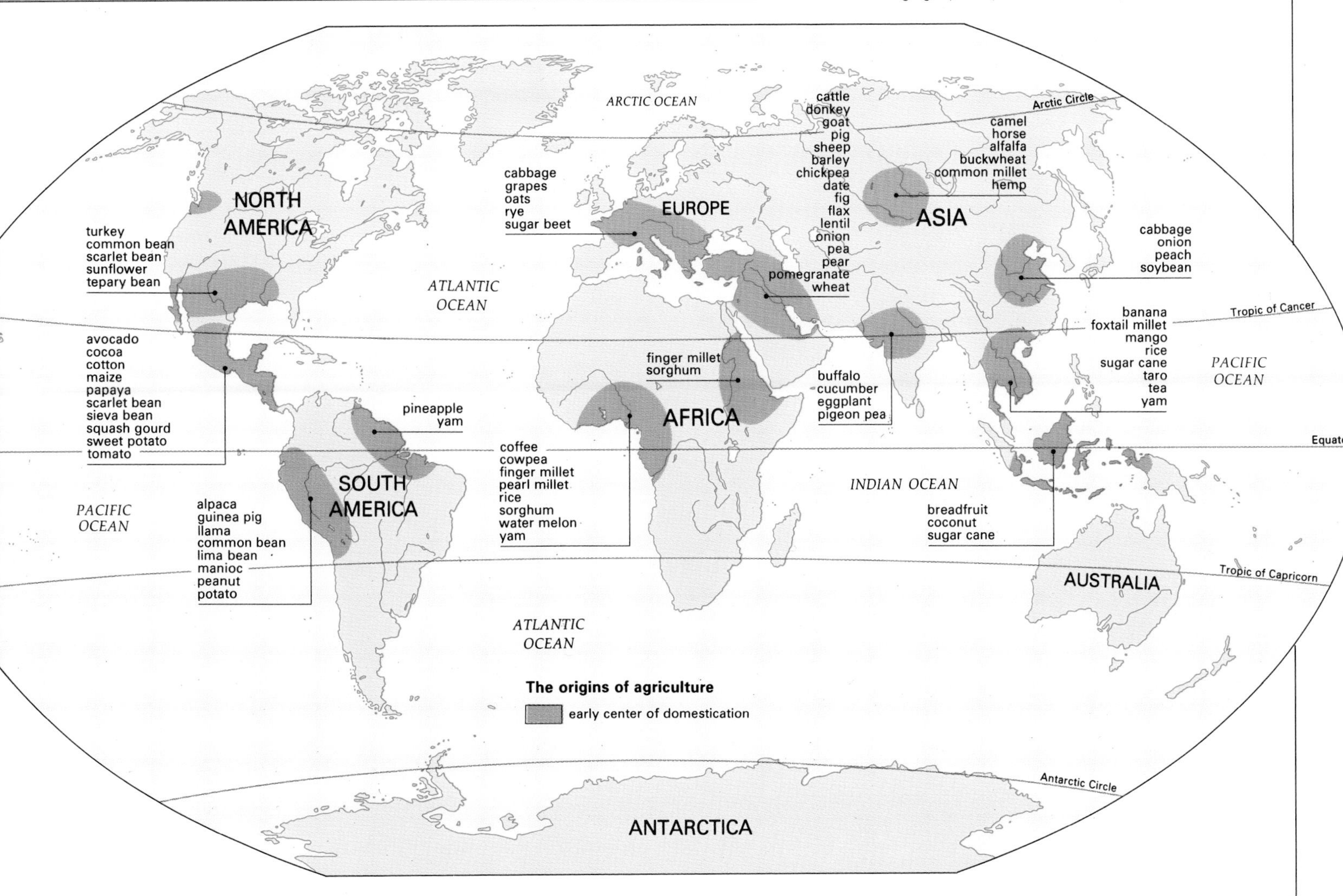

Muscle Power to Mechanization

Agriculture requires tools. Even at its simplest, vegeculture is made easier with the help of a sharpened stick to dig holes for planting. The digging stick, sometimes weighted with a stone, is still used where agriculture remains based on root crops such as cassava and taro.

With the cuitivation of grains came the development of the plow – the world's most important agricultural implement. At first the digging stick was provided with handles to allow it to be pulled or pushed through the soil. This method would only work for the lightest of soils. Later it was shod with iron, domesticated oxen were harnessed to pull it, and later still wheels were added, and then a wooden moldboard – all these modifications allowing more and heavier soil to be cultivated. Finally came the disk and rotary plows in use today.

For most of history the seed for growing crops was scattered by hand (broadcast), as it still is in many parts of the world. In Britain, in about 1700, Jethro Tull invented the seed drill, which allowed the seed to fall onto the plowed ground in straight lines, enabling laborers with hoes, and later machines, to weed between the crop rows.

Early grain crops were harvested – as they still are in parts of southern and Southeast Asia – by cutting the stalks by hand; the first mechanical reapers were introduced about 150 years ago. The grain was detached by beating the stalks against a stone or trampling them on threshing floors. Next it was thrown into the air to allow the wind to separate the grain from the chaff (the outer covering). Machines to do both became widespread in Europe and North America after 1850. Reaping and threshing were combined into one process with the introduction of the combine harvester, which came widely into use after World War II.

These heavy machines required more power to drive them than could be supplied by humans alone. Large animals such as oxen and water buffaloes were domesticated early in human history, and were used to pull heavy agricultural implements. The horse was domesticated more recently; from the Middle Ages until early this century it was the chief provider of motive force for farming in Western Europe and in those parts of the world where a European-based agriculture was established.

The replacement of draft animals by machines began first with steam engines in the mid 19th century, but they were heavy and unwieldy and soon gave way to the internal combustion engine; the tractor was developed in the United States in the 1890s. After it came a train of specialist machines: these were used for separating peas from the vines and pods; for picking cotton and tomatoes; for spraying; for baling; and for milking. Side by side with these developments there came a decline in the need for farm labor, and the trend continues today as computerized technology takes over many functions, particularly in connection with the rearing of livestock.

Modifying the environment

Throughout agricultural history human ingenuity has been applied to the problem of raising water to irrigate cultivated fields. Today most irrigation systems are driven by diesel and electric pumps, but in the past humans and animals supplied the power. Among the first of these,

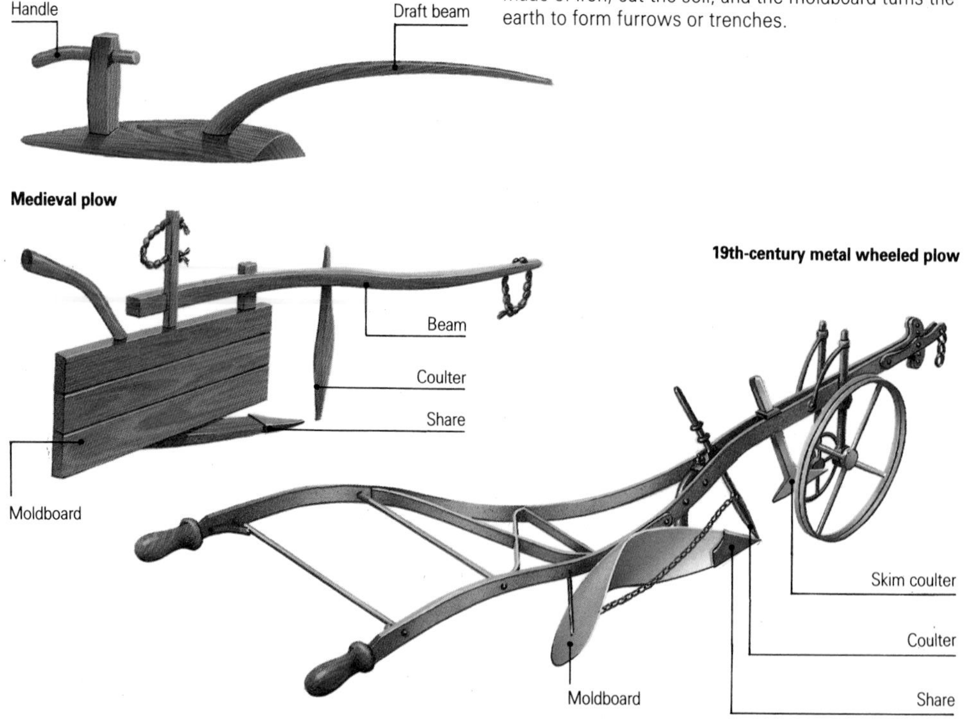

The evolution of the plow The plow is agriculture's most essential tool. It has been gradually modified to make it more efficient. The share and the coulter (later made of iron) cut the soil, and the moldboard turns the earth to form furrows or trenches.

Horse power (*above*) An early combine harvester at work. The advent of the tractor to pull farm implements ended the need for vast teams of horses like this one, thereby releasing land, previously used for growing fodder for farm animals, to raise other crops.

Ancient technology (*left*) Irrigation has long been practiced in dry parts of the world. The Persian wheel, turned by a draft animal, lifts water from underground wells.

Modern reversible plow

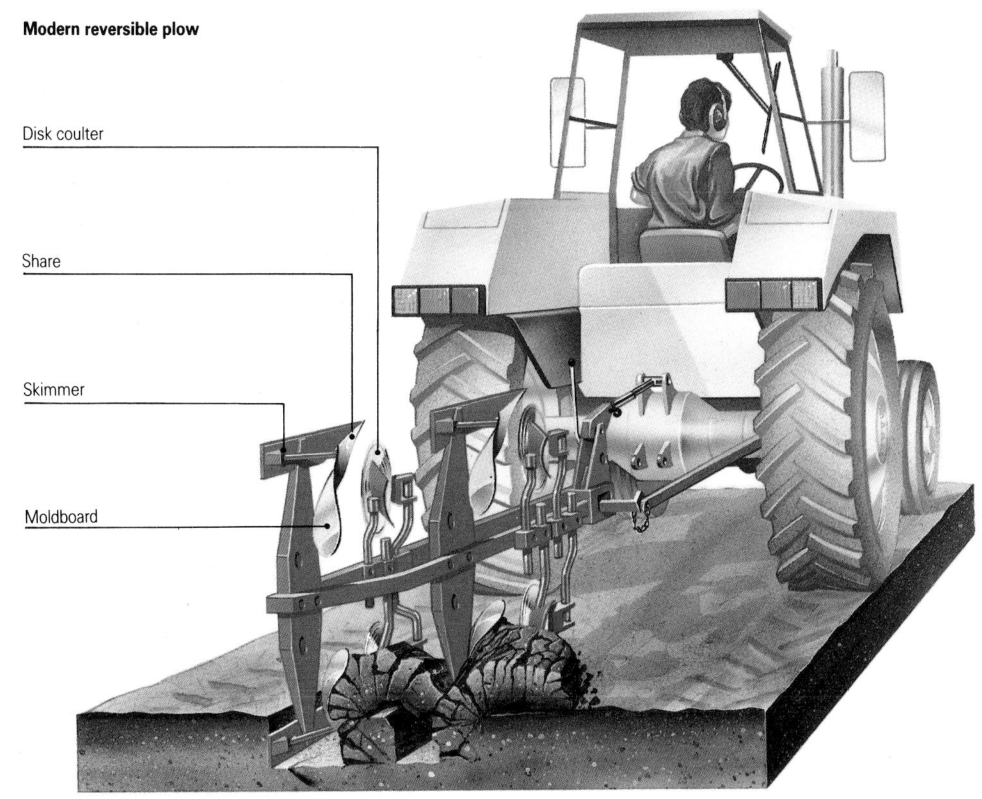

which originated in ancient Egypt, was the shadoof – a pivoted pole with a bucket at one end and a counterweight at the other. It was used to lift the water from the river and empty it into distribution canals in the fields. As the water level in the river drops, a series of shadoofs can be arranged to lift it over greater heights.

In dry areas without rivers techniques were developed that enabled reserves of water lying beneath the surface to be tapped using wheels turned by oxen. Another method used in Asia from Iran to Afghanistan and western China, depended on a series of wells, dug about 100 m (110 yards) apart into the alluvial gravels at the foot of mountains. Tunnels were dug from the bottom of one well to the next, and the water that seeped into them from the gravels was carried along to be used for irrigation. Such water systems, called qanats, are still widely used.

In flat coastal areas, where rivers are slow-moving and soils heavy, the problem is reversed. Rain lies on the surface and is difficult to remove. Engineers in the Netherlands in the 16th century found ways of draining land so that it could be cultivated. Banks kept the sea out while windmills, and later steam power, were used to lift water into the rivers. Electric pumps make such water-lifting methods even more efficient, and in some drained areas, such as the polders in the Netherlands, the peat soils are drying out and shrinking. As a result, even more careful protection from flooding is needed as they sink farther below sea level.

The Accelerating Technology

THE FACT THAT LAND PRODUCES POORER AND poorer crops the longer it is under continuous cultivation must have been noticed from the beginnings of settled agriculture. The early farmers of tropical root crops moved on to another newly cleared plot once the land they had been using was exhausted but grain cultivation used land more extensively, and needed more sophisticated techniques to allow soil fertility to recover. The methods that farmers used – which included fallow periods, the application of animal and human fertilizer, and eventually rotating cereals with root crops and clover to restore soil nutrients – must all have been discovered through experience rather than understanding.

It was not until Sir Humphry Davy (1778–1829) published his work *Elements of Agricultural Chemistry* in Britain in 1813 that the interactions between plant growth and soil chemistry – in particular the role of the nitrates, phosphates and potash – became clear. Saltpeter (potassium nitrate), imported from India in the early 1800s, was the first chemical additive to replace or augment the traditional manuring of fields. By the end of the 19th century the chemical fertilizer industry was firmly in place, and fertilizer use has expanded ever since. In the United States the average use increased from 23.6 kg (52 lb) to 92 kg (202.7 lb) per hectare (2.5 acres) in just over 10 years in the 1960s, and from 104 kg (229 lb) to 231 kg (509 lb) per hectare (2.5 acres) in Europe during the same period.

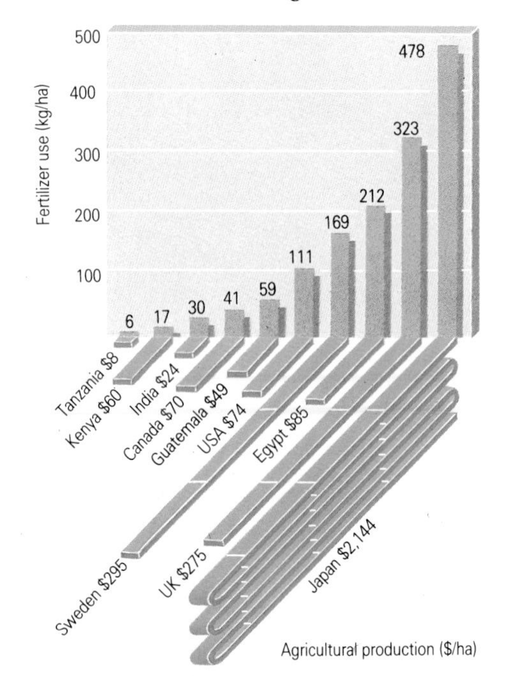

Fertilizer use One effect of the Green Revolution has been an enormous worldwide increase in fertilizer use. It has undoubtedly helped to raise food production, but since oil-based fertilizers are very expensive and have to be imported by most developing countries, they are in short supply.

At about the end of World War II the agricultural chemists turned their attention to finding ways of controlling the weeds and pests that flourished, along with the increased crop yields, wherever fertilizers were applied. They also looked at ways of boosting livestock production. The introduction of growth hormones and antibiotics into their feed meant that a greater number of animals could be reared on a smaller amount of land.

The Green Revolution

When people first started to select the strongest and the best varieties of domesticated crops and livestock, and to spread them into suitable environments around the world, they were applying biological techniques to agriculture. As genetic understanding developed in this century the scientists began deliberately to cross breed varieties of plants and animals to produce new, hybrid varieties that combined the best features of the parent stock. Cereal varieties were crossed to create higher-yielding hybrids, and from the 1940s on this principle was applied to the problem of increasing food supply for the rapidly growing populations of the Third World. Termed the "Green Revolution", this experimental work produced dwarf species of grains that responded to the application of fertilizers not by growing taller, but by developing a larger head of grain. Results in farmers' fields were less impressive than the scientists' tests, which had showed tenfold increases in yield, but the effects on world food production have been dramatic.

Hybridization programs of this kind had relied on the selective breeding of varieties of a single crop species to isolate the genes that determine specific, useful characteristics, such as shorter stems or greater resistance to disease. Now genetic engineers are working at new ways of transplanting useful characteristics across species – probably a more significant development than all the many previous technological changes in agriculture put together. For example, if the genes responsible for producing the specific proteins that are lethal to insect pests in one particular plant species were transplanted into crop species, it could greatly reduce our dependence on chemical pesticides. Genetic resistance to viruses will follow.

The problems surrounding this new technology arise not from the techniques themselves but from their acceptability to the public. However, increasing questioning of the possible harmful effects of agricultural chemicals may help to remove some of the doubts about these new genetic strains. Work on the genetic engineering of animals is likely to proceed more slowly, both because it is intrinsically more difficult and because studies will be needed over a period of several generations of animals to ensure that the manipulation is successful and involves no longterm risks.

THE NEW AGRICULTURAL REVOLUTION

Biotechnology – biological and chemical control over living organisms – is having both profound and far-reaching effects on modern agriculture. For centuries farmers have made use of cross-breeding in order to produce the plants or animals best suited to their needs. Rapid advances in biotechnology have opened up a great spectrum of new opportunities for the management of these natural resources.

Genetic engineering means that new varieties of plants can be developed much more quickly than was possible before. Because genetic engineers can take genes from any source, rather than having to rely on reproductively compatible plants, they also improve the nutritional value of low protein crops such as cassava or potatoes; this has important implications for developing countries where root crops such as these are staple foods.

Livestock farming is also now being revolutionized by the new biotechnology. Sheep that are treated with an epidermal growth factor produce a finer wool; it can be removed by brushing rather than shearing. Fecundity, meat production and milk yields can all be increased by genetic engineering.

Much Western research is directed toward improving profitability, but biotechnology also has the potential to increase food production in developing countries. Research in this area is being focused on improving yields without having recourse to expensive fertilizers or pesticides.

Crop spraying by air (*above*) The application of fertilizers to boost crop yields benefits weeds and animal pests as well. In the past farmers relied on hoeing and mulches to prevent weeds from destroying their crops, but now more sophisticated methods are used.

Agricultural research (*left*) is usually financed and run by the large corporations that supply pesticides, fertilizers and seeds to the farming industry as part of their development strategy to increase the range and profitability of their products. Laboratories and research stations that are responsible for studying the problems and diseases that affect crops are also funded by individual governments and by a number of international agencies.

Bug control One way to reduce dependence on chemical pesticides is to use insects themselves. These larvae prey on an aphid that attacks certain varieties of bean. They could be used to keep the aphid population in check.

Environmental and Market Factors

LAND IS THE FUNDAMENTAL RESOURCE OF agriculture. The intensity with which it is farmed depends on population pressure and on the strength of competing demands for its use. On the Indonesian island of Java, for example, every square meter of available land is cultivated. In the Netherlands competition for land has brought some crop production under glass, and most farm animals are kept indoors, with their food brought to them.

Land alone is not enough for successful agriculture. Soil and climate determine its farming potential. The soil is the reservoir of the water, minerals and nutrients that are needed for plant growth. The three most important minerals are nitrogen, phosphorus and potassium. These are derived from the parent material of the soil (the underlying rocks), the organic matter that is present as a result of earlier plant growth, and the work of nitrogen-fixing bacteria in the soil.

In general, the deeper the soil is, the greater will be its capacity to store water and minerals. The nature of the soil is determined by its particle size. The large particles of sandy soils make light, well-drained land that is easily warmed for early spring planting. Fine particles of clay soils retain water, and are heavy to cultivate. The spring warmth is expended in evaporating excess water from such soils rather than in warming them up; as a result they produce plant growth later in the season than lighter soils.

Humans have domesticated a great variety of plants and animals that are adapted to withstand an equally great variety of climatic conditions. Sometimes the environment is sufficiently benign to allow a wide range of agriculture within a particular region, but more often the climate is too hot or too cold, too wet or too dry for more than a narrow choice of crops or livestock. Plant growth stops when conditions are too cold. The length and reliability of the warm season, when temperatures are above the minimum required for growth, are critical factors in crop cultivation.

The timing of rainfall may also be critical. For example, in Mediterranean regions rain falls in the winter rather than in the main growing season, so plants have to rely on water stored in the soil from the previous wet season. The reliability of rainfall decreases as it becomes sparser; consequently, in semi-arid areas, such as the Sahel in sub-Saharan Africa, the frequency of total crop failure increases. Loss of water through both evaporation and plant transpiration means that more rain is needed for growth in hot climates than in temperate and cold ones. Parts of eastern Britain have less rainfall than much of the Sahel, but lower temperatures there conserve water in the soil, ensuring successful cultivation.

Where the demand for food is high the environment can be modified to allow agriculture to take place or to make it more productive. A shortage of water can be remedied through irrigation; the length of the growing season increased by artificial light in greenhouses; and poor soil enriched by the application of fertilizers. New varieties of plants have been domesticated or bred to produce high yields within a shorter growing season.

Market gardening
Vegetables and dairying
Mixed farming
Grains
Profit (set against higher capital costs closer to center)
Distance from market (higher transport costs away from center)

Zones of specialized agricultural production develop around a central market as a result of a number of factors. Perishable goods such as fruit and vegetables have high transport and production costs and command high prices; they are grown closest to the market. Less perishable crops, such as grain, with lower costs, can be grown at a greater distance.

The global supermarket

The advent of modern transportation, and especially refrigeration methods, provided the means of taking perishable foods to distant markets, thus removing a further serious constraint on what could be produced where, and enabling an international food industry to develop. Previously, circular zones of production had

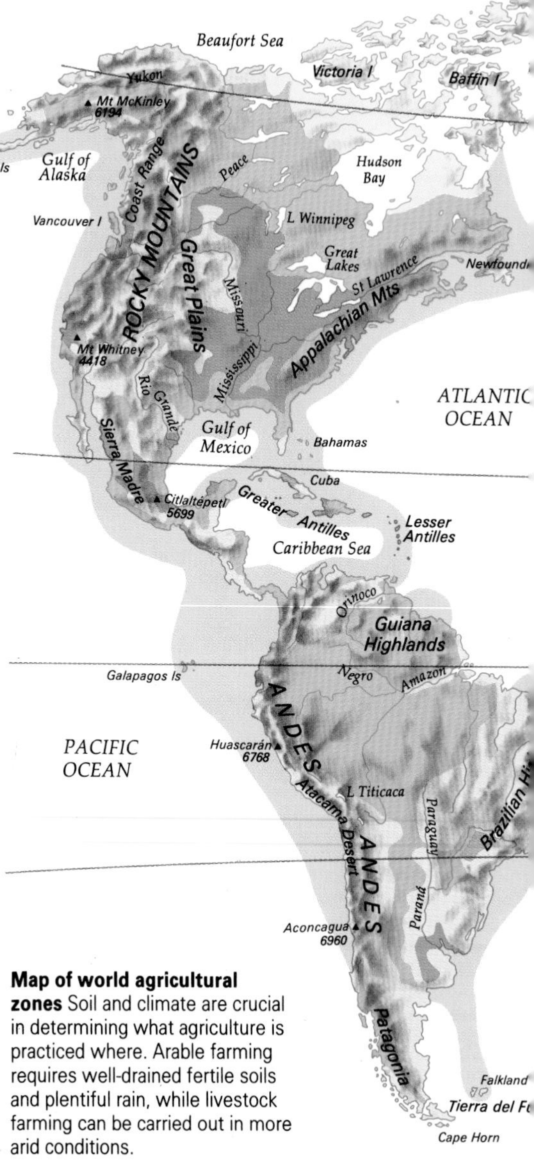

Map of world agricultural zones Soil and climate are crucial in determining what agriculture is practiced where. Arable farming requires well-drained fertile soils and plentiful rain, while livestock farming can be carried out in more arid conditions.

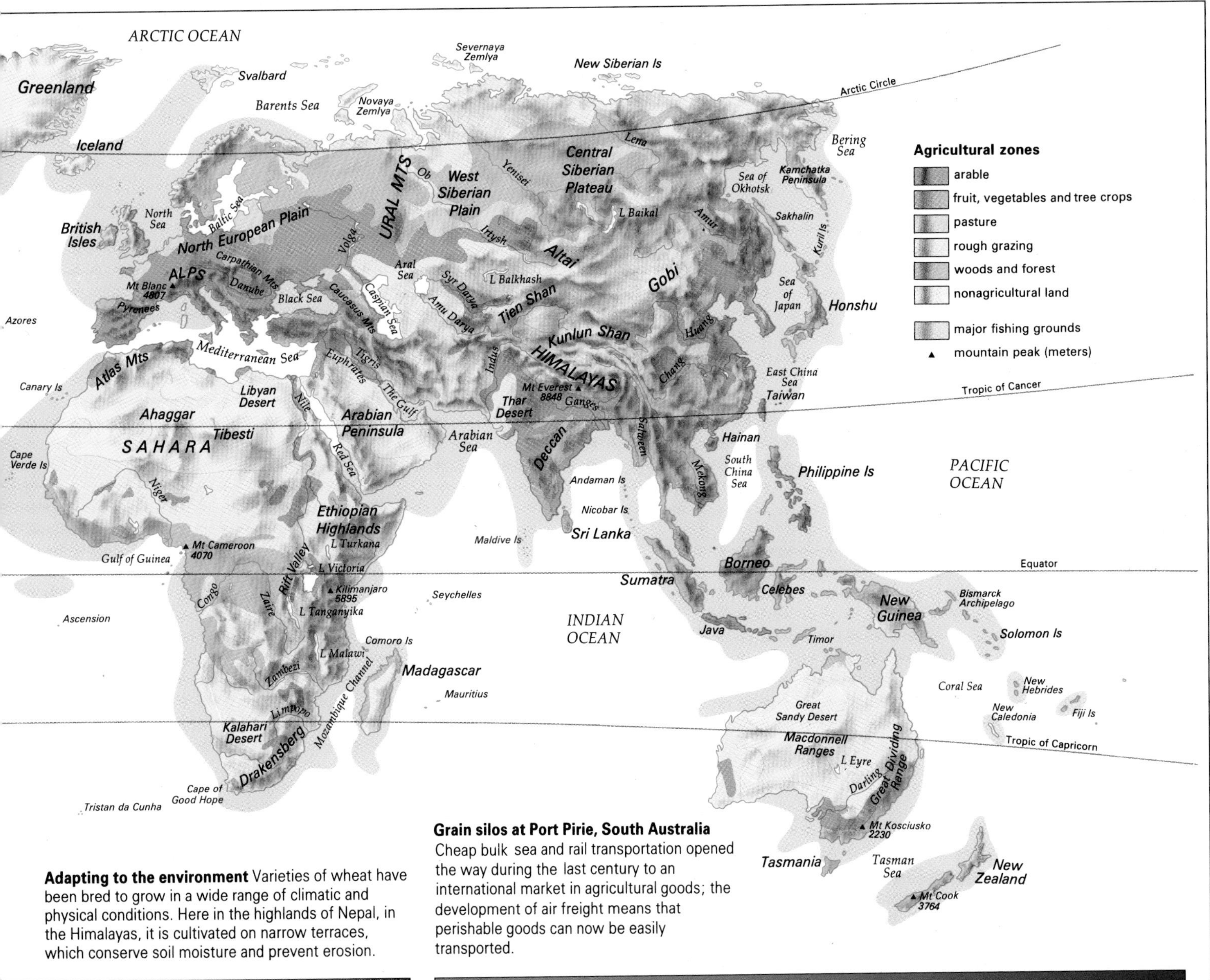

Adapting to the environment Varieties of wheat have been bred to grow in a wide range of climatic and physical conditions. Here in the highlands of Nepal, in the Himalayas, it is cultivated on narrow terraces, which conserve soil moisture and prevent erosion.

Grain silos at Port Pirie, South Australia Cheap bulk sea and rail transportation opened the way during the last century to an international market in agricultural goods; the development of air freight means that perishable goods can now be easily transported.

existed around cities and towns: those goods that could be transported most easily without damage were grown at the greatest distance from the center, while other more perishable foods, such as fresh milk, were produced in the immediate vicinity.

Comprehensive global transportation now allows food to be carried to wherever markets exist in the world. A single visit to the supermarket will yield vegetables from Guatemala, fresh herbs from Israel, cheese from France and Italy, bacon from Denmark, grapes from California, lamb from New Zealand and potatoes from North Africa. This encourages specialized food production, so that, for example, bananas grown in the West Indies supply a world market. The world is now interdependent in terms of food production, freeing it from the constraints of the local market, and allowing consumers, if they can afford it, the choice of foods from far outside the local producing area.

Social and Political Factors

FARMING IS REMARKABLY DIVERSE. IT MIGHT be expected that the sort of crops and the way they are grown would be different in southern India and the Midwest of the United States, but even neighboring farmers, sharing the same climate, soil and market conditions, may choose very different styles of agriculture.

All the same, environmental factors will obviously affect the decisions a farmer makes: the combination of climate, soil and water resources gives a range of choices. The range may be very narrow, as in Iceland, where cereal production is limited by the short growing season, or very much larger, as in northern Europe. Within these constraints, the style of farming will be determined by how profitable the various options appear, and by the individual knowledge, skills and preferences of the farmer.

Farming is risky and unpredictable. Only rarely will conditions in any two years be the same: the weather may be unseasonably wet, dry, or cold; there may be an outbreak of pests or disease; or a sudden fluctuation in the market. The farmer must often decide what crops to grow a year or more before they can be harvested, and the financial returns remain uncertain. For this reason, a farmer with only a small amount of land and limited financial resources will probably choose to practice mixed farming – a few pigs, several dairy cattle and a small amount of cereals – rather than risk investment in a single activity.

A peasant cultivator who grows just enough in an average year for the needs of the family and lives close to the edge of failure, will be less keen to grow a new variety of grain, or a new crop, than will a larger farmer who has the resources to risk innovation. In the Indian subcontinent or parts of Africa, where there are several million such smallscale cultivators, this produces natural resistance to change in methods or production, except by the small minority of farmers who can bear the element of risk attached to doing something new. When higher-yielding grain varieties became available during the 1960s, the larger farmers adopted the new technologies first, giving them such a head start that they were able to dominate the market and buy up the land of their smaller, more vulnerable neighbors. Although the new varieties eventually spread to the smaller farmers, it was too late for the many who had

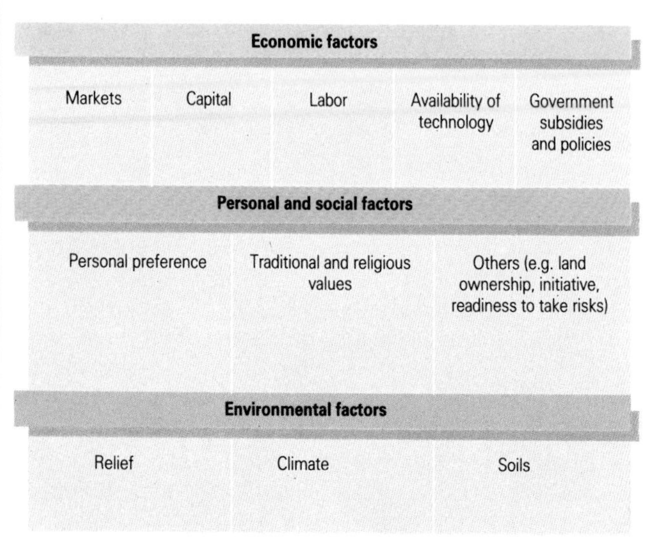

Traditional wisdom (*above*) A peasant rice farmer in India may prefer to trust his own knowledge of local conditions than to take the risk of trying new crop varieties and farming methods.

Inspecting the wheat crop (*right*) In most developed countries governments regulate the prices paid to farmers, helping to stabilize market fluctuations and reduce the risk of failure.

How farming decisions are made (*left*) Many factors affect a farmer's choice of what crop to grow. Which has greatest weight depends on individual circumstances and the region of the world he is farming in, but all will have some influence.

FARMING COOPERATIVES

Each farmer contributes only a minute fraction to total national food production, and is powerless to influence the market. Acting as individuals, farmers are vulnerable to the monopolistic power exerted by a small number of large food-processing companies or by state purchasing agencies. Their bargaining strength is increased if they form organized groups, or cooperatives. Such cooperatives can vary from simple organizations in which harvested crops or meat and dairy production are marketed communally, with each member taking a share of the profits in proportion to the size of their contribution, to more complex arrangements.

Farmers may get together to share all the stages of decision-making, production and marketing, holding land, machinery and labor in common. The proceeds are shared according to the scale of contribution, or according to need, as determined by the cooperative. The former pattern is adopted in many Western countries; it has been particularly successful in France, helping to consolidate many small, peasant holdings into larger, more efficient units. The latter was the practice in the communes of postrevolutionary China, and in the collective farms called *kibbutzim* that were established by pioneering Jewish settlers in the new state of Israel.

Plenty to choose from This patchwork effect of fields is characteristic of Western Europe where the conjunction of temperate climate, fertile soils and access to urban markets supports mixed farming: a wide range of crops can be grown, and there is good pasture for livestock.

already lost their land. As a result rural resources were even less equitably divided than before the change.

Government intervention

Throughout the world governments are increasingly influencing the decisions farmers make. In most developed countries the prices paid to farmers are regulated and many farming costs are subsidized, either directly through controlling the price of fertilizer or by providing grants for land drainage and farm building improvements, or indirectly by exempting farmers from land taxes. Such policies in the United States and Western Europe have encouraged production to rise so rapidly that it is now necessary to limit it by paying farmers to take land out of cultivation. The produce that is surplus to the needs of the domestic market is disposed of by exports, if necessary with subsidies, or by developing extensive food aid programs.

It is an irony that in most developing countries, where a very large proportion of the population are farmers, government intervention is usually directed toward reducing food prices to urban consumers who, though fewer in number, have greater political influence. Even where direct food subsidies are not used, governments often keep farmgate prices artificially low to ensure cheap food and to maximize government revenue from the export of crops grown for surplus. These low prices discourage any change or investment in agriculture, especially among a largely smallscale, risk-sensitive population of farmers.

Farm Organization

Agriculture requires access to land, labor to work it and capital to finance production. Throughout history there have been many systems by which land has been owned and controlled, labor organized and capital acquired, and these help to explain the way agricultural production has developed, and is carried out today, in different parts of the world.

In the earliest styles of agriculture land was not owned but cleared, used and then abandoned, possibly to be returned to once fertility had been restored. With settled agriculture, based on the cultivation of grains, concepts of land ownership, as distinct from simple use, began to arise, and with it came the beginnings of wealth division in society. The Greeks, for instance, had developed large estates on which a wide range of different types of economic activity were carried out by the 5th century BC. The various agricultural tasks were already becoming more specialized, while production was controlled by the people who owned the land. The same system was adopted by the Romans. They used it to bring under production the large areas of territory they acquired by conquest.

In the beginning slaves provided the labor for these large estates, but as the costs of feeding, housing and clothing them rose, tenant farmers were established on small plots within the estates, which they farmed for their own use in return for labor. This system of tenancy, known as *latifundia*, survived in Italy and southern France until this century, and in Spain, from where it was taken across to South and Central America. It remains a very common system of land holding in that continent.

The practice of large landowners employing peasant labor also spread into northern Europe. Here the land was divided into narrow strips, groups of which were cultivated by a single family who kept a proportion of the produce for themselves. The reform of this strip field system at the end of the medieval period created consolidated units of farmland, each of which had its own farmhouse standing apart from its neighbors, rather than clustered in villages. This pattern of land organization was then carried from northern Europe to the parts of the world colonized from there.

Many of these farms are family-owned, but others are worked by tenant farmers. In the simplest form of tenancy agreement the landowner provides the land, and the tenant the necessary labor and the majority of the working capital to farm it. In return for use of the land the tenant pays rent in cash.

A form of tenancy that is encountered in some parts of the world, such as India, is sharecropping; it was also frequently used in the southern United States following the abolition of slavery. The sharecropper rents the land in return for a proportion of the harvest – usually about half. Unlike cash-based tenancies, where the rent has to be paid regardless of the success of the harvest, sharecropping allows the risks to be shared between the landowner and the tenant. However, there is little incentive for the tenant to develop and improve the land, since any resultant increases in production have to be shared with the landlord.

State ownership

In 20th century China, the Soviet Union and many other communist regimes, the concept of community land ownership was adopted. The state, representing all the people, owns all the land, which is cultivated by workers as tenant farmers-grouped in communes, state farms, cooperatives or other forms of collectivized labor. Production is shared between the workers and the state.

There is no simple correlation to be made between a particular style of landownership and the level of agricultural productivity achieved. While the family farms of Europe and North America are among the most productive in the world, until recently their success was matched by the communal agriculture practiced in China. However, agricultural productivity on the *latifundia* of South America is very low; the landowners often have little interest in farming, giving their workers little incentive to increase output – in marked contrast to the levels of productivity they achieve on the small areas of land (*minifundia*) that they cultivate for their own use. This pattern is repeated in the politically opposite system of the Soviet Union, where the productivity of unofficial private plots far outstrips that of the state farms.

Medieval farming (*above*) A 15th-century French manuscript showing winter wheat being sown. The manorial system of landholding that predominated in northwestern Europe has had a longlasting influence on the pattern of land organization.

The Chinese collective system (*right*) A team leader supervises the threshing of rice. The collectives were quite effective in raising productivity, but were disbanded in the economic and political turnabout that followed the death of Mao Zedong in 1976.

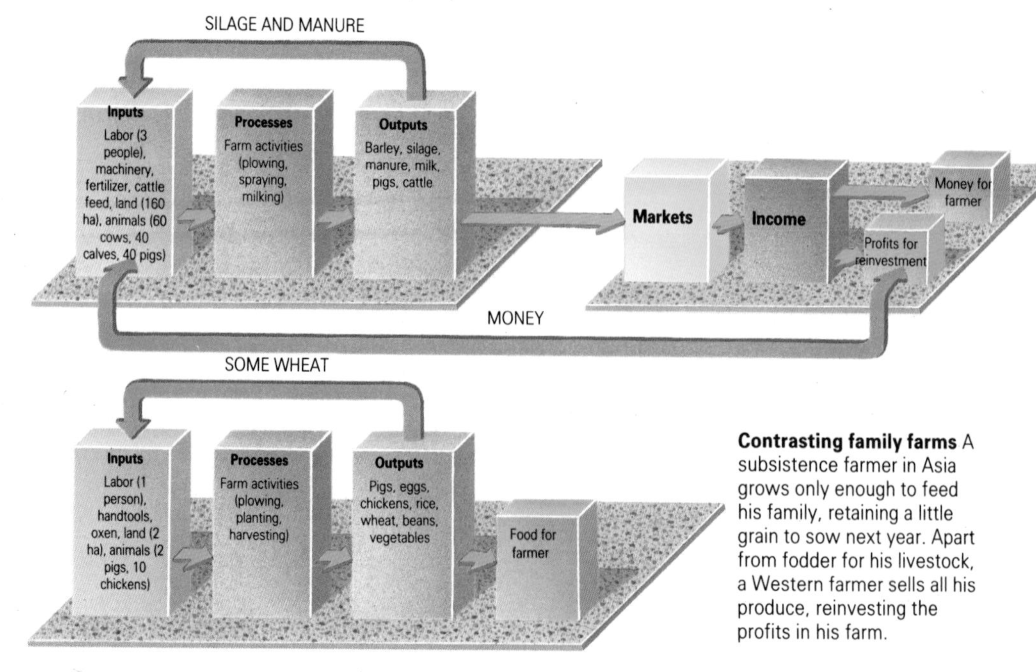

Contrasting family farms A subsistence farmer in Asia grows only enough to feed his family, retaining a little grain to sow next year. Apart from fodder for his livestock, a Western farmer sells all his produce, reinvesting the profits in his farm.

Principal Food Crops

MOST AGRICULTURAL ACTIVITY IS CONCERNED with growing the staple crops that provide the principal source of food for local populations. Usually these food crops are cereals (grain-yielding domesticated grasses), and nearly all of the populated regions of the world have at least one important locally grown cereal. The most important, in terms of worldwide production, are wheat and rice, the former grown in temperate regions and the latter in tropical and subtropical parts of the world.

Wheat can be cultivated in a great variety of environments, from sea level to 3,000 m (10,000 ft) and where levels of rainfall range from 300 to 1,800 mm (12 to 70 in) a year. Cultivation takes two distinct forms. Winter wheat is planted in the fall, germinates and grows to about 12 cm (4.5 in) tall before lying dormant through the winter. Where winters are severe the plants need a layer of snow to protect them from frost. If the winters are too cold, or there is insufficient snow cover, wheat is planted in the spring when the ground has warmed sufficiently to allow germination. Winter wheats have a higher yield than spring wheats.

In contrast to most other grains, wheat is nearly always ground to make flour before it is used. The protein content of wheat is generally high, and nearly all is used for human consumption; only a little is grown as fodder for livestock.

As much as half the world's population depends on rice for its staple diet. It is grown on land that is covered by 50 to 100 mm (2 to 4 in) of water throughout the growing season. Oxygen is moved to the roots of the plant along hollow stems. Rice is generally planted in seed beds and the young plants transplanted later into the main paddyfields. Only when the paddyfields are deeply flooded is the seed sown directly; the yield from such rice is much lower than transplanted rice.

Rice is almost without exception used as a human food. Before it is cooked, by boiling it in water, the grain has to be separated from the husk by milling. Sometimes the layer of bran surrounding the seed is also removed, and the grain is polished during a final processing stage. The residue from milling the rice is used as animal feed, for fuel and as a fertilizer.

Maize (known in the United States as corn) is a native crop of North and South America, introduced into Europe in the 16th century. It is now grown around the world, its area of cultivation extending from 58°N in Canada to 40°S in South America. As a staple food for humans, maize is most important in Central and South America, where it is ground to make flour, and in Africa, where it is boiled to make porridge. Its protein is of relatively low value. It is widely grown as a fodder crop: in more northerly regions, where the growing season is too short to allow the grain to ripen, the whole plant

Planting out rows of young rice plants (*above*) In tropical parts of the world, where population densities are extremely high, rice is the leading staple crop. Unlike other cereals, rice is grown exclusively for human consumption.

Winnowing grain (*right*) Ladakh farmers in Kashmir, northwest India, using wooden handtools to separate the grain from the rest of the plant. The straw residue will be used as animal feed. Much of the world's grain production is grown by traditional farmers.

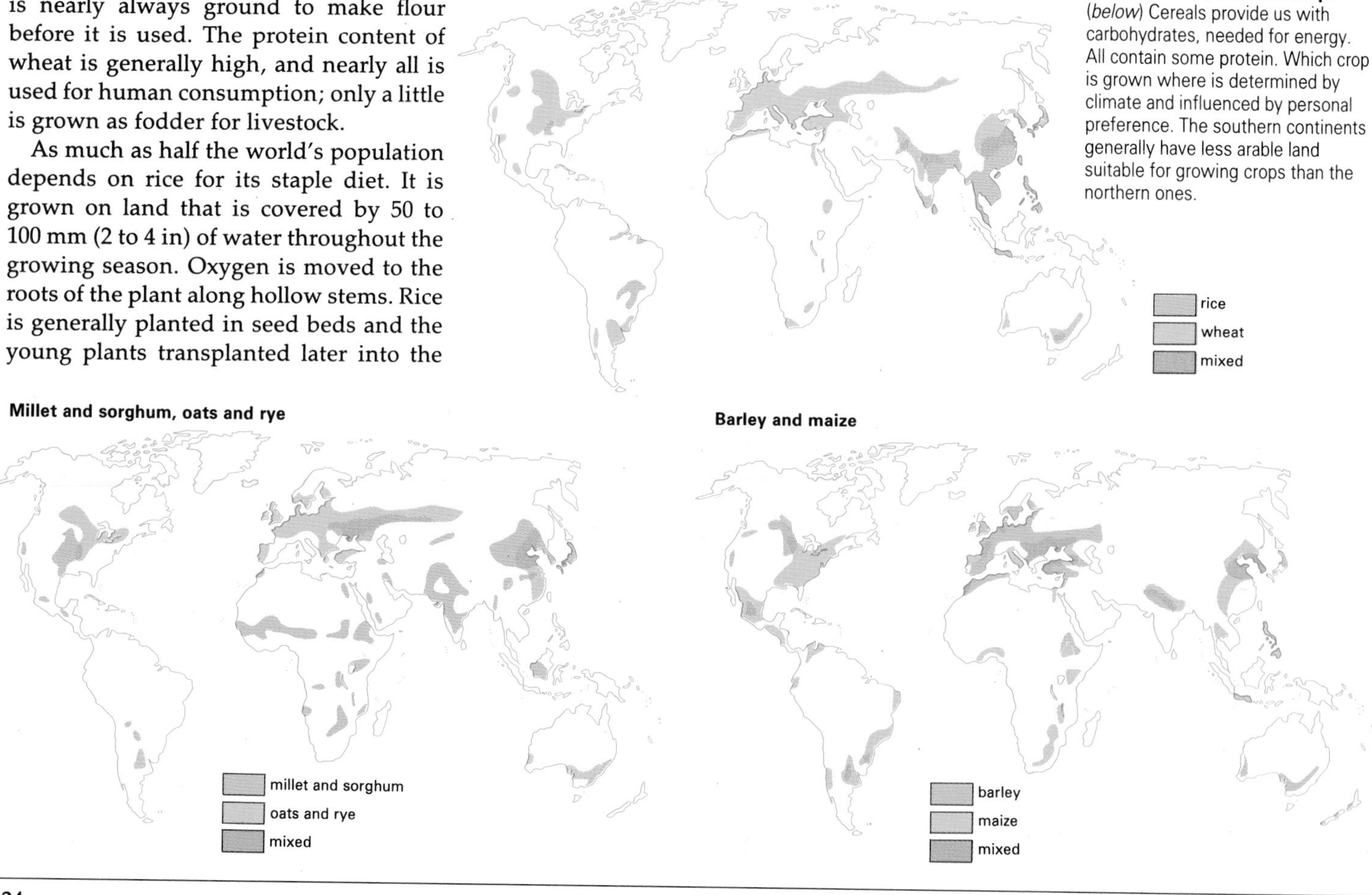

The distribution of food crops (*below*) Cereals provide us with carbohydrates, needed for energy. All contain some protein. Which crop is grown where is determined by climate and influenced by personal preference. The southern continents generally have less arable land suitable for growing crops than the northern ones.

is used to make silage, which is fed to livestock during the winter.

Millet is an important staple grain in much of Africa and India. The plants grow between 1.5 and 3 m (5 and 10 ft) high, and tolerate drought and poor soils. The grain may be eaten like rice, or milled into flour. Sorghum (sometimes called large millet) also grows in dry conditions. In the United States it is grown for use as animal feed. Rye, oats and barley are all temperate grain crops grown for human and animal consumption. Barley is cultivated for use in brewing beer.

In areas where conditions are too cold and wet for grain production, potatoes become a staple food crop. They are the edible tubers (swollen root organs) of a plant that is a native of the Andes in South America, and can be grown at altitudes of about 2,500 m (8,800 ft) and as far north as central Alaska. Although high in water content, they are a very valuable source of carbohydrate and protein.

Sugar is an energy-producing food that adds palatability and enhances the flavor of other foods. Sugar is refined from two commercially grown plants: sugar cane, a perennial grass native to New Guinea and now cultivated throughout the tropical regions of the world, and sugar beet, a root crop that is cultivated in temperate regions, too cold for sugar cane.

Fruits, nuts and vegetables have always been used to add variety to the human diet, as well as vitamins and essential minerals. An enormous variety of species is grown in every region of the world, according to soil and climate. Fruit growing is commercially important in warm parts of the world. Vines, grown for wine production, are particularly important in areas of Mediterranean climate.

Sugar beet and sugar cane

sugar beet
sugar cane

Fruit and potatoes

fruit
potatoes
mixed

Plantation and Economic Crops

CROPS ARE NOT JUST GROWN AS FOOD. SOME provide textiles for clothing, fiber for ropes and sacking, and oils for food processing and other industrial uses. In tropical and subtropical regions of the world, many of these crops are grown on plantations – the legacy of 19th-century colonial rule when the agriculture of the countries in these regions was developed in order to provide cheap food for sale in European markets.

Most climatic regions of the world grow a crop that is farmed for its fiber. In tropical regions it is sisal, a plant of the agave family, whose leaves produce a flexible fiber that is used to make ropes and binding twine. Tanzania and Brazil are the world's largest producers of sisal.

In subtropical regions cotton and jute are grown. Cotton is the world's most important fiber crop. The fiber is derived from the hairs that surround the seeds within the seed pods (bolls). They are separated from the seeds by a process known as ginning, and are spun together and woven into fabric. The seeds are crushed as a source of oil, and the residue is used as cattle fodder.

Jute is second only to cotton as a fiber crop. It is an annual plant that grows to a height of 3 m (10 ft) or more in warm, humid climates. The stems of the plant are soaked in water until the fiber can be stripped away from the center, a process known as retting. The fiber is used for ropes, sacking or carpet backing. Production of the crop is concentrated in India and Bangladesh around the Bay of Bengal.

Hemp is grown in more temperate conditions. The production of the crop originated in China. When it is grown for its fiber, the plant reaches a height of up to 5 m (16 ft); the fiber is produced and processed in a similar manner to jute. Oil is obtained from the seeds of smaller, branchier varieties, which also yield the narcotic drugs marijuana and hashish.

In cooler, more temperate regions flax was formerly a much more important fiber crop than it is today. The fiber is obtained by soaking the stems in water. It is very flexible and is woven into fine linen cloth. Its importance as a textile crop declined with the rise of cotton production in the United States. Smaller plants may be grown for their oil, commonly known as linseed.

A number of other plants are grown as a source of oil, the most important being the groundnut, or peanut, the oil of which is used in the manufacture of margarines and cooking oil. The seeds of sunflowers, rape and soybean also yield edible oils, and are used for various industrial

The distribution of economic crops Many crops were established around the world after European colonization. World demand for most fiber crops, as for natural rubber, is declining in the face of competition from synthetics, but production of oilseed crops is rising, both for human and for animal consumption.

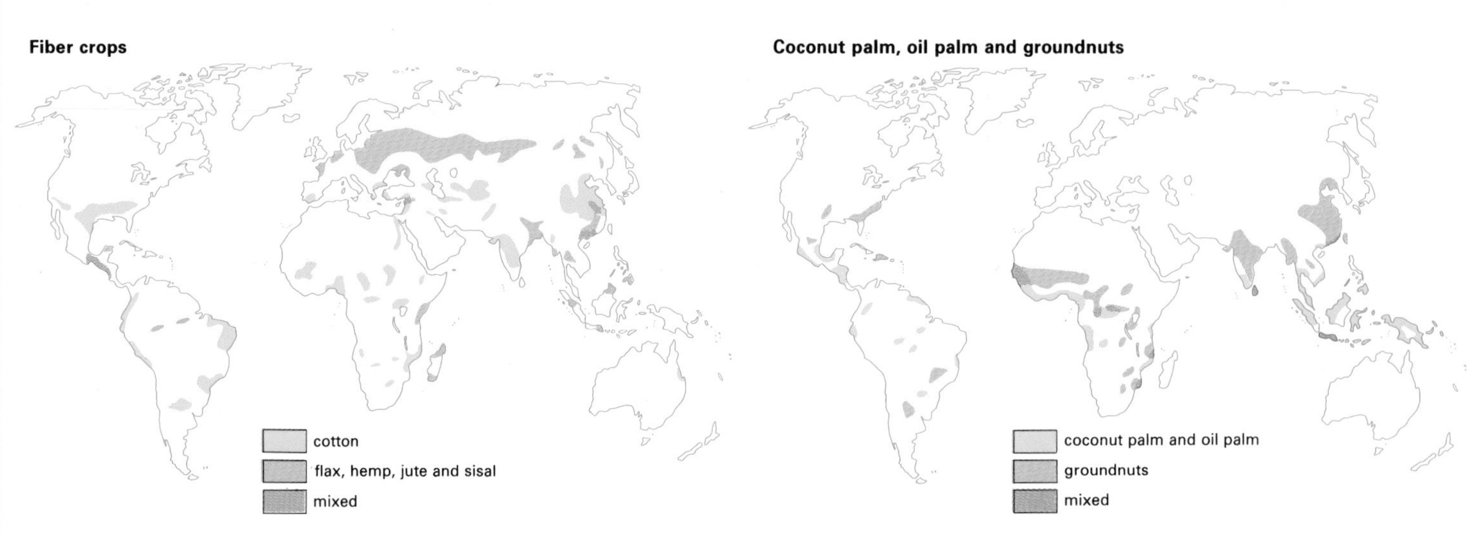

Sisal drying in Kenya (*above*) The leaves of the sisal agave are crushed in rollers to extract the fibers, which may be more than 1 m (40 in) long. They are used for rope making.

Tobacco plants ready for harvesting (*left*) Tobacco is either air-dried or smoke-cured before it is processed for use as pipe tobacco or cut to make cigarettes and cigars. Some tobaccos are chewed; others, powdered and flavored, are taken as snuff.

products, including soap and plastics. The soybean, a member of the pea family, has many other uses: its roots fix nitrogen in the soil, restoring fertility; its beans are an excellent source of protein, and can be processed as "textured protein" meat substitutes and as soya "milk". It is also grown for use as animal feed.

The oil palm is cultivated as a source of oil in West and Central Africa and in Southeast Asia. The oil obtained from the outer parts of the fruit is used to make soaps, candles and lubricating greases; the oil from the kernels is used in food processing. The coconut palm is another multipurpose crop. The husk of the nut furnishes a fiber called coir; the dried white meat, known as copra, is a major agricultural product of Pacific countries.

Natural rubber (latex) is obtained from a tree, *Hevea brasiliensis,* native to South America. Although demand has declined as a result of the increasing use of synthetic rubbers, it is still widely grown on plantations in Sri Lanka, Indonesia and Malaysia. Tobacco is obtained from another native plant of South America. After its discovery as a stimulant by the Spanish, its use spread rapidly around the world, and it was soon being grown on plantations in the southern United States, which remains a major producer of the crop.

Three tropical plants have worldwide importance as beverage crops. Tea, a native of China and India, is grown in plantations in most tropical regions of the world. Coffee is obtained from a shrub that originated in Africa and Arabia but is now grown around the world. Cocoa or chocolate is obtained from the bean of the cacao tree, a native of South America, from where its cultivation has spread to make it a major plantation crop of west Africa and parts of Southeast Asia.

Animals in Farming

Wherever agriculture is practiced in the world there is likely to be some livestock farming – the breeding and raising of domesticated animals for meat, milk and hides or fleeces. However, the role it plays within a particular agricultural system may be minor and incidental – as when a farmer in a predominantly arable area keeps a few dairy cows for personal preference – or central, as cattle herding is to the Masai people of Kenya. The reasons determining the importance of livestock, and the kind of animals kept, depend on a whole range of environmental and social factors. Cattle, sheep, goats, pigs and poultry are the most commonly farmed animals, but a large number of other species have local importance; for example, guinea pigs are kept for meat in South America.

The domestication of animals began very early in human history. As settled agriculture developed, livestock farming became fully integrated within it. Land in continuous cultivation soon loses its fertility. This can be restored by letting it lie fallow, or uncultivated, for a period. If animals are grazed on the fallow land, not only do they provide a source of meat but most importantly, they manure the land, speeding up the return of fertility.

All domesticated farm animals are herbivores, converting energy from the grasses and other plants they consume to meat and passing it on to the humans that consume them in their turn. They are usually set free to graze the grasses and herbs of the pasture or rangelands on which they are kept. This system of livestock farming requires the extensive use of land. Where farmland is at a premium, and populations from towns provide a market for meat, animals are kept intensively in sheds or yards, and all their food has to be specially grown.

A variety of uses

The meat of every domesticated animal is eaten in some part of the world, though religious or social taboos may restrict consumption of a particular animal by some groups of people. Pigs are forbidden to Jews or Muslims; cattle are considered sacred in India; horses are thought to be unfit for eating by many Western people. Most domesticated animals also provide nutrition in the form of milk, which may be drunk fresh or processed into butter and cheese. Poultry (hens, ducks, turkeys and geese) are kept both for their meat and for their eggs, which provide a valuable source of protein that is well suited to humans.

Both wild and domesticated animals have always been valued for their skins. Sheep are multipurpose animals, reared both for their meat and their fleeces, which provide wool for clothing. Llamas, alpacas and goats also have valuable long-haired fleeces. The treated hides of cattle and pigs provide leather. Some animals, such as mink and fur foxes, are farmed for their pelts.

Many of the very large domesticated animals were, and frequently still are, used as beasts of burden and to provide power in agriculture. Before mechanization, horses pulled plows to turn the heavy soils of Europe and North America. They do less well in hot and humid climates, and in tropical parts of the world oxen, water buffaloes and even elephants have been domesticated to provide pulling power for farming and forestry.

The most valuable farmed animals are cattle. They have been bred for survival in many different kinds of environment. However, a good growth of grass is necessary for reliable milk production, so dairying is limited to areas of plentiful rainfall. Where winters are severe cattle have to be housed indoors because of the lack of pasture and because they do not thrive in cold conditions. Pigs are the mainstay of subsistence farmers around the world, kept as single animals on a small patch of ground where they will scavenge for scraps and vegetable matter. In many Western countries they are intensively farmed for meat.

Sheep and goats can tolerate a wide range of conditions, and are able to thrive in areas that are too dry for successful cattle rearing. Goats – particularly hardy – can even survive in arid areas on rangeland that is able to support little other life. Unfortunately, overgrazing in such extreme conditions can lead to severe soil erosion. The most hardy domesticated animal of all is the camel, which can survive for weeks without water and is used by the desert peoples of Africa and the Middle East as a beast of burden. Camels also provide a valuable source of milk, meat and hides.

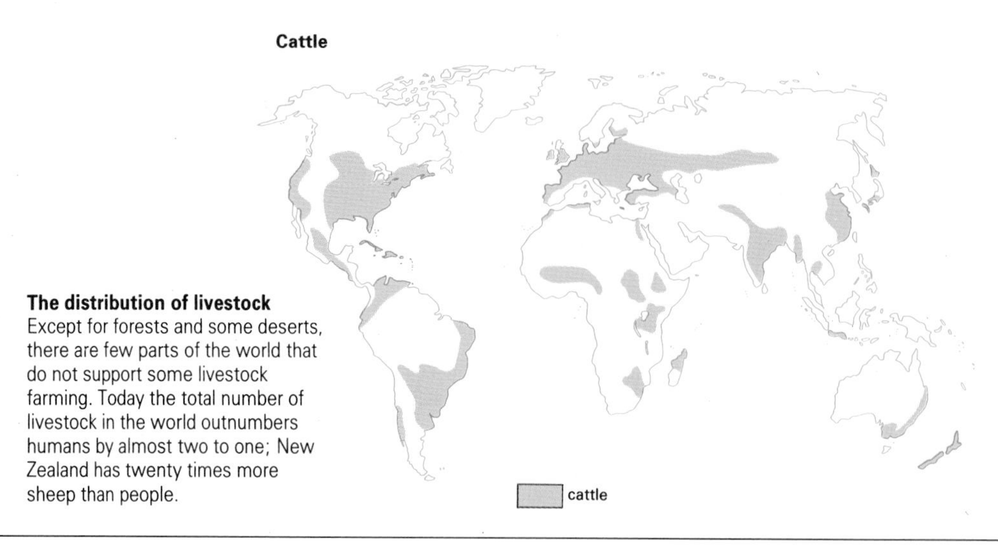

The distribution of livestock
Except for forests and some deserts, there are few parts of the world that do not support some livestock farming. Today the total number of livestock in the world outnumbers humans by almost two to one; New Zealand has twenty times more sheep than people.

Cattle ranching in California (*above*) Cattle are the principal livestock of the United States, whose population has the highest meat consumption in the world. India has the lowest, but supports most cattle – 15 percent of the world total.

Sheep in the desert (*right*) In arid areas of the world sheep, goats and camels are the only animals that can survive. Herders are often nomadic, moving animals considerable distances in search of areas of pasture for grazing.

Sheep and goats

sheep and goats

Buffalo, camel and llama

buffalo
camel and llama
mixed

Supply and Demand in Agriculture

FARMERS GROW FOOD TO MEET DEMAND. Food supply and population size need to be held in broad balance, though it is not always clear whether population growth in the past was made possible by the agricultural improvements that increased food production, or whether population pressure on existing food supplies stimulated the innovation needed to increase food supply. In reality, it was probably a combination of both.

Collapse of the food supply brought catastrophe to societies that had no external sources of food to tide them over periods of shortage. The Mayan civilization of Central America, at its height in about 900 AD, had up to 40 cities with populations of between 5,000 and 50,000; archaeologists estimate that there was a total population of about 2 million people. They were supported by an extensive and quite sophisticated agriculture that included techniques for irrigation, drainage and terracing. After that time the number and size of the cities declined rapidly, and there was a return to a village-based society. Many reasons for this collapse still remain unexplained, but there is evidence of extensive malnutrition in the population at this time, suggesting that the return to the land was triggered by a collapse of the food supply.

The industrial revolution that took place in the Western world early in the 19th century created a huge demand for unskilled labor, which was met by drawing people from the land into the rapidly growing towns and cities. This could not have taken place without parallel developments in agriculture, which not only freed it from its reliance on large numbers of laborers, but also increased its productivity. At the same time advances in transportation opened up agricultural trade with the rest of the world, enabling this new urban population to be fed. For the first time in history the majority of the population did not have to concern itself with the daily labor of producing its own food.

Rising demand

For all the many people who do not grow their own food, what they eat, and how much, depends on whether they can pay for it. Their age, sex and the amount of physical activity they take will determine what they need, but their level of income determines how much they get. Poor people spend a larger proportion of their income on food than the rich do. By far the greater part of any increase in the earnings of impoverished workers in the Third World will be directed to feeding themselves and their families better. As basic incomes begin to rise with a country's increasing economic success, there will be a greater demand for both more and better food. This means that people in the Third World are likely to spend as much as 80 percent of any income increase on food, in comparison with less than 10 percent in the richer economies, where most of this will be spent on "trading up" to more expensive foods, rather than buying more.

Increased demand for food is not linked simply to population growth; economic growth is an additional factor. The more successful a country is in raising basic incomes, the faster demand for food will grow. This is not because population growth rises faster if people are better off (though this may be the case for a short while), but because people who are better off can afford better diets.

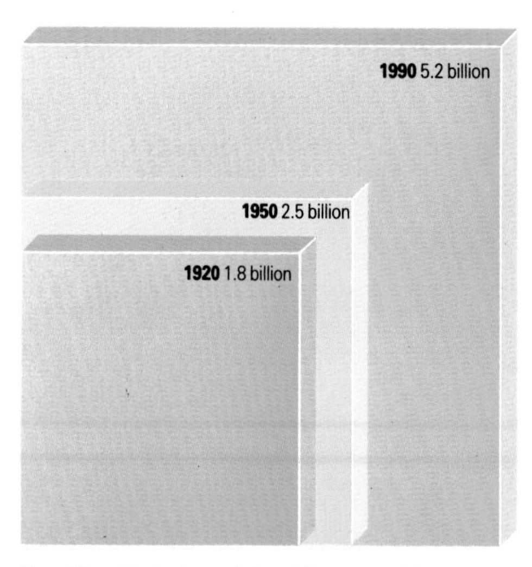

Upsetting the balance (*above*) For most of human history population growth was broadly in line with rising food production, but the 20th-century population explosion has put the food supply in jeopardy.

Squeezing the land (*right*) Terraced paddyfields fill every available square centimeter of land on Luzon in the Philippines. Southeast Asia has one of the fastest rising populations in the world, and there is little new land to bring into cultivation. Consequently existing land has to be farmed as intensively as possible. But as industrial success increases the spending power of the urban population, the demand for food will rise further.

Why People Starve

WE ARE WHAT WE EAT. THE FOOD THAT we consume enables our bodies to grow and develop, and provides fuel for our mental and physical activity. The building blocks of the body, the major constituents of all cells, are proteins. Proteins are made up of amino acids, 11 of which have to be in our diet as they cannot be manufactured within the body: the proteins we take in are broken down in digestion and the amino acids are then reassembled into the different proteins that our bodies require.

The amount of proteins in different foods varies greatly – fish has the most – and some proteins have greater nutritional value than others, according to how closely they approximate to human proteins. Different foods, eaten together, make up for each other's protein deficiencies. Thus mixed diets, such as the rice and pulses that are commonly eaten in India, and the pork and beans, which was the typical diet in the southern United States before and after the Civil War, have great dietary importance.

A number of other substances are essential for a healthy diet. Certain fats are used in the body's manufacture of prostaglandins, which regulate a host of functions from blood pressure to the working of the large intestine. Vital minerals include calcium and phosphorus to build bones, magnesium to process proteins and iron to utilize oxygen. A lack of vitamins, which are present in many natural foodstuffs, particularly fresh fruit and vegetables, can lead to a number of different health problems – the shortage of Vitamin A (retinol), for example, causes blindness in thousands of children in the Third World every year.

Food as fuel

We also eat food to give us energy, which is measured in calories. The principal sources of energy are the starchy foods (known as carbohydrates) derived from grain cereals and root crops such as cassava or potatoes. Energy is needed for the processes that keep our bodies ticking over (basal metabolism), and to fuel activity. The amount varies according to sex, body size, and occupation. A man weighing 65 kg (143 lb) may use as many as 4,000 calories a day if he is involved in an active occupation, but only 2,700 calories if he spends his day working at a desk. About 1,700 of this daily requirement is for basal metabolism.

If energy intakes are insufficient to support the body's basal metabolism and rate of activity, then fat stored in the body is raided – the person loses weight. Once this is exhausted, the proteins present in cell tissue are consumed; the person becomes wasted, and suffers from starvation. Two of the recognized conditions caused by malnutrition are marasmus, which is the result of calorie deficiency, and kwashiorkor, caused by lack of protein in the diet. However, as proteins can be used to make up a shortage of calories, but not the other way round, protein deficiency also appears when all the available proteins have been used to make up a deficit of calories to supply the body's energy needs. Consequently, it is now common to refer to protein/calorie malnutrition, or PCM, to take account of this interaction.

Malnutrition and hunger are prevalent in large areas of Asia, Africa and Latin America. They lead to premature death, high mortality rates in infants and hold back physical and mental development.

The effects of drought An American farmer during the drought of 1988 examines his ruined crops and parched fields, where erosion is already starting. In the Third World a scene like this might be an indication of famine to come; in the West it spells personal hardship, but people will not starve.

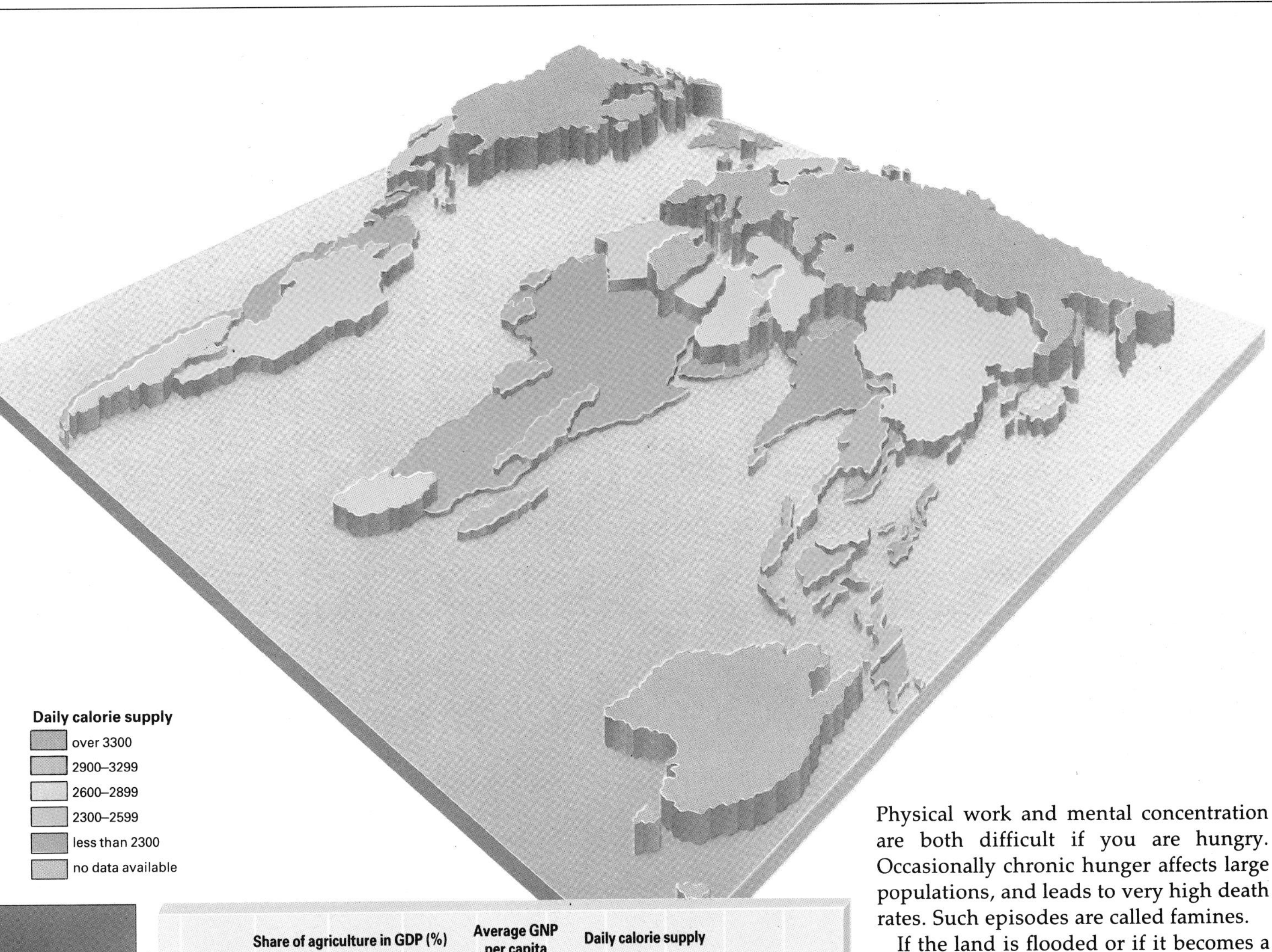

Share of agriculture in GDP (%)
Average GNP per capita
Daily calorie supply
300
890
2090
4790
14790
40 30 20 10 0
0 800 1600 2400 3200

Physical work and mental concentration are both difficult if you are hungry. Occasionally chronic hunger affects large populations, and leads to very high death rates. Such episodes are called famines.

If the land is flooded or if it becomes a theater of war, or dries up through lack of rain, farmers cannot cultivate it, or crops fail in the field and cannot be harvested. While climatic vagaries may cause extensive crop failure at any time anywhere in the world, they do not always lead to famine. Rich countries have stored surpluses to draw on, or can buy food once farmers are unable to grow it, or – in exceptional cases – because government policies have disrupted the food-supply system or work to prevent the supplies reaching particular groups, even though food may be locally available.

The daily calorie intake throughout the world (*above*) is calculated by totaling the calorie content of the food a country produces and imports, and dividing the sum by the size of population. This is then expressed as a daily amount. As a country's gross national product rises, agriculture's share in it decreases, yet people have more to eat.

Aid for the hungry (*left*) When drought hit the sub-Saharan countries of Africa in the 1980s a massive relief program was launched in the West. But international food aid brings only short-term solutions; it may bring relief, but it does not break the cycle of famine.

MALNUTRITION AND WEANING

Mothers' milk is the most suitable food for babies, providing protein intake in sufficient quantities to sustain rapid growth and mental development. In parts of the world where diets are generally poor in proteins, infants are at serious risk when weaned. In western Africa, for example, they are transferred from breast milk to foods made from cassava, which is low in protein. Moreover, what protein content it has is not of high value. The term kwashiorkor, applied to protein malnutrition in children, derives from a west African dialect word meaning "deposed child". It happens when the firstborn child is weaned on the birth of a sibling.

Similarly, among poor families in the West Indies weaned children are given sugar cane to suck. This provides more than sufficient energy, but no protein. Such malnourished children are called "sugar babies".

The consequences of severe protein deficiencies in the very early stages of life can never be overcome. Many infants will die, and those that do manage to survive will suffer the effects of stunted physical and often mental development throughout their lives.

The Issue of Healthy Food

IN MOST OF THE THIRD WORLD THE PRIORITY IS to produce enough food to eat. In richer countries concern is now shifting to the quality, convenience and safety of food. In the drive to make farming more and more productive, increased amounts of fertilizers, herbicides and pesticides have been applied to the land. Encouragement has come from the companies that manufacture the agrochemicals, from governments anxious to increase food self-sufficiency, and from the retail food suppliers who demand cheap, consistent and high quality produce for the consumer. A few caterpillar holes in a lettuce or cabbage may lead to a whole batch being rejected by a supermarket chain.

In animal production, the same pressures for cheapness and consistency of product have led to intensive rearing conditions for broiler chickens, calves and pigs. Eggs are produced by hens that are kept in automated battery houses. Animals reared in these intensive conditions are prone to infection, and this encourages the widespread use of antibiotics. These promote growth, which is further encouraged by the injection of hormones and by high protein feeds.

A very rapid rise in the consumption of convenience foods, ready cooked, frozen or chilled, to be prepared in seconds using a microwave cooker, has increased pressure on the farmer. Such foods are expensive to manufacture; some of the processing costs are passed on in high prices to customers who are willing to pay them, some are met by using cheap raw ingredients. A consequence is that a few very large food manufacturing companies are coming to dominate the industry, demanding cheap produce tailormade for their particular production processes.

In North America and Western Europe an increase in public awareness about diet and health, fueled partly by concern over these recent trends, is having a marked influence on the food industry. For example, consumption of saturated fats (found in animal fats, milk, cheese and butter) is now associated with the high levels of blood cholesterol that are known to increase the risk of heart disease. As a result of health warnings, many Western countries have seen a clear movement toward greater use of butter substitutes containing unsaturated fats. Many of these are made with sunflower oil, and production of this crop in Europe has risen sharply.

There is some concern about the safety of foods, particularly about the possible effects of residues from pesticides, the use of food preservatives and the high levels of hormones used in animal production. Some of these fears need to be weighed against the undoubted benefits that some treatments bring. For example, increased use of food preservatives to prevent the development of botulism and other toxic poisoning has saved many lives. Modern methods of preserving food are almost certainly much safer than traditional methods of smoking and salting. Adding chemicals that do nothing more than boost color and flavor is less justifiable.

Fields of sunflowers (*right*) are a common sight in Europe: in France alone production has increased a thousandfold since the 1960s – partly due to a rise in consumer demand for healthier foods.

An intensive pig unit (*below*) The feed the pigs are given contains antibiotics and other additives to prevent infection and promote growth; traces remain in the meat that humans then eat.

ORGANIC FARMING

Increasingly, people in the West are looking for healthy food that is free from chemicals. This has led to the growth of a small but thriving organic farming sector. Organic produce is grown without the use of agricultural chemicals; manure from livestock is used for fertilizer; and crops are rotated to keep down crop pests and to restore soil fertility. The people who want to enjoy their food in the knowledge that it has been produced naturally are prepared to pay higher prices for it. This compensates for the lower yields generally achieved on organic farms.

Organically grown produce is beginning to find its way on to the supermarket shelves. Consumer pressure will certainly continue to be for cheap and convenient foods, but it is becoming more diverse. Consumer groups have been encouraged by better, more informative labeling of foods to allow people to make informed choices about what they eat, and they will continue to encourage the trend toward healthier, safer foods as well. Many people argue that this will mark a return to fuller flavor and texture too.

The risks from meat

In affluent countries human health problems stemming from the intensive rearing of animals are also a matter of concern. Salmonella – bacteria that live in the intestinal tracts of many animals – are widespread among broiler chickens, and may also be present in eggs. If they are passed on in food to humans they can cause acute gastroenteritis. The bacteria are killed by proper cooking, but the careless handling and preparation of poultry and egg dishes has led to a rapid rise in cases of salmonella food poisoning.

The spread of the salmonella organism among hens was certainly encouraged by the practice of feeding the waste products of poultry processing, including droppings, back to the birds. Similar efforts to boost growth in other livestock by feeding them on high-protein compound foods, including animal waste products, have increased the risks of transferring diseases between species. There are fears that they may be passed on to humans through the food chain; these, allied to concerns for the welfare of animals reared in intensive conditions, are leading more and more people to turn to vegetarianism.

The Global Market

MOST OF THE WORLD'S FARMERS ARE STILL subsistence farmers who are concerned with little beyond their farmgate. They keep the greater part of their crops to feed their family, and either sell the surplus to dealers visiting the farm or at the local market. Animals are kept to provide power and transportation, manure for fuel and fertilizer, and food.

Beyond the farmgate, however, the world food system is becoming increasingly integrated. Vast quantities of food are moved daily around the globe by bulk carrier; perishable goods are refrigerated and transported by ship or, increasingly, by air. This means that seasonal goods are now available all the year round. Fruit such as apples, pears and grapes grown in the southern hemisphere, in Australia, New Zealand, South Africa or Chile, can be coated with wax to prevent them from spoiling, or sprayed with chemicals to retard ripening. They are then shipped to northern markets in Western Europe and North America, to supply shops and supermarkets throughout the summer with fruit that is cheaper and in better condition than any local products, which will have been kept in storage since the previous fall.

The rise of world trade

There has been international trade in food for thousands of years, but when overseas journeys were long and costly, and risked loss from shipwreck or brigandage, small quantities only of luxury items were carried: spices, silk, tea, coffee and sugar. The history of European colonialism is the history of a growing international trade in food – as the Portuguese, Spanish, British and Dutch established and extended their territorial empires in Asia, the Americas and Australasia, new and exotic produce began to find a market in Europe.

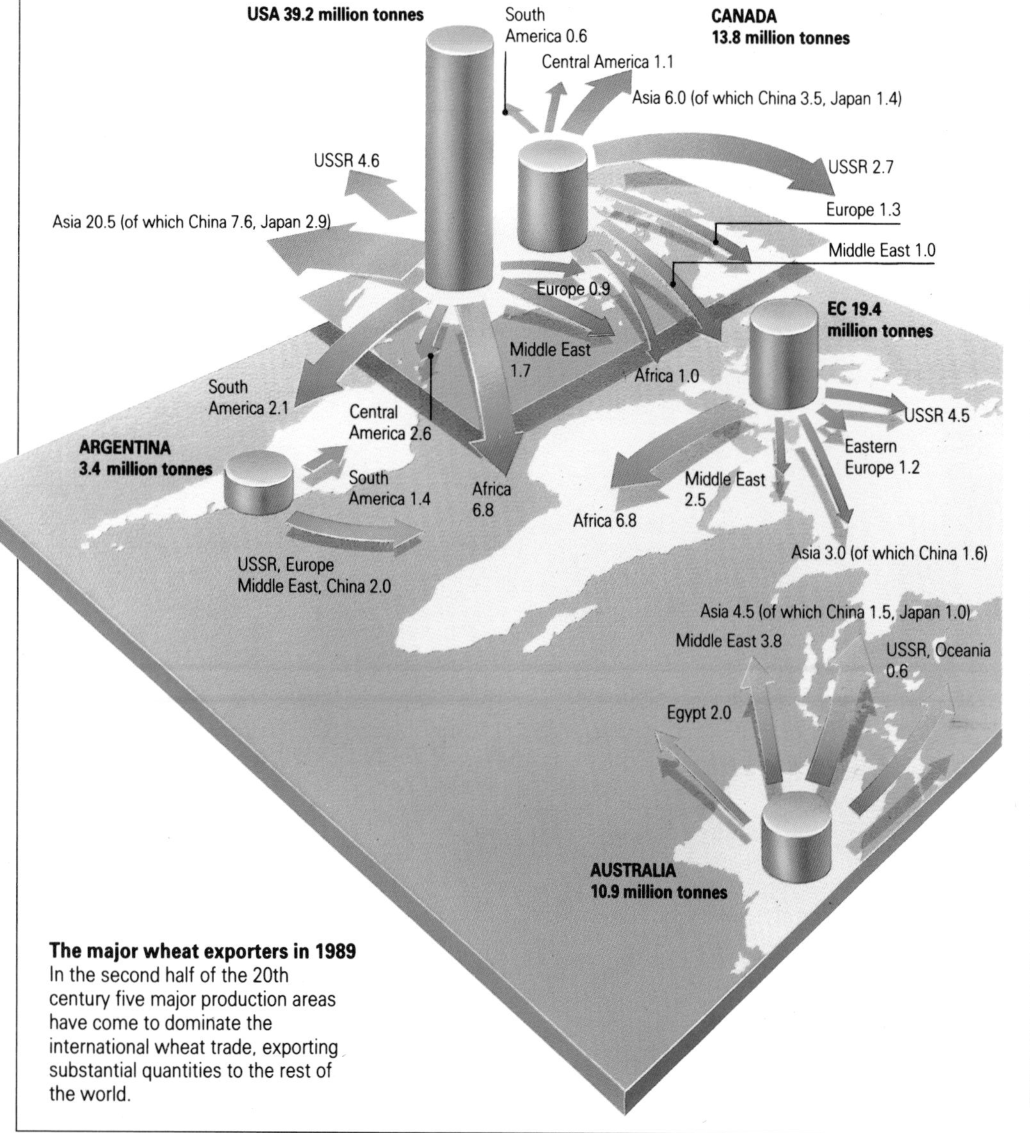

The major wheat exporters in 1989
In the second half of the 20th century five major production areas have come to dominate the international wheat trade, exporting substantial quantities to the rest of the world.

Moving food around the world (*left*) Australian beef is being loaded onto a Swedish freighter. Food grown at one end of the world frequently finishes up on people's dinner plates at the other end, having passed through the hands of a series of agents and middle men and others involved in the international food industry.

The local market place (*above*) Many of the world's farmers look no farther than the local town as a market for their surplus produce. On the Caribbean island of St Vincent small amounts of vegetables and fruit have been brought from farms and smallholdings nearby to be sold in this lively street market.

CHOCOLATE: A WORLD COMMODITY

A drink known as *xocoatl* (chocolate) made from cocoa beans was popular at the court of Montezuma, the Aztec ruler of Mexico. He served it to the Spanish invader Hernándo Cortés in 1519. Returning home, Cortés introduced the drink, sweetened with both vanilla and sugar, to the Spanish court, and from there its popularity spread throughout the countries of Europe.

Chocolate houses were all the fashion in mid 17th-century England, and were opened alongside the coffee houses in which coffee, recently introduced from the Middle East, was drunk. During the 19th century the habit of chocolate-drinking was promoted by teetotal manufacturers such as John Cadbury as an alternative beverage to gin, then a cheap, popular drink. Today it is much more widely consumed as candy or is used to cover buns and cookies. The chocolate is obtained by pulverizing the roasted beans to make a paste, and adding sugar, additional cocoa butter and condensed milk.

The cocoa (or cacao) tree grows naturally in tropical America, but was spread by the Spanish from Mexico throughout Central and South America and the West Indies, and later to the Philippines; the Dutch established the tree in their colonies in Indonesia and Ceylon (now Sri Lanka). Cocoa became a principal crop of western Africa only about 100 years ago; the region is now the producer of over half the world's total crop. Very little is used locally; the bulk of it is exported for processing and consumption, mostly in Europe and North America.

Tea, for example, was first introduced into Europe from China (where it had been cultivated since the 3rd century AD) in the 17th century. The East India Company, a British company formed to exploit trade with the Far East, soon had a monopoly on the world tea trade, and under its auspices tea cultivation was introduced into India and Ceylon (Sri Lanka). Specialized ships, the tea clippers, were built to transport the product as quickly and cheaply as possible to both Europe and North America, and as the trade developed tea ceased to be a luxury. By the beginning of the 20th century tea growing had spread from Asia and had become established as far afield as southern and central Africa, South America and Australia.

In order to meet the world demand for particular commodities, agriculture in some regions has become highly specialized. Of the 42 million tonnes of bananas grown in the world, 7.5 million tonnes enter world trade. Of these, 2.8 million tonnes are grown in the Caribbean, and agriculture on some of the islands is limited to little except bananas for export to the United States and Europe. Coffee is even more reliant on world trade. Of a total world production of some 6 million tonnes, 4.4 million are exported.

The international grain market

Nearly every country in the world grows some grain. Even so, as much as a fifth of the 510 million tonnes of wheat produced enters world trade. Before 1930 no single region had real dominance in the grain market, though Europe was the only substantial importer. By 1960, however, the situation had changed. The population increase in Asia had outstripped local food production; Eastern Europe and the Soviet Union were no longer exporting grain; and the United States and Canada between them provided about 80 percent of world exports. Since 1982 the dominance of North America has declined considerably. As a result of subsidies to farmers the countries of the European Community now have substantial grain surpluses; and since the introduction of new Green Revolution higher-yielding crop varieties, India has been able to achieve self-sufficiency in grain production in most years.

The Politics of Food Supply

GRAIN IS THE MOST IMPORTANT TRADED food commodity. The few countries which have grain surpluses to export have potential power and influence over those that must rely on food imports. Except in times of war, however, this power may be more apparent than actual.

Most attempts to exert influence over another country by withholding grain exports have ended in greater damage being inflicted on the country with the food surplus. In 1980, for example, following the Soviet invasion of Afghanistan, the United States imposed a boycott on its cereal trade with the Soviet Union in order to express its disapproval of the action. The American grain companies, which had contracted to purchase large quantities of grain for onward shipment to the Soviet Union, found themselves unable to sell, and the United States' government was forced to purchase the contracts, at a high cost to the American taxpayer. In addition, the embargo failed as other grain-exporting countries were not persuaded to join in, and the Soviet Union was able to look to them for alternative sources of supply.

Food aid is a potentially more effective way of wielding political influence, since the normal conditions of the open market do not apply. The recipient countries are too poor to buy enough food and, as most food aid is either given away, or sold on very generous terms, the donor countries can obviously afford to be selective about where its food aid is sent.

Picking tea (*above*) on a plantation in Assam, northeast India. The founding of India's tea industry stems from the years of British rule. Many plantations are now owned by multinational companies; they have the reputation of paying their workers very low wages.

The food paradox (*right*) While the countries of North America and Western Europe are regularly producing grain in excess of their needs, others chase the illusory target of food self-sufficiency, though this will not necessarily bring them closer to feeding their people.

Food security or self-sufficiency?

The aim of most countries, consequently, is to achieve food self-sufficiency. In reaction to the high world food prices in the early 1970s, and with the encouragement of the Food and Agriculture Organization (FAO) of the United Nations, most developing countries created food plans to help them to do so. However, though grain self-sufficiency has strong emotional and political appeal, it makes little economic sense when there are grain surpluses elsewhere in the world and prices are low: in 24 out of the last 30 years, the world cereal trade has been characterized by surplus rather than by shortage. Food security can be achieved without struggling for self-sufficiency. Self-sufficiency can also be an entirely artificial concept. India achieved it in the 1980s and even exported a small surplus, but a large proportion of its population

THE PLANTATION ECONOMY

Most cash crop production in the Third World is a legacy of European colonization; plantations were established in agriculturally productive tropical areas – from South and North America to Southeast Asia, India and western and eastern Africa – as a way of supplying the factories of Europe with cheap raw materials. The production, originally carried out with the use of slave labor, was quite distinct from the economy of the rest of the colony. The plantations grew food for their own work force, and the input of senior staff, management ideas and equipment was supplied from the home country.

Large tracts of territory were given over to the production of a single cash crop. Tea, coffee, cocoa, sugar, tropical fruits, oil palm and rubber were all typical plantation crops. The colonizing powers were responsible for extending their cultivation around the globe into regions with similar growing conditions; for example, rubber trees, native to Brazil, were transplanted to Malaysia and Indonesia, while sugar cane made the journey the other way round, from Southeast Asia to the Caribbean.

The plantations have survived independence, and are now mostly run by multinational companies. The largescale "colonization" of the Amazon Basin by United States' owned cattle ranches has much in common with the plantations of colonial days, as does largescale state-run rice production in western Africa.

did not get enough to eat as food prices were higher than they could afford. As much as 40 percent of its population remains undernourished.

When so many people in, for example, Brazil are without sufficient food, the expanses of land devoted to the production of over 16 million tonnes of soybeans may be hard to accept, particularly as most of it will be exported to feed livestock in the United States and Western Europe. As a result of their former colonial status many Third World countries have been tied into a cash crop economy. One-fifth of the arable land in Senegal, for example, is devoted to the production of groundnuts for export, increasing its need to import staple foods. Thus the economically weak Third World countries are made doubly dependent on the economically strong First World.

Every country needs to import some goods. To do so it requires foreign exchange. No matter how unsatisfactory it may be for a country like Bangladesh to be at the mercy of the world jute market, jute is the country's most important agricultural export, and its only substantial non-aid source of hard currency.

A balance needs to be struck between the importance of secure food supplies and the need for export earnings from cash crops. A major consideration in this balance is how the benefits of cash crop production are distributed. Any rise in the number of landless people in the developing countries means there will be more people without enough to eat. So if large amounts of land are concentrated in fewer and fewer hands in order to raise cash crops, the poverty gap will be increased between the rich elite and the mass of rural and urban poor. Furthermore, if most of the profits are repatriated to the home base of a multinational company, the advantages become even smaller. However, cash crop production need not be like this. Jute production in Bangladesh, so vital to the economy, is in the control of a multitude of small producers who are free to choose whether to grow rice or jute on their flooded land rather than submit to decisions made by companies whose interests lie outside the country of production.

The soybean phenomenon There has been an enormous expansion of the cash crop cultivation of soybeans, grown mainly for cattle fodder. Brazil is now the third largest producer of the crop, but its largescale cultivation does little to benefit local people.

The Future for Farming

TOWARD THE END OF THE NEXT CENTURY, THE population of the world will have reached about 10 billion, double its present size. It is believed that population levels will then remain more or less stable. Will the world be able to produce enough food to allow the population to grow to this size?

Our present growth in food production is based on technological developments that are already damaging the basic raw material of agriculture – the soil. Increased cultivation of the soil means that more is lost through erosion. It is important not to exaggerate the scale of the problem. For example, if erosion in the United States continues at its present rate for the next 100 years, agricultural production might be reduced by between 3 and 10 percent. Losses of this kind can easily be compensated for by technological improvements in crop yields. However, erosion is far more serious elsewhere in the world, particularly in Nepal and parts of India, eastern Africa and some parts of South America.

Between 1960 and 1980 perhaps as much as 60 percent of the world's increased food production was achieved through additional irrigation. Future sources of irrigation will be more expensive and difficult to exploit. It will bring increased salinization, the silting of reservoirs and the depletion of groundwater reserves. New methods of water management will be needed to cope with the increasingly serious problems of environmental degradation already associated with major irrigation schemes such as the Aswan Dam in Egypt, and the drying up of the Aral Sea in the Soviet Union.

It is important that present problems of surplus food production in Europe and the United States do not lead to a reduction in worldwide agricultural research. The development of more efficient nitrogen utilization by plants, of low dose, highly specific pesticides, better ways of using limited irrigation water, the better management of dryland margins of agriculture, and methods of increasing the yields of the staple crops of Africa are all essential. Above all, the agriculture we develop over the next years must increase food production in a sustainable manner for the future.

Sustainable agriculture

What is meant by sustainable agriculture? It is the obligation to produce enough food for ourselves without reducing the stock of resources for future generations, so that we hand on at least as much as we enjoyed ourselves. This stock includes not only the raw materials for farming – productive soil, energy, water resources – but also the knowledge and the new technology to make more efficient use of them and to find ways of renewing or replacing those that become scarce. Technology is vital – without it the future populations of the world will starve.

Sustainable agriculture means finding new ways of applying technology to make more efficient use of existing resources to increase food production. This water treatment plant in San Diego, California, is being used for experiments in aquaculture.

Some people argue that the world should stop using fossil fuels altogether, in order not to harm the environment. This means that future generations will never be able to use them either, for the same reason. Moreover, it will be impossible to increase food production fast enough without them. People fortunate enough to live in Europe and North America are able to propose a return to low-technology agriculture because they already produce excess food. No policy that reduces productivity is practical in countries already short of food and where the population is growing fast.

The goal should not just be to produce enough food to feed the world's growing population. The imbalances that already exist between the rich and poor countries, and between rich and poor people, will rapidly increase unless there is united political agreement to concentrate our technological and development efforts on those most in need.

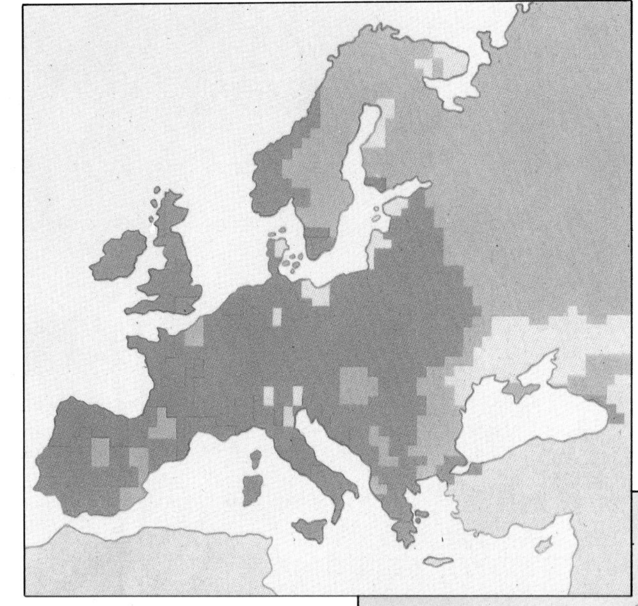

Growing season (months)

- more than 9
- 7–9
- 3–6
- less than 3

The potential effects of climate change (*above and right*) Should carbon dioxide levels double, as is likely, there will be great changes in the length of the growing period around the world as temperatures rise. Northern Europe is likely to increase its cropping potential, while the growing season around the Mediterranean could shorten.

Soil degradation (*left*) in Western Australia, where the removal of vegetation has exposed the soil to extreme temperatures during the summer months, running the risk of serious erosion. Better farming practices could prevent a good deal of unnecessary land loss.

AGRICULTURE AND CLIMATIC CHANGE

The "greenhouse" gases – carbon dioxide and methane – associated with global warming are accumulating so rapidly in the atmosphere that by the year 2030 levels are expected to be twice what they were before the Industrial Revolution. Most of this increase results from carbon dioxide released into the air through the burning of fossil fuels. However, agriculture is a major contributor as well. The clearing of the tropical rainforests removes a "sink" for carbon dioxide (plants absorb carbon dioxide), while the burning of the trees adds further amounts to the atmosphere. Levels of methane, produced by decomposing vegetable matter, rise as the amount of land used for rice paddies and the worldwide numbers of grass-eating livestock increase. As these gases encourage the heating up of the atmosphere, global temperatures may rise by as much as 2°C (3.6°F) by the year 2030.

Most agricultural crops grow better in a carbon dioxide-rich atmosphere. It seems likely, however, that the effect of global warming will dramatically alter the regional pattern of crop production and trade. The warming is expected to be greater toward the poles, especially in the continental interiors – important grain-growing areas. If rainfall levels here remain the same, or decline, yields will drop considerably. The degree of warming will be less in places closer to the Equator, so monsoonal India may even expect an improvement in growing conditions since precipitation levels are expected to rise.

Changes in climate will have other environmental consequences that will affect agriculture. As the oceans become warmer the frequency and intensity of tropical cyclones is expected to increase. Figures for the estimated rise in sea level by the middle of the next century vary from 0.5–1 m (16–36 in). These two factors in conjunction will spell further disasters for many low-lying fertile agricultural areas such as the Ganges Delta in India and Bangladesh, already subject to frequent widespread and devastating flooding.

REGIONS OF THE WORLD

CANADA AND THE ARCTIC
Canada, Greenland

THE UNITED STATES
United States of America

CENTRAL AMERICA AND THE CARIBBEAN
Antigua and Barbuda, Bahamas, Barbados, Belize, Bermuda, Costa Rica, Cuba, Dominica, Dominican Republic, El Salvador, Grenada, Guatemala, Haiti, Honduras, Jamaica, Mexico, Nicaragua, Panama, St Kitts-Nevis, St Lucia, St Vincent and the Grenadines, Trinidad and Tobago

SOUTH AMERICA
Argentina, Bolivia, Brazil, Chile, Colombia, Ecuador, Guyana, Paraguay, Peru, Uruguay, Surinam, Venezuela

THE NORDIC COUNTRIES
Denmark, Finland, Iceland, Norway, Sweden

THE BRITISH ISLES
Ireland, United Kingdom

FRANCE AND ITS NEIGHBORS
Andorra, France, Monaco

THE LOW COUNTRIES
Belgium, Luxembourg, Netherlands

SPAIN AND PORTUGAL
Portugal, Spain

ITALY AND GREECE
Cyprus, Greece, Italy, Malta, San Marino, Vatican City

CENTRAL EUROPE
Austria, Germany, Liechtenstein, Switzerland

EASTERN EUROPE
Albania, Bulgaria, Czechoslovakia, Hungary, Poland, Romania, Yugoslavia

THE SOVIET UNION
Mongolia, Union of Soviet Socialist Republics

THE MIDDLE EAST
Afghanistan, Bahrain, Iran, Iraq, Israel, Jordan, Kuwait, Lebanon, Oman, Qatar, Saudi Arabia, Syria, Turkey, United Arab Emirates, Yemen

NORTHERN AFRICA
Algeria, Chad, Djibouti, Egypt, Ethiopia, Libya, Mali, Mauritania, Morocco, Niger, Somalia, Sudan, Tunisia

CENTRAL AFRICA
Benin, Burkina, Burundi, Cameroon, Cape Verde, Central African Republic, Congo, Equatorial Guinea, Gabon, Gambia, Ghana, Guinea, Guinea-Bissau, Ivory Coast, Kenya, Liberia, Nigeria, Rwanda, São Tomé and Príncipe, Senegal, Seychelles, Sierra Leone, Tanzania, Togo, Uganda, Zaire

SOUTHERN AFRICA
Angola, Botswana, Comoros, Lesotho, Madagascar, Malawi, Mauritius, Mozambique, Namibia, South Africa, Swaziland, Zambia, Zimbabwe

THE INDIAN SUBCONTINENT
Bangladesh, Bhutan, India, Maldives, Nepal, Pakistan, Sri Lanka

CHINA AND ITS NEIGHBORS
China, Taiwan

SOUTHEAST ASIA
Brunei, Burma, Cambodia, Indonesia, Laos, Malaysia, Philippines, Singapore, Thailand, Vietnam

JAPAN AND KOREA
Japan, North Korea, South Korea

AUSTRALASIA, OCEANIA AND ANTARCTICA
Antarctica, Australia, Fiji, Kiribati, Nauru, New Zealand, Papua New Guinea, Solomon Islands, Tonga, Tuvalu, Vanuatu, Western Samoa

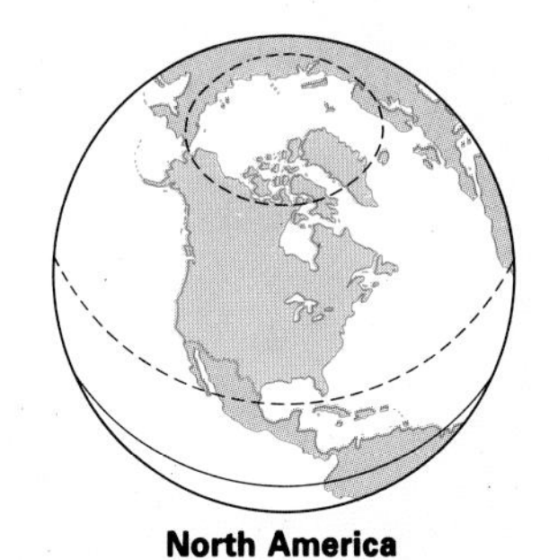
North America

CANADA AND THE ARCTIC

THE UNITED STATES

CENTRAL AMERICA AND THE CARIBBEAN

SOUTH AMERICA

Central and South America

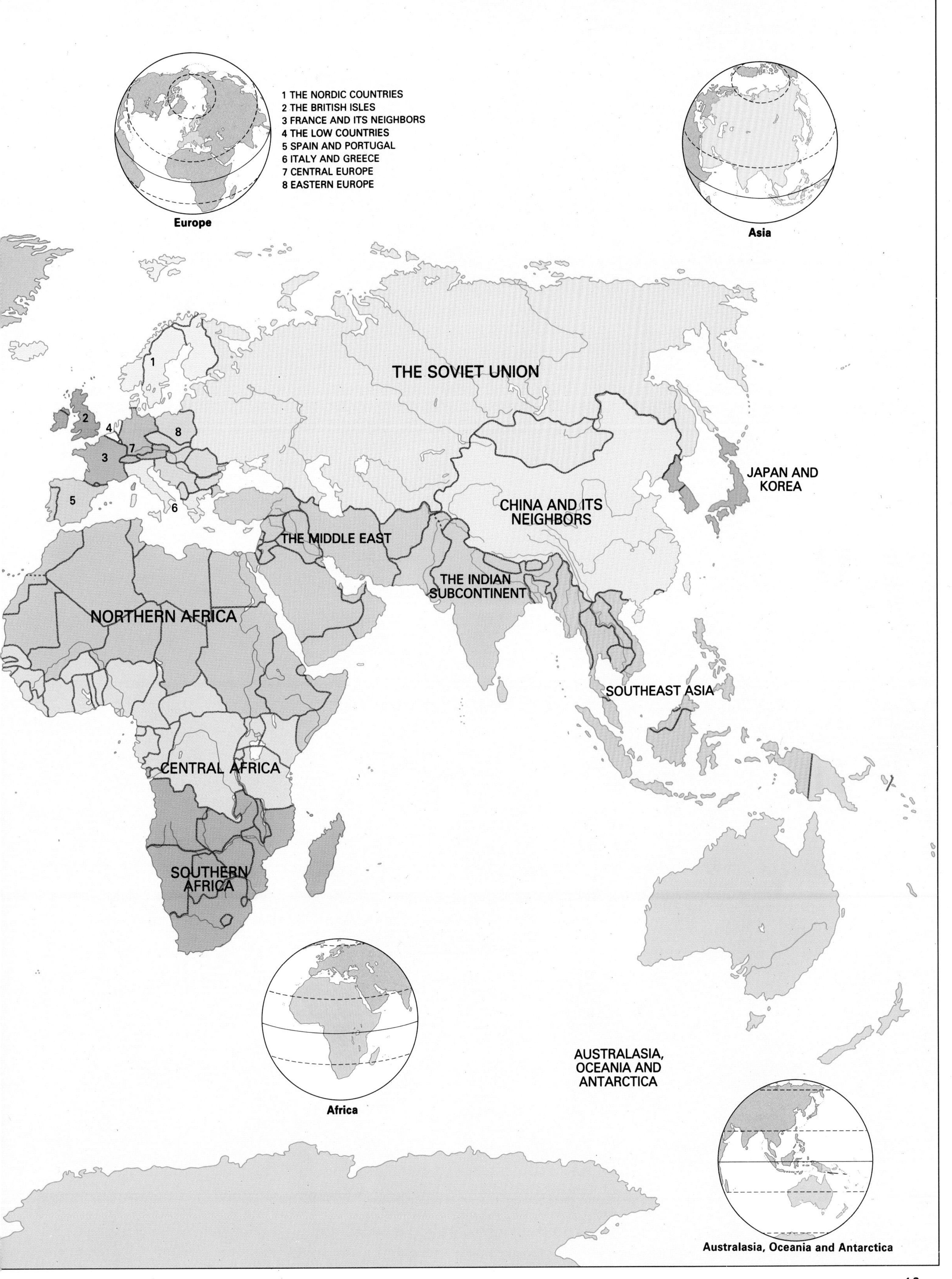
1 THE NORDIC COUNTRIES
2 THE BRITISH ISLES
3 FRANCE AND ITS NEIGHBORS
4 THE LOW COUNTRIES
5 SPAIN AND PORTUGAL
6 ITALY AND GREECE
7 CENTRAL EUROPE
8 EASTERN EUROPE
Europe
Asia
THE SOVIET UNION
JAPAN AND KOREA
CHINA AND ITS NEIGHBORS
THE MIDDLE EAST
THE INDIAN SUBCONTINENT
NORTHERN AFRICA
SOUTHEAST ASIA
CENTRAL AFRICA
SOUTHERN AFRICA
AUSTRALASIA, OCEANIA AND ANTARCTICA
Africa
Australasia, Oceania and Antarctica

NORTHERN WHEATLANDS

WESTWARD HO! · FAMILY FARMS, WORLD MARKETS · SUPPORTING AGRICULTURAL GROWTH

Although Canada is the second largest country in the world, its agricultural lands are limited. Farming occupies just 8 percent of the total land area of nearly 1 billion ha (2.5 billion acres), and only a tiny proportion of this farmland has high agricultural capability. Almost half the land is under forest; the rest is covered by lakes, swamps and high mountains, or is in areas with Arctic or subarctic climates that are unsuitable for agriculture. The best agricultural land is in the Great Lakes/St Lawrence lowlands, close to the border with the United States. To the west, the wheat-growing prairies – the northern part of the Great Plains of America – have been Canada's "bread basket" since the 1880s. Forestry predominates on the Pacific coastlands, with intensive agricultural production occupying the fertile valleys.

COUNTRIES IN THE REGION

Canada

Land (million hectares)

Total	Agricultural	Arable	Forest/woodland
956 (100%)	78 (8%)	46 (5%)	354 (37%)

Farmers

472,000 employed in agriculture (4% of work force)
97 hectares of arable land per person employed in agriculture

Major crops
Numbers in brackets are percentages of world average yield and total world production

	Area mill ha	Yield 100kg/ha	Production mill tonnes	Change since 1963
Wheat	13.5	19.3 (83)	26.0 (5)	+69%
Barley	5.0	27.9 (120)	14.0 (8)	+262%
Rapeseed	2.7	14.4 (101)	3.8 (17)	+1,284%
Oats	1.3	23.7 (129)	3.0 (7)	−51%
Maize	1.0	70.2 (193)	7.0 (2)	+554%
Linseed	0.6	12.3 (216)	0.7 (29)	+42%

Major livestock

	Number mill	Production mill tonnes	Change since 1963
Cattle	11.7 (1)	—	+4%
Pigs	10.5 (1)	—	+101%
Milk	—	8.0 (2)	−4%
Fish catch	—	1.6 (2)	-

Food security (cereal exports minus imports)

mill tonnes	% domestic production	% world trade
+23.5	45	11

WESTWARD HO!

French settlers first introduced European-style agriculture to eastern Canada in the early 17th century, and further waves of immigrants from France, Britain and other European countries then progressed westward across the country during the course of the 18th and 19th centuries. The prairies were not really opened up to cereal farming until the passing of the Homestead Act (1870), which offered free land to prospective settlers. Many different ethnic groups took part in the surge of pioneer settlement during the last quarter of the 19th century, giving rise to distinctive local communities: Ukrainian, German and Scandinavian settlers were among those who benefited from the emerging export trade in grain. Settlement continued into the 20th century, with mining and forestry opening up new areas north of the established agricultural frontier. Even parts of the extreme north developed a limited amount of agriculture to provide food for mining settlements.

The influence of climate

From colonial times, Canadian agriculture has mainly been determined by the widely different physical conditions across the country. The cold, harsh climate of the north generally discouraged European settlers, so that is one area where the indigenous Inuit people's traditional way of life, based on hunting and fishing, has been preserved. Elsewhere in the country indigenous Indian peoples, mostly hunter–gatherers, were displaced by the European colonists and were eventually confined to designated reserves.

The early settlers in the Great Lakes/St Lawrence lowlands found extremely favorable conditions for agriculture. The fertile soils, the ample rainfall evenly distributed throughout the year, and the hot summers that compensated for cold winters were all factors that encouraged

Fish drying (*above*) The Inuit, the indigenous Indian people of the north, remain largely unchanged by colonial influences. Isolated by the harsh climate, they follow a traditional way of life based on hunting and fishing. Fish is an important source of protein and is preserved by air drying.

Opening up the wheatlands (*right*) The vast expanses of the prairies began to be exploited after 1870, when pioneer settlers were offered free land. The introduction of a drought-resistant variety of wheat and of dry farming techniques caused a massive increase in the area of land under wheat cultivation over the next few decades.

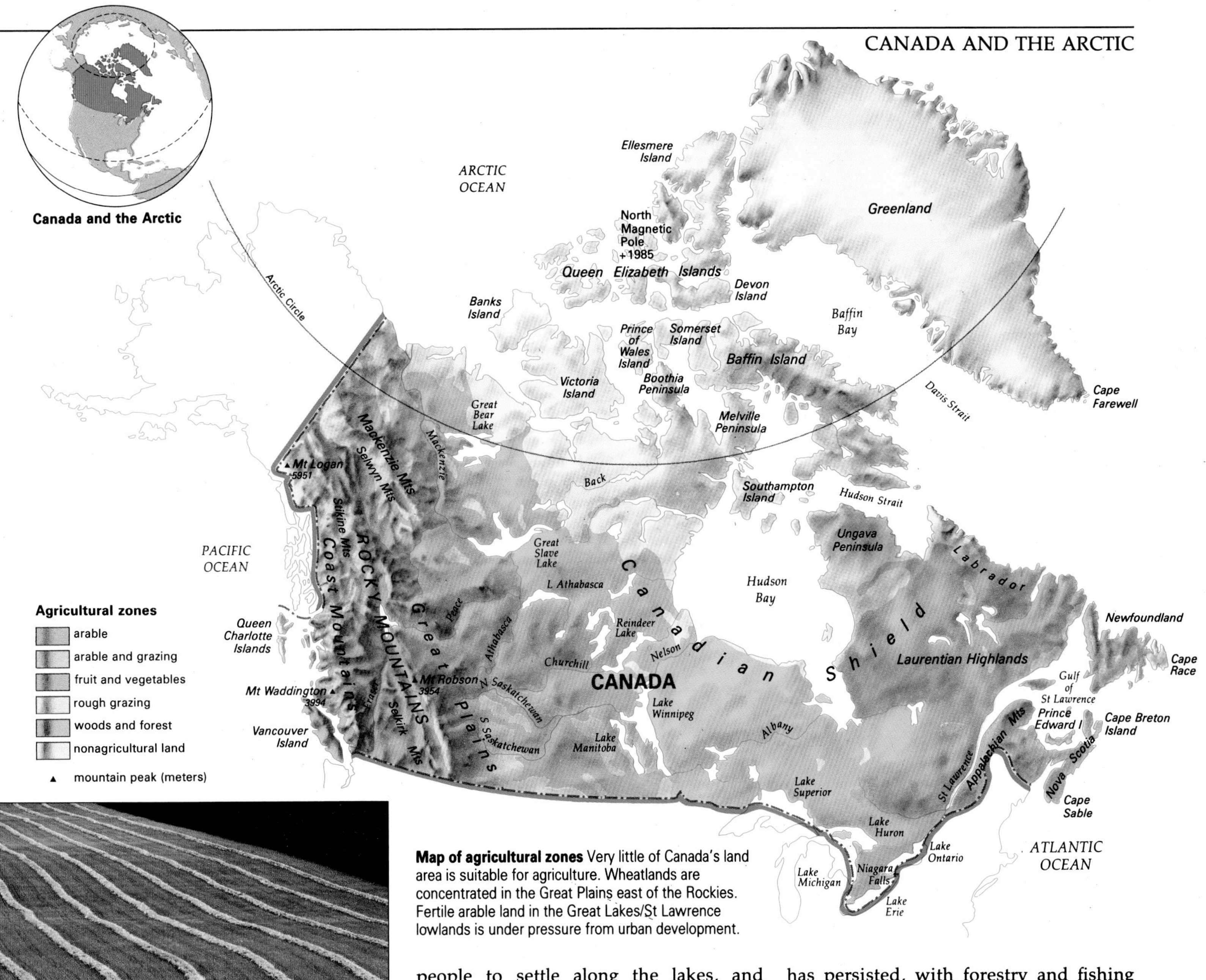

Map of agricultural zones Very little of Canada's land area is suitable for agriculture. Wheatlands are concentrated in the Great Plains east of the Rockies. Fertile arable land in the Great Lakes/St Lawrence lowlands is under pressure from urban development.

people to settle along the lakes, and enabled the development of both arable and livestock farming to take place. Fruit growing was introduced on the Niagara peninsula, which is free of frost for 190 days of the year.

Farther west, the harsher winters of the prairies reduce the length of the growing season, and rainfall in the south of the region is less than 400 mm (15 in) per year. Nevertheless, settlers were able to produce good cereal crops thanks to the timely summer rains and humus-rich chernozem soils. To the west of the Rocky Mountains, the mild, damp climate gives a frost-free period of over 250 days, supporting mixed farming and fruit growing in the valley lowlands. By contrast, the Maritime Provinces of eastern Canada have severe conditions, with harsh winters and cool springs that restrict agricultural activity.

Natural riches

From the earliest European settlement of eastern Canada, farming was only a part of a diverse rural economy. This diversity has persisted, with forestry and fishing both playing a major role in Canada's economic development. The physical and climatic restrictions on agricultural activity have been offset by favorable conditions for forestry, particularly on the Pacific coast in British Columbia and in northern parts of Ontario and Quebec. Forest products are now the single most important component of Canada's international trade, accounting for nearly a third of the country's exports.

Fishing played an important part in the European colonization of the Atlantic seaboard, and continues to be an important industry in the region. About 1,000 communities along the coast are wholly or mostly dependent on income derived from fishing; it is particularly important to the local economy of Newfoundland, in the east of the region, where the waters of the Grand Banks to the southeast of the island provide rich catches of cod, haddock, flat-fish, mackerel and herring. The fishing industry in British Columbia yields valuable exports of processed salmon and tuna.

FAMILY FARMS, WORLD MARKETS

Grains, notably wheat, provide nearly half of Canada's agricultural revenue. Cultivation is concentrated in the three prairie provinces of Alberta, Manitoba and Saskatchewan, which were settled in the years after Canada's Confederation in 1867. A special variety of wheat – Marquis – was developed to withstand the dry climate and short growing season, and the wheat economy boomed with the coming of the railroads, particularly the Canadian Pacific Railroad in 1886. Rapid transportation to eastern ports provided access to world markets, and large grain elevators were built to store grain beside the railroads.

A system of "dryland farming" was first introduced into the prairies in the 1880s, thereby facilitating the spread of large-scale cereal farming. It involves the systematic use of summer fallow to limit the effects of low annual rainfall. Once a crop has been harvested, the land is allowed to remain fallow throughout the following year, enabling it to store moisture and yield a good crop the next year.

The adoption of dry farming methods, as well as the introduction of fast maturing grain varieties, helped Canadian farmers to bring about an increase of 15 million ha (37 million acres) in the area of land under cultivation in the first three decades of the century. Although dryland farming gradually reduces soil fertility over a long period of time, this can be restored by the use of fertilizers, and the system is still widely used as a means of conserving moisture.

Wheat still dominates farming in the prairies, except in the driest areas where it has been replaced with cattle ranching. Between 1900 and 1930 the area under wheat rose from 610,000 ha (150,700 acres) to 10 million ha (almost 25 million acres). This phenomenal expansion was halted during the 1930s by drought and world economic depression, and subsequently by the disruption of World War II. But by the 1950s output and exports were again increasing. Longterm contracts to supply wheat to China and the Soviet Union opened up new markets, in addition to

Ontario cereals Wheat gives way to maize and rye in eastern Canada, where the ample rainfall reduces reliance on dry farming. The farmhouse reflects the north European influence in the province.

Specialist fruit production Canada supports a wide range of fruit growing in areas where the climate permits. Most of the fruit comes from specialist farms in the southeast, but there is also an important fruit area in British Columbia. Apples are a major crop, and are sold either as fresh produce or for processing.

the traditional ones in Britain and Japan. The 1960s saw an expansion of beef production in the region at the expense of some wheat production, but since then higher grain prices have led to a new resurgence in wheat growing, and grains have now overtaken livestock as the largest single source of revenue in the Canadian agricultural economy.

Unlike the grain monoculture of the prairies that stretch across central Canada, the agriculture of the eastern provinces has tended to be more diverse. Commercial dairy farming became locally significant, supplying the growing cities of Ontario and Quebec; horticultural farming developed in the Niagara peninsula and southwestern Ontario; and greater reliance was placed on livestock production in mixed farming systems.

The tendency toward specialist dairy and livestock farming grew stronger after World War II as demand for most meat and livestock products increased. By 1981 the number of cattle had risen by 60 percent, pigs by 50 percent, and poultry by 53 percent. On the other hand, the size and number of sheep flocks fell in the face of cheaper competition from Australia and New Zealand.

While a number of long-established crop specializations, such as potatoes on Prince Edward Island, have been overtaken by a greater reliance on livestock, in general the high prices available during the 1980s for crops such as maize, soybeans, tobacco and vegetables have stimulated greater production. There has been a particularly rapid expansion of the cultivation of oilseed rape (canola) in the east of the country, as well as on the prairies, where the area devoted to it rose from 5,000 ha (12,350 acres) in 1960 to 2 million ha (almost 5 million acres) in the mid-1980s. Higher cereal prices, and the practice of using grains with rapemeal and silage for livestock feed, also brought about a significant increase in grain production in Ontario.

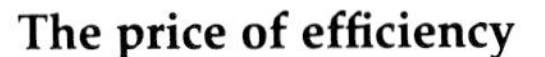

The price of efficiency

The move away from mixed farming in favor of specialist production of a single crop or livestock animal demands high capital investment. Very often this brings greater financial returns, but it also makes producers more vulnerable to fluctuations in market prices and has led to a high rate of farm business failure.

The need to use costly inputs of fertilizers and machinery to sustain high productivity has meant that many farms have had to amalgamate to achieve economies of scale. The postwar period has consequently witnessed a sharp fall in the number of farm holdings: from 623,000 individual farms in 1951 to 293,100 in 1986. In the eastern Maritime Provinces the drop has been enormous – as much as 80 percent – and fishing and forestry here are now much more important in the rural economy.

Despite the highly commercial nature of Canadian agriculture, the vast majority of farms are still family run. Nearly two-thirds of farms are entirely family owned, and many others constitute a mix of owned and rented land. This high level of owner occupation has its roots in the early pioneer days, when colonists were able to buy land outright from the colonial, federal or provincial authorities.

THE NORTHERN LIMITS FOR FRUIT

Harsh climates, particularly the length and severity of winter frosts, limit the areas where fruit and vegetable production is possible in Canada. However, there are significant pockets where the climatic conditions of the region are modified by the natural features of the landscape to allow earlier spring warming and a longer maturing season. One of the most important of these is the Okanagan valley of British Columbia in the extreme west of the region.

The valley lies in the central plateau to the west of the Rocky Mountains and in the rainshadow of the Pacific coastal range. Fruit growing here is therefore dependent on irrigation, and this means that a wide range of fruit can be grown. Apples are the major crop, however, accounting for two-thirds of the area of production.

Most of the orchards are situated on terraces on the top and sides of the slopes above Okanagan Lake. The terraces are divided by deep gullies running from the high slopes down to the lake. These enable cold air to "drain" down the hills to the valley bottom, thereby preventing damage from frost. Nevertheless, this can be a hazard if there is a severe frost in late spring. Very low winter temperatures can also damage the apple trees; sometimes as much as a fifth of the orchard stock has been destroyed.

Most of the apples grown are sold as fresh produce, but there is also a large local fruit processing industry that makes juice concentrates, apple sauce, pie fillings and frozen apple slices.

SUPPORTING AGRICULTURAL GROWTH

The importance of agriculture to Canada's national economy is recognized in the government's record of supporting water and land conservation measures and of exercising price controls to help maintain farm incomes. After Confederation, the government's agricultural policy was designed to encourage farming on the prairies, and freight rates and the grain handling system were both regulated. The Prairie Farm Rehabilitation Act (1935) was passed to counter the effects of drought and economic depression, which had led to the abandonment of farmland. It introduced better management of water resources, improved farming methods, and some resettlement schemes. Irrigation has been substantially extended as a result of government policy, especially in Alberta and dry southern Saskatchewan. Nevertheless, no more than 1.5 percent of Canadian land is irrigated, indicating the continued reliance on dryland farming on the prairies, and the plentiful rainfall in other agricultural regions.

An agricultural revolution
More recent government measures have helped to stabilize the farming industry by raising commodity prices, making agricultural credit easier to obtain, and encouraging mechanization and the development of more efficient farms.

To meet the demands of the expanding Canadian and world markets, farmers have turned more and more to the use of expensive inputs to generate higher yields. There was a fivefold increase in the use of chemicals in the 1960s and 1970s, a doubling in the area of land sprayed with pesticides, and a near doubling in the real value of investment in farm machinery. The rising costs of these inputs in the 1970s prompted an extension of both federal and provincial income maintenance programs for farmers. Intervention in the marketing of eggs, poultry and dairy produce meant that the producers were able to have more control over the price of commodities.

In effect, what has taken place has been a rapid process of agricultural industrialization, with high capital investment and more sophisticated methods of business management. The average size of farms has more than doubled since 1951, and despite government measures to protect them, many smaller, less efficient farmers have been driven out of business. In order to remain in farming many have resorted to off-farm work to supplement their incomes; about 40 percent of Canadian farmers are now engaged in paid employment off their own farms for an average of 171 days a year.

Urban sprawl
The loss of agricultural land to urban development is a serious concern in a country where prime agricultural lands are limited. In many cases this conflict is inevitable, since urban centers in the past were established to serve good agricultural areas, and they have expanded as farming and its related industries have grown. A highly efficient monitoring

Reaching the world market The construction of railroads to connect the prairies with eastern ports was vital for the wheat economy. Grain destined for Europe was stored in vast grain elevators beside the tracks. Many are no longer in use following the reorganization of the railroads.

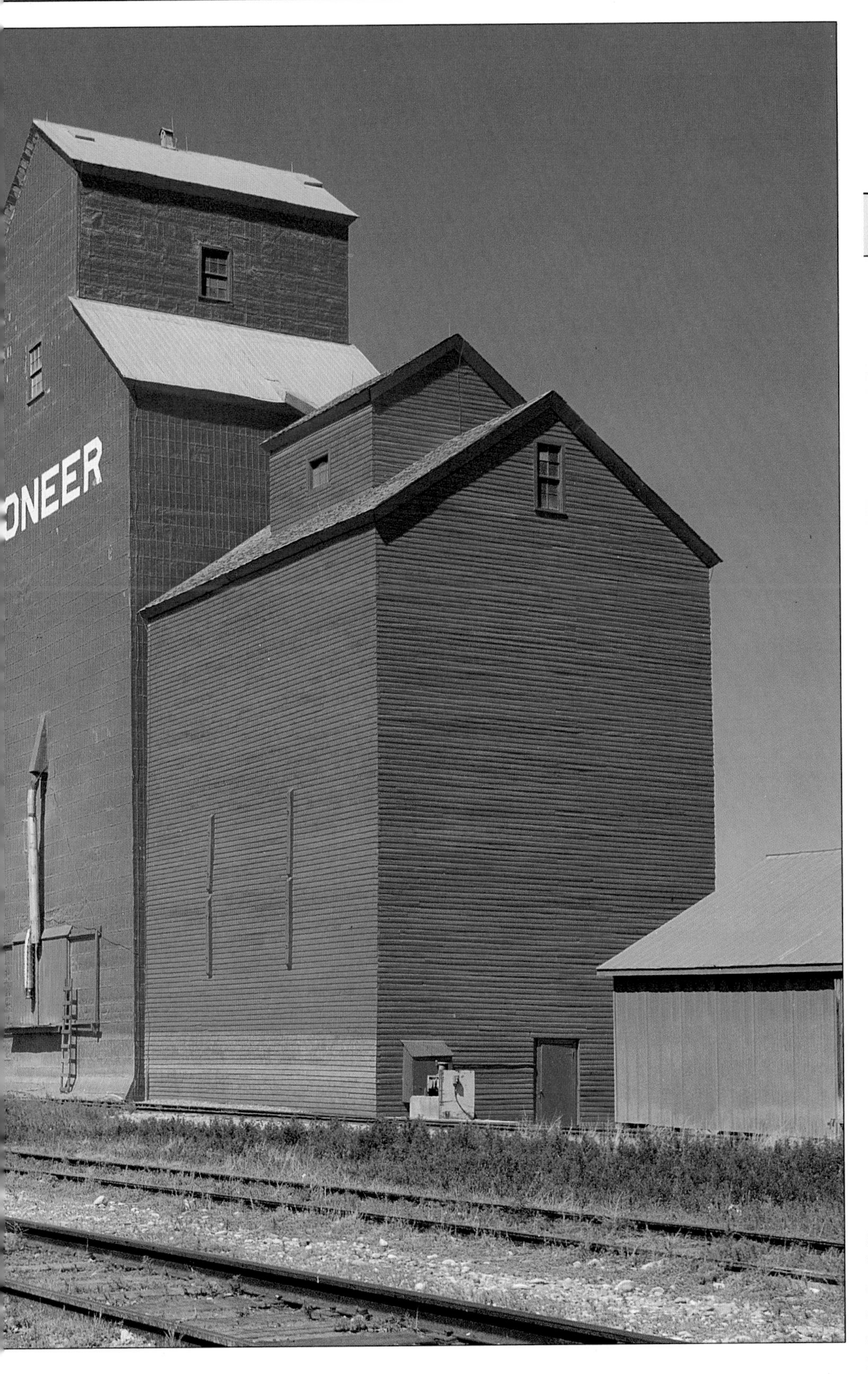

FARMING CHANGE IN THE PRAIRIES

When the prairies were opened up to agriculture in the 19th century, the surveys used by the pioneers to allocate land divided the territory into 259 ha (640 acre) farm "sections". Usually the land given out to individual farmers measured one-quarter of the size of these original sections, or 65 ha (160 acres). Rationalization of farming in recent years has seen the disappearance of many of these small, family-owned farms as land is amalgamated into larger, more efficient units that can be farmed using huge machines rather than human and animal labor. At the same time, the reorganization of the railroad network led to the closure of more than 2,500 grain elevators between 1953 and 1983. These huge structures, left empty and abandoned in the flat, bare prairie landscape, symbolize what many regard as the death of oldstyle farming in the prairies.

Many people have had to abandon farming to find employment in the towns and cities, and in some cases whole farming communities have been broken up. Those that remain have been forced to change their traditional way of life: scores of small local schools have been closed down, along with other rural services. This movement away from the country is typical of a dramatic decline in rural populations throughout Canada. Between 1951 and 1990 – a period when the total population rose from 14 million to 25 million – the number of people employed in agriculture fell from just under 3 million (21 percent of the population) to slightly less than 1 million – a mere 4 percent.

system was established in 1966 to chart the transfer of farmland to nonagricultural uses; in some provinces there has been legislation to protect farmland. British Columbia, for example, has created Agricultural Land Reserves that strictly control the extent of urban expansion in order to protect its valuable fruit growing areas, and similar controls in the Niagara fruit belt in Ontario have helped to maintain production there.

Since monitoring began, losses have been greatest in Ontario. More than a third of the total amount of agricultural land lost to other uses is in this province – more than 17,000 ha (nearly 50,000 acres) of prime land was surrendered to urban use in the early 1980s. A triangular area that extends northward from Toronto and contains most of the province's best farmland is now threatened by the rapidly encroaching city. By contrast, in the northern prairies the agricultural frontier has expanded in recent years.

In response to the continually expanding world market for cattle fodder, the federal government has supported long-term research on plant breeding to adapt certain crops – most notably wheat and oilseed rape – to withstand Canadian climatic conditions. It was the development of the Marquis variety of wheat in the 19th century and early in the 20th that helped push the frontier of cultivation on the prairies northward. More recently new varieties have been bred that are able to reach maturity even more rapidly, and this has resulted in the expansion of grain production in areas like the Peace river area of Alberta. In particular, a completely new cereal crop, triticale, has been developed by crossbreeding wheat and rye. Triticale is very high in protein, and is grown as a food grain for humans as well as for livestock feed.

Forestry and the logging industry

Forests cover well over a third of Canada's land area; they account for 7 percent of the world's forests. The exploitation of this vast natural resource has played a critical part in the country's development, and today forestry generates jobs for over three-quarters of a million workers. A substantial export earner, it makes a greater contribution to the balance of trade than any other commodity group.

Three-quarters of Canada's forests form part of the cold climate boreal forest that extends across the high latitudes of the northern hemisphere into Alaska, Siberia and Scandinavia. The main species of trees are white and black spruce, but other conifers such as balsam fir, jack pine and tamarack are also widely distributed. Both inaccessibility and the unfavorable climatic conditions of large parts of the forest, especially in the far north, means they are not economically productive; it is in the eastern part, in a belt stretching from Manitoba east to Newfoundland, that Canada's principal area of pulpwood production lies. In parts of Ontario and Nova Scotia the conifers give way to deciduous forest. These were once much more extensive; they are all that survive of the forests cleared by the original European settlers. West of the Rocky Mountains along the Pacific coast there is temperate, mainly coniferous, forest. It is more productive than the boreal forest.

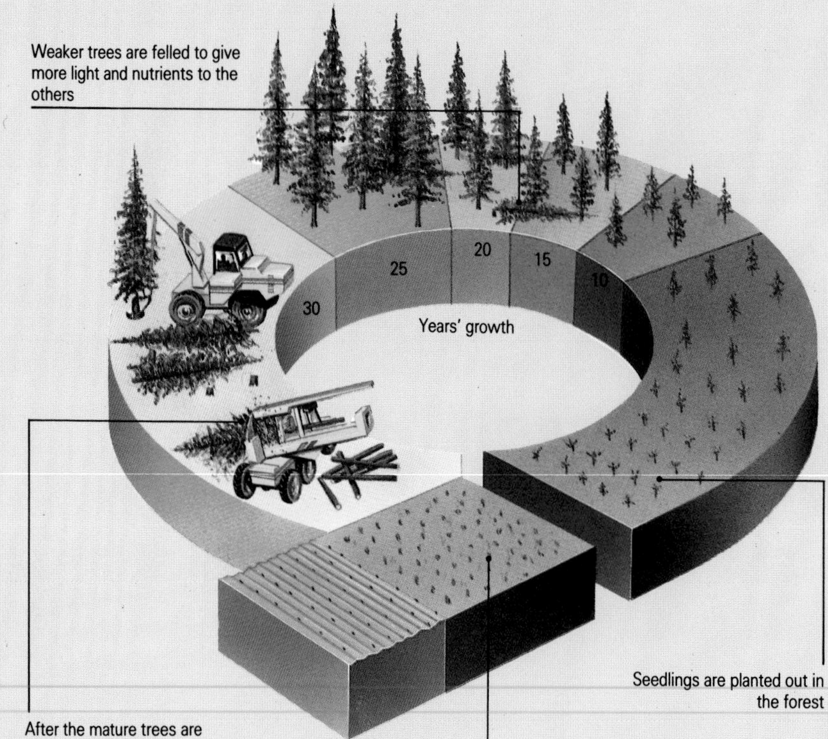

Forests for the future (*above*) Forests – unlike fossil fuels – are a renewable part of the Earth's resources. Responsible forestry can fulfill both commercial and environmental aims.

A plentiful resource

The forests yield a wide range of products. From the logging industry come logs, plywood, pulp chips and poles; from sawmills, a variety of boards and milled wood; from the paper industry, newsprint, pulp, paper and tissue. In addition

A valuable cargo (*below*) Timber forms a significant part of Canada's export trade. Logs are transported from the forest to mills that convert them into a range of products from lumber to plywood and paper.

there are specialist ventures such as the supply of Christmas trees and the manufacture of maple syrup.

From the earliest days of the colonial period, the forests of eastern Canada became the major supplier of timber to Britain. They still provide a range of forest products for export, but since 1917 British Columbia has been the leading export producer in the country. The favorable climate, with cool winters and warm, moist summers, contributes to the forests' high productivity, especially of western hemlock, western red cedar, sitka spruce and Douglas fir. In the first half of this century too much forest was felled for any natural regeneration to take place, but efforts are now being made to manage the forests as a renewable resource.

Rivers were the only way of moving the logs out of the forests in pioneer days, and water transportation is still used in British Columbia to carry logs from isolated islands or coastal inlets. Logs are taken by truck to the rivers and then formed into rafts of log bundles for transportation to the mills. Bundling the logs reduces losses from sinkage and prevents logs from breaking away.

British Columbia's forest industry supplies two-thirds of Canada's softwood lumber and nearly half its woodpulp, which it exports in great quantity to the United States. The unrestricted market that was created by the Free Trade Agreement concluded between Canada and the United States in 1988 was consequently greeted with enthusiasm by British Columbia's timber exporters, though other interest groups are concerned about the industry's encroachments in one of the world's last remaining areas of temperate rainforest.

Commercial forestry operations are dominated by private companies, 40 percent of which are foreign owned. In recent years the trend has been for smallscale saw mills or pulp and paper mills to be replaced by large complexes that produce a range of products under single corporate control. Although the number of individual processing plants has fallen, both output and the influence of foreign investment have increased.

River of wood The transportation of logs by water was the standard method in pioneer days, and is still widely used in some areas. Water transportation is cheap, and allows huge quantities of wood to be moved at one time. It is still the only way of moving timber from inaccessible areas.

THE FOOD GIANT OF THE WORLD

THE FIRST EUROPEAN SETTLERS · THE EXPANSION OF FARMING · FARMING IN THE SOUTH
FARMING IN THE NORTH · BIG BUCKS, BIG BUSINESS · FARMING AND GOVERNMENT

The United States plays a special role within the world's agricultural systems. Not only does it possess a tremendous diversity of farm types and produce – as befits the world's fourth largest country by land area – but it is also the leading producer of beef, maize, soybeans and citrus fruits, and is in addition a major producer of dairy and poultry products, pork, cotton and tobacco. It dominates the world's export markets in cereals, particularly in wheat, holding most of the world's grain reserves. This success came through the opening up of the Great Plains west of the Appalachian Mountains to agriculture in the second half of the 19th century. The United States' rise to world trade preeminence since the 1950s has contributed in no small part to its strategic political role as a world superpower.

COUNTRIES IN THE REGION

United States of America

Land (million hectares)

Total	Agricultural	Arable	Forest/woodland
917 (100%)	431 (47%)	188 (20%)	265 (29%)

Farmers

3.1 million employed in agriculture (3% of work force)
61 hectares of arable land per person employed in agriculture

Major crops

Numbers in brackets are percentages of world average yield and total world production

	Area mill ha	Yield 100kg/ha	Production mill tonnes	Change since 1963
Maize	24.0	75.0 (206)	179.6 (39)	+88%
Soybeans	23.1	22.7 (125)	52.3 (52)	+168%
Wheat	22.6	25.3 (109)	57.4 (11)	+74%
Sorghum	4.3	43.8 (297)	18.8 (30)	+35%
Barley	4.1	28.3 (122)	11.5 (6)	+33%
Cotton lint	4.1	7.9 (144)	3.2 (19)	−1%
Vegetables	—	—	28.0 (7)	+46%
Fruit	—	—	25.7 (8)	+50%

Major livestock

	Number mill	Production mill tonnes	Change since 1963
Cattle	102.0 (8)	—	−2%
Pigs	51.2 (6)	—	−11%
Milk	—	64.7 (14)	+13%
Fish catch	—	5.7 (6)	—

Food security (cereal exports minus imports)

mill tonnes	% domestic production	% world trade
+74.2	24	34

THE FIRST EUROPEAN SETTLERS

The diversity of farming in the United States reflects the country's range of physical environments. The wide array of climates ranges from subtropical humidity in the south, and a milder Mediterranean climate, with warm summers and winter rainfall, on the southwest coast, to more temperate climates farther north, characterized by marked seasonal extremes in the continental interior. This means that all but tropical crops can be grown. East of the Rocky Mountains, and in their rainshadow, cattle ranching is carried out in the drier parts of the southwest and west; irrigation has increased the amount of cultivable land here. Agriculture in the center and north is dominated by grain, with mixed farming in the northeast.

As a result of these environmental differences, and of varying market demand, different parts of the country have developed specializations. Dairying predominates in Wisconsin in the central north, for example, and maize (corn) cultivation in Iowa in the center. Beef cattle are reared in Texas in the central south, cotton grown in Georgia in the southeast, and fruit in California on the southwest coast and in Florida, on the other side of the continent.

The extinction of a way of life

Virtually nothing remains of the agricultural pattern that existed before European colonization of North America in the 17th century. The indigenous Amerindians subsisted primarily through hunting and fishing, though several eastern tribes grew maize, beans, melons and tobacco.

Some of their crops and cultivation practices were adopted by early European settlers along the Atlantic seaboard. They quickly displaced the Indians, forcing them into more inhospitable territory in the country's interior, where hunting rather than farming was dominant. Many

The Old World and the New Immigrant groups from many different parts of Europe have left their distinguishing marks on the American farming landscape. This farm in Pennsylvania belongs to the austere Amish religious sect, of German origin, who still use traditional methods to work the land.

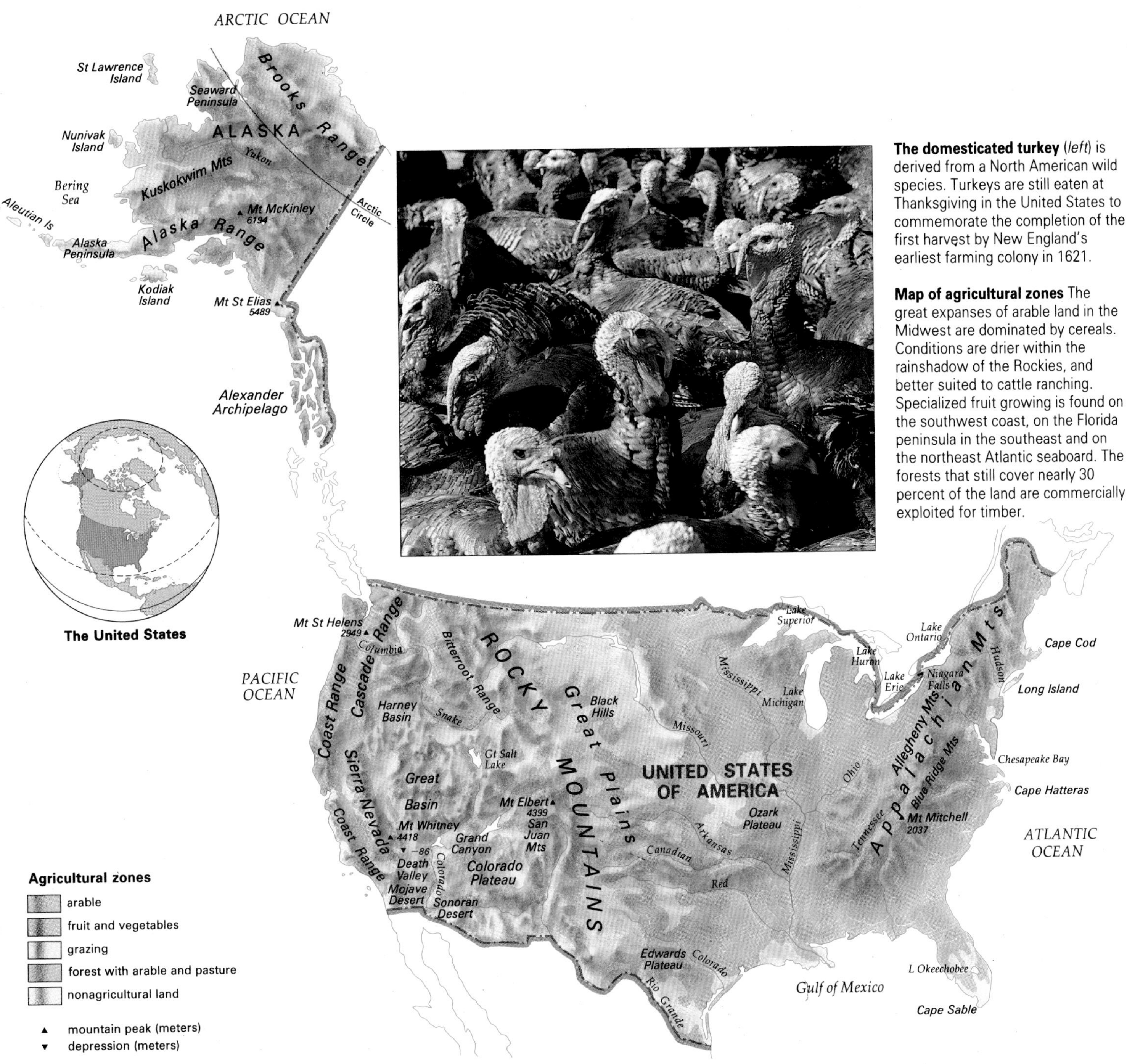

The domesticated turkey (*left*) is derived from a North American wild species. Turkeys are still eaten at Thanksgiving in the United States to commemorate the completion of the first harvest by New England's earliest farming colony in 1621.

Map of agricultural zones The great expanses of arable land in the Midwest are dominated by cereals. Conditions are drier within the rainshadow of the Rockies, and better suited to cattle ranching. Specialized fruit growing is found on the southwest coast, on the Florida peninsula in the southeast and on the northeast Atlantic seaboard. The forests that still cover nearly 30 percent of the land are commercially exploited for timber.

Indians now live on reservations in the southwest and center, but are largely unable to sustain their traditional way of life as the reservations are too small or lack wild game, or the land is too marginal to support them.

The European settlers who took over the land and farmed it introduced new crops, animals, implements and farming methods as well as alien concepts of property rights and of individual land ownership. These concepts were critical in shaping the nature of rural settlement in the new colonies, and played a central role in molding the country's economic and social institutions. Even in the earliest colonial settlements, ownership of land was the key to social status.

New colonists established individual farmsteads on "virgin" land on the edge of the first European settlements, and this process was carried further into the interior by successive waves of immigrants. Initially, food for the expanding urban settlements was grown by small independent farmers. However, even by the early 18th century a group of larger landowners had emerged, seeking to own more land as a means of gaining more wealth. Subsequently, American agricultural production has been dominated by large, well-capitalized farms benefiting from economies of scale.

In the southern states, landowners grew rich on farming tobacco for export, using the plantation system of production in which slaves, shipped to the American colonies from Africa, provided the labor force. By the end of the 18th century cotton began to rival tobacco as the main export crop grown on the plantations of the south, also using slave labor.

To limit private land acquisition, large tracts of land were declared "public", but this policy was reversed after the United States gained its independence from Britain in 1776. The new government encouraged disposal of these public lands to facilitate further settlement, and it was the large landowners with sufficient capital to purchase sizable blocks of land who gained most from the change. Some landowners subsequently obtained income by splitting their land into smaller units, which they rented to tenant farmers. This process was accelerated by the breakup of many southern plantations after the Civil War; as a result, by the 1880s a quarter of all farmers in the United States were tenants.

THE EXPANSION OF AGRICULTURE

In 1803 the United States government purchased a vast tract of land west of the Mississippi (the Louisiana Purchase). The interior of the continent, across as far as the Rocky Mountains, was consequently opened up to colonization. As settlers moved in to these areas squatting on public land became increasingly common. To regulate this process the Homestead Act was passed in 1862, granting rights to any citizen to claim up to 65 ha (160 acres) of public land if he or she agreed to settle and farm the land for a minimum of five years.

The chief effect of this measure was to promote rapid colonization of the western interior during the next 30 years. Small family farms and dispersed settlements came into conflict with the capital-rich large landowners over the "open ranges". The competition was fiercest on the Great Central Plain in South Dakota, Nebraska and west into Colorado, where vicious range wars were fought between large ranchers and small "homesteaders".

As ranchers had no legal claim to the public open range they were compelled to buy land or illegally fence the range. Some of them purchased land that contained the only available watercourses, so prompting further conflict with homesteaders. The patenting of barbed wire in 1873 provided cheap fencing to enclose rangeland, accelerating the end of the public open range. Even so, ranchers in the Rocky Mountain states could still graze cattle on the summer pastures of the public domain, free of charge, until well into this century.

The westward movement of the agricultural frontier brought settlers into new environments, often at considerable distances from the main urban markets and ports. This promoted new forms of farming activity with different mixtures of crops and livestock than those used by farmers on the Atlantic seaboard. These still followed the traditional small-scale, labor-intensive systems brought to the country by early European settlers, or belonged to the plantation economy of the south. Gradually, however, farmers in the northeast, particularly in New England, changed from mixed farming to dairying and to the more intensive production of vegetable crops.

The growth of major industrial towns in the north and east encouraged the highly profitable production of wheat, maize and meat in the rich agricultural land of the Midwest, around and west of the Great Lakes. The vast cattle ranges in drier areas to the west were sometimes

Wheat-growing country (*right*) in the Palouse area of eastern Washington, west of the Rocky Mountains. The agricultural frontier was continually being pushed westward throughout the 19th century, and settlers moving into fresh areas developed new techniques to adapt to unfamiliar farming environments.

The cattle kingdom (*below*) Ranchers had their heyday on the open range of the southwest in the years before 1862, when the Homestead Act encouraged farmers to move in. This caused competition over control of the grasslands and water supplies. Ranching eventually settled in drier western areas.

extended over thousands of hectares. In the southwest, pioneer farmers introduced wheat and sheep into California; these were later replaced by the intensive cultivation of fruit and vegetables as the population grew.

Forests and fish

West of the Rockies the enormous forests of the mild and moist northwest coast proved a valuable economic resource: today nearly three-quarters of Washington State is under forest and almost half of Oregon. The United States is now the leading producer of industrial wood, providing a quarter of the world's total. About three-quarters of this comes from coniferous trees. Nevertheless, the United States is still a net wood importer because of its very large domestic market. To meet this high demand, during the 1980s there was a steady rise in output from forests in the southern states, stretching across into eastern Texas. Nearly half the timber harvested now comes from this area.

There is a sharp conflict between the need to exploit available timber for commercial development and the necessity of renewing forest resources. Nearly 2 million ha (5 million acres) are reforested every year; nevertheless, in forests controlled by the federal government tree cutting is at an all-time high. Ancient stands of forest in the Pacific northwest are under particular threat. However, nearly 20 percent of all forest land (a total of some 66 million ha/163 million acres) has been set aside for protection by the US Forest Service, and a further 32 million ha (79 million acres) are protected in national parks.

Fishing has also been of considerable value in the northwest. In the late 19th and early 20th centuries salmon fishing from inshore waters was very important, and stimulated the development of a substantial freezing and canning industry; their products became major exports. Overfishing, however, eventually led to a greater focus on deep-sea fishing in the Pacific, concentrating on different fish species, notably herring and halibut.

In the east, fishing played a major role in the development of New England. Large stocks of cod and other white fish are to be found in the offshore waters of the Atlantic. This area now accounts for 15 percent by weight of the country's total catch, and its contribution is surpassed by only the Pacific fishery.

DEFORESTATION IN THE SOUTH

The belt of forests extending across the southern states remained virtually untouched until the second half of the 19th century. The removal of land ownership restrictions in 1876 led to the rapacious acquisition of cheap virgin forest by lumbermen and speculators. Hundreds of saw mills sprang up, and the large-scale felling of timber resulted in the exposure of fragile soils, and widespread erosion on unprotected slopes.

This degradation of the landscape received scant attention until the 1930s, when the Civilian Conservation Corps and the Tennessee Valley Authority (TVA) were established. The TVA was given broad powers over land within the Tennessee river system, extending across seven states. Its main roles were flood control, electricity generation, maintenance of navigation channels and resource conservation, including timber reserves. After 1945 these activities were supplemented by those of the wood-using industries themselves, which were concerned not just to exploit the forests, but to maintain them.

This new attitude prompted widespread reforestation, giving it the name of the "Greening of the South". Between 1933 and 1985 the TVA's reforestation programs alone added more than 1.5 million ha (3.7 million acres) of forested land. However, increased deforestation by some private corporations of woodlands not under government control reflected the continued high demand for timber during the 1980s.

FARMING IN THE SOUTH

European colonization had far-reaching effects on the way in which the southern states were settled. For example, the plantation system, established to grow both tobacco and cotton, made use of black slaves, whose descendants represent 12 percent of the United States' population.

Between 1500 and 1870 more than 10 million Africans were transported to the Americas, though only one-twentieth of them went to the United States. However, the United States was one of the few countries where slave numbers grew by natural increase, and this expansion helped to maintain the plantation system.

This system was European owned, financed and managed, but relied on heavy inputs of slave labor for sowing, weeding, harvesting and processing the crop. One cash crop was grown per plantation, but often in rotation with woodland to preserve soil fertility. Plantations near the coastal swamps specialized in sugar and rice.

The importance of cotton

In the 18th century tobacco was the main plantation crop of the south, but cotton gradually rose to dominance, encouraged by the growth of both the British and American textile industries and by the invention of the cotton gin in 1793 by Eli Whitney. This mechanically removed the seeds from the fiber, greatly speeding up production. As soil in the southeastern states became depleted by overuse, the plantation system and the use of slave labor spread west of the Mississippi as far as Texas in search of new and better land. By the outbreak of Civil War between the north and the south (1861–65), cotton was "king" in the south.

The dominance of cotton persisted after the Civil War and the abolition of slavery, but the defeat of the Confederate states had far-reaching consequences for agriculture. Many plantations were either subdivided and sold to smallholders or kept intact and operated by sharecroppers. These were often former slaves, who were provided with mules, implements, seeds and fertilizer, and were directed by the plantation owner. A typical sharecropper had 10–25 ha (25–60 acres), with 6–8 ha (15–20 acres) under cotton – the most a family could pick. Maize was the staple grain, usually occupying 4–8 ha (10–20 acres).

About a third of these sharecroppers were black, and they derived very poor returns from this type of farming. Many suffered malnutrition from poor maize-based diets. In the 20th century large numbers of sharecroppers left the land to seek more remunerative work in northern cities. Many smallholdings were amalgamated, and there is now strong polarization between the remaining sharecroppers and the larger farmers.

King cotton (*above*) The southern economy was based on cotton production, and it is still a major export. The geometric efficiency of today's mechanical harvest is a far cry from the early slave plantations.

Harvesting sugar cane (*right*) Sugar cane is a major plantation crop in the humid areas of the lower Mississippi and Gulf coast. This mechanized plantation is near Brownsville in the Rio Grande valley of Texas.

Tobacco leaves drying in the southern sun (*left*) Unlike the other plantation crops of the south, the production of tobacco proved slow to modernize – mechanical harvesters were not introduced until the 1970s.

In 1910 just over 12 million ha (31 million acres) of land in the south was under cotton, but by the 1920s its cultivation was in decline, affected by labor shortages, infestation by the boll weevil (an insect that destroys the cotton boll, which entered the United States from Mexico in the 1890s), and competition from the southwestern states. The imposition of price controls promoted greater diversity, encouraged by the government in an attempt to produce a healthier rural economy and to preserve soil fertility by promoting the cultivation of nitrogen-fixing legumes.

Southern farmers now grow sorghum, groundnuts, soybeans and maize as well as cotton and tobacco. Arable land itself has increasingly given way to forestry in areas where cotton was once supreme. Cotton from the south and southwest of the United States, however, still comprises nearly a third of the world's total exports of this crop.

Ranching – as in the movies

As a result of the rise of cotton, ranching – one of the earliest forms of farming in the south – was gradually pushed into the drier areas of the southwest, such as New Mexico, Arizona and Colorado. There it took on some of the characteristics of cattle rearing used by the Spanish in Mexico. These included open grazing in areas of little rainfall and poor pasture, mass roundups to collect the animals, cattle branding to identify ownership, and long drives following the cattle trails to distant markets and the large meat-packing centers. Of all the farming systems in the United States, ranching is perhaps the one most associated with that country, immortalized in Hollywood's Western movies.

From the 1880s cattle were moved by railroad rather than along the trail drives, and grain and cotton cultivation was introduced into the ranges, using new dry-farming practices that left land fallow for periods to conserve moisture in the soil. These techniques, however, proved inadequate during prolonged droughts such as that of the 1930s, when the failure of crops led to numerous bankruptcies and evictions.

Irrigation has subsequently improved pastures, permitting the more intensive rearing of beef cattle in the southwest. It has also enabled vegetables to be grown here, as well as grain and cotton. Ranching is still practiced in the unirrigated parts of the southwest.

CITRUS SUCCESS

Farming in the subtropical coastlands of the southern states, from Texas through to Florida, has recently been transformed. A variety of factors have been involved – the ability of farmers to capitalize on local physical advantages, to respond to changing market demands and to invest capital in development, as well as government intervention.

Early Spanish and French settlers had found these lowlying, swampy coastlands inhospitable. Agriculture did not become established here until the beginning of the 19th century, when slave labor was used to work sugar cane and rice plantations. But poor drainage and foreign competition always limited production, it was only with large-scale federal and local funding for drainage, after 1945, that the full potential of the rich, black soils of the coastal swamps began to be exploited – and with spectacular success – by the citrus fruit industry.

Oranges are unharmed by occasional light frosts in winter, when they are semidormant, but early frosts in the late fall, before they have finished their annual growth, will damage them. They are therefore well suited by the generally frost-free winters experienced in Texas, Louisiana and Florida, where they are principally grown. By focusing on the production of frozen juice concentrate, which now accounts for nearly 40 percent of the United States crop, growers have helped to create a new market, transforming relatively limited sales of fresh oranges into a multimillion dollar industry.

Growing appreciation of the dietary value of Vitamin C, which is abundant in oranges, helped to increase consumption. The result is a product that can be marketed year-round. As with any other types of farming, production and processing have been integrated, and large-scale operations are the norm.

Truck farming

When California, in the southwest, was annexed by the United States in 1848, the Spanish had been settled there for less than a century. Within thirty years of its becoming part of the United States it had more than 1 million ha (2.5 million acres) under wheat and the largest sheep population in the country. However, much of the wheat was grown in fairly dry conditions, which resulted in rapid losses of soil fertility and reduced yields.

The new settlers who poured into the state from east of the Rockies in the 1870s and 1880s brought with them new farming methods – among them, most importantly, irrigation techniques. This transformed Californian agriculture, greatly increasing productivity, most notably in the central valley formed by the Sacramento and San Joaquin rivers.

Just as importantly, the new settlers brought the railroads. By 1876 California had a rail link through to the markets on the east coast of America. To encourage freight shipments the railroad companies offered cheap rates, and after 1887, when refrigerated trains were introduced, intensive fruit and vegetable growing for distant urban centers quickly began to replace more unreliable wheat and sheep production in the state. Later on farmers were quick to use motorized transport to carry perishable goods to nearby urban centers or to rail shipment points. Thus the term "truck farming" was coined for California's intensive fruit and vegetable production.

Developing a competitive edge

Despite all these advantages, Californian farmers could only be competitive in the eastern markets by developing special organizational methods. A number of these proved crucial in turning California into the country's leading producer of fruit and vegetables.

Cooperatives were formed by small growers to ensure high standards of produce. Processing was undertaken in order to reduce the weight of produce transported as well as to add value, and canning and the production of fruit concentrates were introduced. Heavy investment was made in mechanizing cultivation but not in harvesting methods; labor costs have always been kept low by using cheap migrant Mexican workers for picking and packing produce on the farm.

California's mild Mediterranean-type climate means that it has a longer growing season than many other areas that are closer to the main markets in the northeast. This climatic advantage has been maximized (as it has in Florida, in the southeast, where a similar truck farming industry has grown up) by specializing in early and late varieties of crops, using irrigation, fertilizers, pesticides, selective crop breeding, and greenhouses or plastic coverings. Thus the growers are increasingly able to meet the constant demand for fresh produce.

Many farms, and even whole districts, now concentrate on one crop: raisin production around Fresno, for example; table grapes around Sacramento; wine production in the coastal valleys near San Francisco; citrus fruit in the Los Angeles basin and the San Joaquin valley; and lettuce and other salad vegetables in the Salinas valley. Irrigation has also enabled intensive cultivation of field crops such as cotton, sugar beet and rice in the San Joaquin valley. Indeed cotton has now become the single most valuable crop produced in California.

The term truck farming is also applied to milk production in California, where drylot dairying is practiced in the Los Angeles area. Through lack of space, some 250,000 dairy cattle are kept in small "farmyards" and are fed in stalls on manufactured foodstuffs. The milk and dairy products are sent by refrigerated bulk carriers to centers throughout the United States.

Fresh produce for the cities *(left)* Lettuces being trimmed, wrapped and packed on the field for immediate shipment. Truck farming developed after the introduction of irrigation, particularly in the central Sacramento–San Joaquin valley, allowed Californian farmers to take advantage of their fine climate for intensive fruit and vegetable production. Speed is of the essence in transporting the crops to their urban destinations. Although cultivation methods are sophisticated, many farmers rely on cheap immigrant labor for harvesting and packing.

High-technology farming *(right)* Hydroponic methods of cultivation, in which plants are grown without soil in specially prepared nutrient solution, were first developed in California in the 1930s. Tomatoes and a salad crop are being grown in the highly controlled environment of this advanced greenhouse. Production methods have been improved and refined over the years to result in this type of capital-intensive, high-value farming. Yields have also been improved through scientific advances in plant breeding.

FARMING IN THE NORTH

The first Europeans who settled on the Atlantic coast of North America were compelled to be farmers. All the food needs of the precarious young settlements had to be met from what they could themselves produce, supplemented by what little they could trade with the indigenous peoples. This established a tradition of mixed farming that has long continued in the northeast, though the ready market provided by the growth of major northern cities later encouraged greater agricultural specialization.

Before the building of the railroads, milk could only be produced very close to the cities, as it was too perishable to transport far. Improved refrigeration and expansion of the railroads, however, widened the area of fresh milk production into districts that previously had been able to produce only less perishable cheese and butter. Two main dairying regions have subsequently emerged in the north of the country: one in the east, centered on New England, New York state, New Jersey and southeast Pennsylvania; the other farther west, around the upper Mississippi valley, in Wisconsin, Minnesota and part of Illinois and Iowa. In the former area production is almost entirely of fresh milk, while in the latter most of the fresh milk is destined for the local Chicago market, and butter and cheese are the main products for wider markets. Wisconsin is the most important state for dairy products. Climatic conditions here give good summer pastures, though they require the cattle to be housed indoors in winter.

Largescale grain production

The growth of urban populations, which assisted the rise of dairy specialization, also encouraged largescale grain production in the new lands of the country's vast interior. This type of arable farming proved to be of lasting significance to the United States' economy. Not only were ready markets to be found within the country, but surpluses were successfully exported to Western Europe, helped by reductions in oceanic freight rates.

Land was cheap, the virgin soils of much of the interior were fertile, and the rapid mechanization of wheat farming led to rapidly expanding production, from the 1860s, on the Great Plains. Only large

Huge silos (*above*) containing fodder for the winter feeding of cattle dominate a dairy farm in Wisconsin. The state has become the country's major producer both of milk and of cheese products.

Grain-growing areas (*right*) There are four major areas where grain is produced in the United States. Three of these concentrate on wheat and the other – the so-called Corn Belt – on maize.

Major grain-growing areas
- spring wheat
- winter wheat
- maize

farms were economic in the dry interior, where annual rainfall was less than 500 mm (20 in), and farm amalgamation often occurred soon after initial settlement.

Largely as a result of the expansion of cereal production, the area under cropland in the United States more than doubled between 1860 and 1910. Thus, like ranching, largescale grain production was a product of the great economic and technological changes of the 19th century. Output was later increased even further by inventions such as automatic reapers and binders, combine harvesters, seed drills and more efficient plows. New crop varieties were also developed and further helped to improve yields.

The United States today has four major areas of largescale grain production: the Corn Belt producing maize (corn) in the center of the country, in Iowa and Nebraska; a spring wheat area in the central north, in the Dakotas and eastern Montana; a smaller spring wheat area farther west, in part of Washington; and a winter wheat area south of the Corn Belt, centered in Kansas. The three wheat areas account for three-quarters of American output for this crop.

One important physical division in

CHANGES IN NEW ENGLAND

New England's farmers, particularly in the mountains, have to contend with three main physical limitations to agriculture: the growing season, which may be as short as 90 days in some upland areas, so that maize cannot be grown; accessibility, which is limited by heavy winter snowfall – more than 250 mm (100 in) a year in the mountains; and soils that are thin, stony and susceptible to leaching.

Faced with growing competition from farmers in the west who were sending produce in increasing quantity to the northern cities, many New England farmers gave up the struggle to maintain their farms against these physical odds. Just over half the region was under farmland in the early 1900s; some 60 years later the proportion had fallen to under a fifth, with nearly two-thirds of the land on farms being turned over to woodland.

Much full-time farming has been reduced to part-time farming in recent years, and many farmsteads have become summer houses for residents of nearby cities such as New York and Boston. Such changes are part of a growing trend throughout the United States toward part-time, or "suitcase", farming. Full-time farming still persists on the better land, where dairying predominates; as much as 90 percent of commercial farmers in Vermont are involved in dairying. Poultry, fruit and vegetables are also grown throughout New England for urban markets.

Profitable fruit specialization in New England. Cranberries – a native plant – are being processed for manufacture into juice or sauces. Formerly limited to the North American market, export demand for these vitamin-rich products is rising.

these wheat belts is between those areas receiving more than 640 mm (25 in) of annual rainfall and those areas – to the west and northwest, in the eastern foothills of the Rockies – receiving less. These drier areas need careful management of soil moisture to produce crops, while the wetter areas have fewer problems with droughts and soil loss.

In the 1930s prolonged droughts caused great problems for farmers in the drier areas of the Great Plains. As the dusty, dry topsoil blew away in huge quantities, the most seriously affected areas in northern Texas, Oklahoma, Kansas and Nebraska earned the nickname of the "Dust Bowl". Since that time production of arable crops has been discouraged in the very dry areas, and cattle grazing has been reestablished there.

Elsewhere wheat has persisted as the dominant crop, though some important changes have occurred since 1945. Better cultivation practices and new varieties of crop species have led to huge increases in yields, resulting in surpluses and government grants to farmers to cut output. In the more humid parts of the winter wheat belt, federal government aid has supported crop rotation and cattle rearing in an attempt to reduce the monoculture of wheat in these areas.

BIG BUCKS, BIG BUSINESS

The number of people employed per area farmed has always marked one of the crucial differences between farming methods in the United States and those in the countries from which the first European settlers came. From the beginning of colonization, land in North America was abundant and labor in short supply. Landless laborers were uncommon, because they could always join the pioneer trail westward and claim land for themselves.

Family labor provided the workforce in most American farms. There were some exceptions, however. Black slaves were essential on southern plantations, and their descendants are equally important on the large cotton-producing estates that succeeded the plantations. Hired hands worked on the cattle ranches during the roundup and accompanied the long drives to the railheads. Migrant laborers, originally drawn from among the unemployed of the northern cities, were employed at harvest time, moving northward through the grain belts as the season progressed. Today large, mechanized contract teams perform the same function. However, 90 percent of work on American farms is still carried out by the farmer and his family.

The lack of available cheap labor was a significant factor in developing agricultural mechanization, particularly in the cereal-growing areas, where tractors and combine harvesters reduced dependence on labor. However, the need for capital to buy machinery favored the larger and wealthier farmers at the expense of the many smallholders.

Economies of scale

Farm growth has characterized American agriculture from pioneer days, particularly in the grain areas. The standard homesteading of 65 ha (160 acres) did not last long, as farms were enlarged through amalgamation. This process has speeded up throughout the 20th century as farming has become increasingly commercialized and industrialized. In 1990 there were 2 million farms, compared with almost 7 million in 1935, and the average farm size had risen from 117 ha (290 acres) to 188 ha (465 acres).

Most of the largest farms – some of them extending to well over 1,000 ha (2,470 acres) – are in the Midwest and Great Plains, from the Canadian border in the north through to west Texas in the south, specializing in cash grains or livestock. They are highly capitalized and mechanized, and are frequently closely linked to major retail outlets, wholesalers, food processors and farm suppliers. There are also a number of large corporations, such as Boeing Aircraft and Dow Chemicals, that have extended their activities into farming.

This industrialization of American agriculture has been accompanied by a steady fall in the farm population. Today only 3 percent of the country's workforce is employed in agriculture, but the number of smallholdings has grown with the rise of part-time farming. Hobby farmers – people without an agricultural background, who have acquired land to farm at weekends, during vacations and after their retirement – are becoming increasingly common.

The transformation of farming into big businesses has not been without problems. Many farmers have fallen into debt,

Conveyor-belt harvest Irrigation and mechanized production methods have substantially increased yields from Hawaii's sugar cane and fruit plantations. Pineapples are being harvested for 24 hours a day on this plantation.

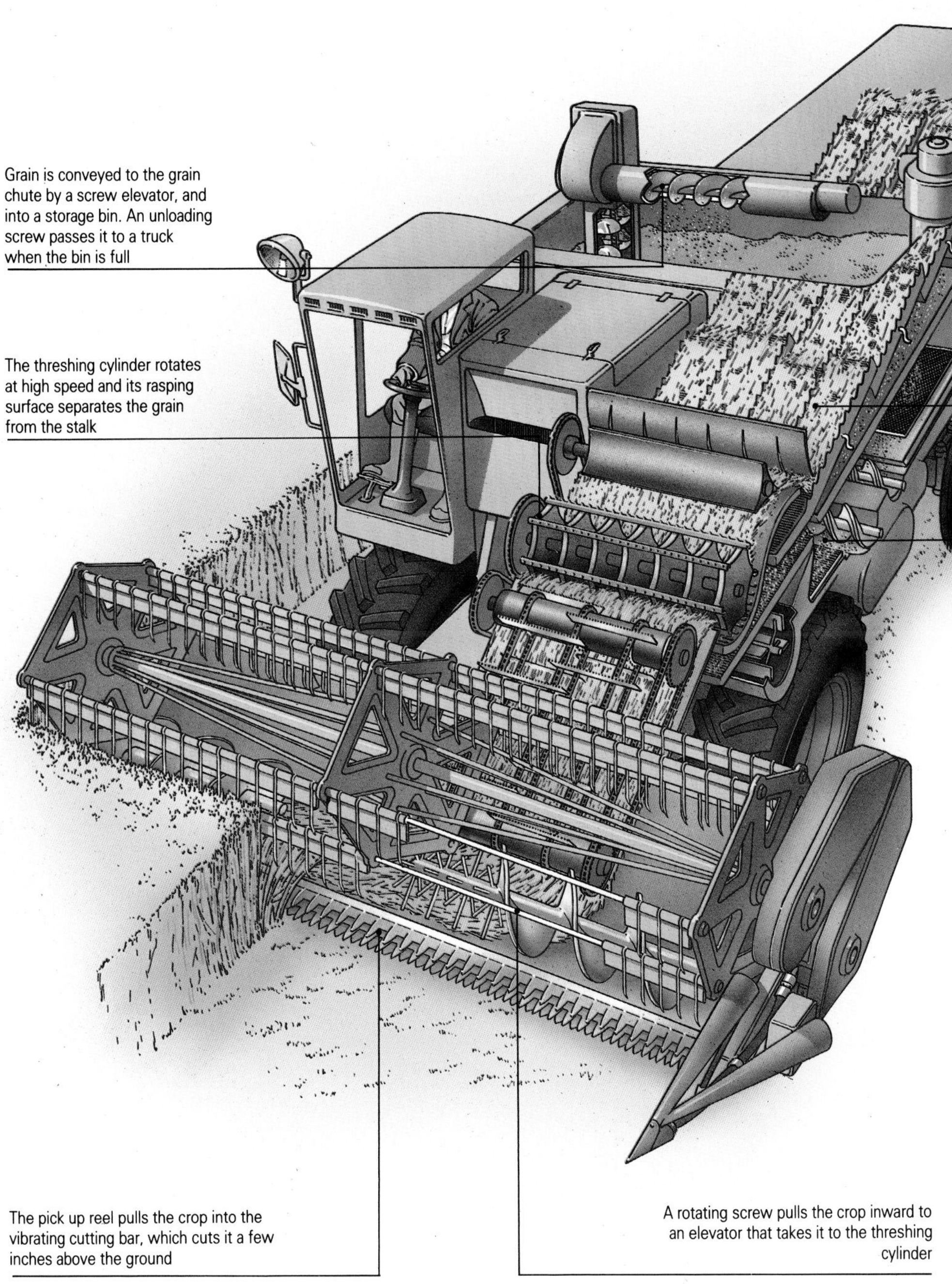

Mechanized power A combine harvester is so called because it combines all the tasks of harvesting a cereal crop that would otherwise have to be carried out by large gangs of hired labor. Gasoline-driven combines were developed in the United States in the 1930s.

increased productivity has at times brought large surpluses of grain, milk and beef, and intensive farming methods have caused environmental damage. Water runoff from fertilized fields, for example, leads to the concentration of nitrates in watercourses. Wildlife habitats are destroyed by the intensive use of land. Huge, sprawling, unsightly farm buildings disfigure the landscape.

Farming where water is scarce

The Dust Bowl of the 1930s provides a classic example of innovative farming techniques bringing environmental disaster in their wake. Early attempts at farming the dry areas of the Great Plains made use of a fallow period, in which soil was formed into a finely powdered "dust mulch" in order to reduce loss of water through evaporation. Unfortunately, the prolonged drought of the 1930s led to this dusty, dry topsoil blowing away in huge quantities, causing widespread and damaging soil erosion, and great human misery. As a result better "dry-farming" techniques were developed, which preserve soil moisture by planting crops with low water requirements, and growing a crop only in alternate years.

This system is still practiced in some places, but the use of irrigation since 1945 has transformed farming in the Dust Bowl and in dry areas to the west. Unfortunately, however, this irrigation is based on "mining" the groundwater that is held in natural reservoirs (aquifers) in permeable rock. These supplies are now known not to be inexhaustible, so additional problems may loom on the horizon for these farmers.

DIVERTING THE RIVERS

The state of Utah is sometimes referred to as the "cradle of American irrigation". It was in the Great Basin containing the Great Salt Lake that, in 1847, the Mormons – a religious sect that migrated in mass to Utah – first directed the waters of City Creek into irrigation channels, and planted out maize, potatoes, beans, buckwheat and turnips to feed their newly established colony. Irrigation has since been extended elsewhere in the state, and federal irrigation schemes now support the production of sugar beet, potatoes, wheat, vegetables, fruit and forage crops such as alfalfa.

To increase the irrigated area, large dams – the Roosevelt Dam on the Salt river, the Hoover Dam on the Colorado river, and the Imperial Dam above the Imperial valley – have all been constructed, transforming the semiarid deserts. The neighboring states of Idaho, to the north, and Arizona, to the south, have also gained substantial blocks of irrigated land. These have enabled new areas to be brought into cultivation as well as bringing security to existing agricultural activity that was very close to the survival margins.

This extension of irrigation has been heavily subsidized by federal government. There has been much flouting of the government's regulations controlling the volume of water allowed per farmer. Cheap water for agriculture, combined with urban water shortages and lower returns from farming, are leading some farmers to sell their water to the cities in California.

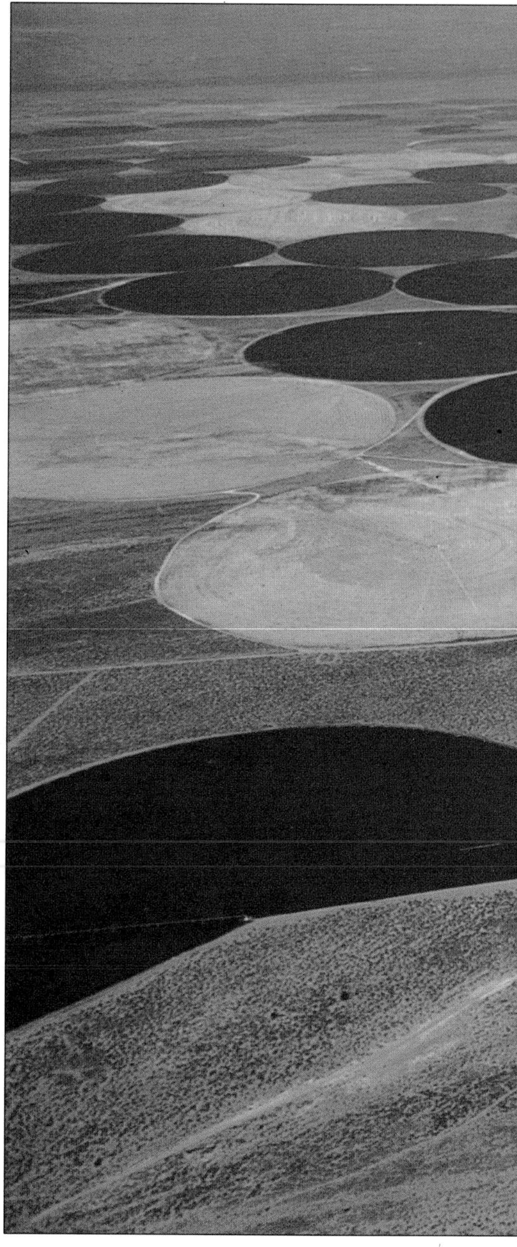

A devastated landscape Dust piles up over what was once good farmland in Colorado in the west. The farm was abandoned in the 1930s after successive years of drought resulted in crop failure and erosion of topsoil. The Dust Bowl spurred the government to help farmers during the Depression.

FARMING AND GOVERNMENT

From the early years of independence, farming has traditionally enjoyed some support from the federal government, but since 1945 its role has become a major influence on agriculture. The change of emphasis was brought about by the need to regulate productivity – the government's earlier role having been confined to one of controlling land dispersal and the use of public lands.

The dramatic rises in productivity and the ability to produce substantial surpluses, especially of grain, reflected tremendous developments in plant and animal breeding, allied to the scientific application of fertilizers, pesticides and herbicides as well as the use of irrigation and mechanization. Many of these major advances in agricultural techniques were developed in state- and federal-supported research centers, and were then exploited by commercial companies.

As a result of such research, fertilizer usage has increased fiftyfold since 1940, often permitting farmers to practice continuous cropping of their main crops instead of rotating different crops to preserve fertility and kill weeds. High-yield crops and rapidly maturing animals have been bred. The development of new varieties of maize, wheat and barley, giving rise to the term Green Revolution, saw yields increase by well over a half since 1960 (as much as 90 percent in some years in the case of maize). Poultry, pig and cattle production has intensified, as exemplified by battery hens, penned pigs (hogs) and "feedlot", or drylot, cattle. Hormone-injected cattle now mature in 11 months, compared with the 18 months that was common during the 1980s.

Mechanization – leading to increased efficiency and reduced labor needs – has had other benefits. In 1940 there were 35 million working farm horses in the United States; these have been replaced by 15 million tractors. This change has released large amounts of arable land previously used to grow feed for these traction animals.

The need for price support

The federal government first sought to regulate domestic production after the collapse of farm prices in the Depression of the 1930s. It introduced price support for agricultural commodities and has continued ever since to act as an insurance for farmers against unfavorable market conditions. Price support operates by paying a farmer a proportion of the difference between the market price and an agreed "parity price", so that, should the market price drop significantly, a farmer is cushioned from the worst effects of the fall.

The goals of this support are to ensure the continued existence of the family farm, to maintain a reasonable level of income for farmers and to secure an adequate food supply at reasonable prices for the consumer. However, as large farms produce more and therefore receive most subsidies, the support system has tended to encourage large commercial farmers, or agribusinesses, rather than small family farmers. The government has, therefore, limited the maximum payment that can be made to an individual farm or farmer.

Price support policies have enabled regions that are disadvantaged for certain types of farming to remain competitive.

THE ULTIMATE ABSURDITY?

The measures adopted by the federal government to stem the rapid rises in cereal production in the 1960s and 1970s included the Payment-In-Kind (PIK) program. Introduced in 1983, it encouraged farmers to take land out of grain production; payments for this action were to be made not in dollars but in grain, which could be sold when the market improved.

More than 1 million farm holdings, containing a third of the country's cropland, participated. The maize harvest fell by 38 percent, the government made a substantial saving on storage costs, and farm incomes actually rose. However, it is debatable whether paying farmers not to produce a certain crop is a rational use of public funds.

In 1985 a new Conservation Reserve Program (CRP) was introduced, setting aside 18.2 million ha (45 million acres) of land for a 10-year period at an average payment to farmers of $119 per hectare ($48 per acre). This may help limit surplus production, but there are certain drawbacks to the CRP: the amount of land available for young farmers, just starting their careers, to rent is reduced; farm incomes may suffer in the long term; and depressed demand may affect the farm supply industry.

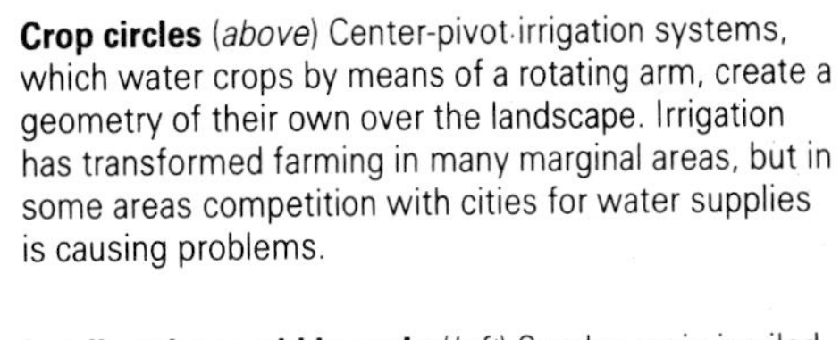

Crop circles (*above*) Center-pivot irrigation systems, which water crops by means of a rotating arm, create a geometry of their own over the landscape. Irrigation has transformed farming in many marginal areas, but in some areas competition with cities for water supplies is causing problems.

Leading the world in grain (*left*) Surplus grain is piled up next to silos that are already full to the brim after a bumper harvest. Government policies encouraged overproduction; the increase in exports brought diminishing returns for American farmers.

Such policies have, however, encouraged surplus production. Some of this surplus has been exported: in the 1950s farm products totaled 10 percent of the country's exports, but by 1990 they were responsible for more than 30 percent – 60 percent of which was wheat. This increased reliance on foreign markets has made farmers more vulnerable to fluctuations in world markets. The rising strength of the US dollar in the 1980s made American agricultural commodities less competitive in world trade, so more produce was directed toward the domestic market, depressing farm prices. At the same time rising farming costs further limited farmers' incomes, and there was a sharp rise in farm sales and bankruptcies, especially among those who had overcapitalized in the days of high prices in the 1970s and early 1980s.

The government, therefore, decided to subsidize exports – by selling grain to the Soviet Union at a reduced price, for example – and to reduce overproduction by introducing quotas. (Food aid programs also in effect subsidized exports.) The quotas program set a limit on the amount of land that could be put down to each crop, and authorized payments to farmers to take land out of cultivation, or to put it to a specified use such as woodland or pasture. Quotas without accompanying regulations on output, however, have simply encouraged farmers to increase productivity on their remaining land; taxation benefits and price supports, too, have often continued to work against production controls.

Corn and beef: an integrated system

One of the most highly organized, best run, most economic and highest-yielding farm areas for its size in the world is the Corn Belt – the agricultural heartland of the United States. Here the fattening of cattle and pigs on locally grown fodder (originally maize – or corn, as it is generally called in the United States – but now often replaced with soybeans) has been turned into a money-making science. The animals are kept in penned-in areas known as feedlots, and their feed brought to them. This highly intensive system contrasts with the type of production farther west, where cattle are grazed on extensive rangelands.

Intensive methods of livestock production were first developed by pioneer homesteaders in the Midwest, south and west of the Great Lakes, who established a three-crop rotation – corn, wheat, hay – which still survives as the basic crop assemblage in Ohio, Indiana and Illinois. Integral to this system was the rearing of pigs, which are efficient converters of cereal surplus and hay or silage into meat, and could easily be managed on the small family farms of 65 ha (160 acres) established under the Homestead Act.

A similar system was followed by farmers opening up the lands to the west. Here corn soon became the key crop, grown for grain and for converting to silage for animal fodder. Farmers quickly adjusted to fluctuating market conditions by converting corn into pork when pork prices were high and selling grain for cash when corn prices were high. The advent of the railroads, which provided vital links with Chicago in the north and with the cities in the eastern United States, also helped to form the basis for the Corn Belt's prosperous rural economy.

Feedlot production

Although pigs and cattle were both reared in the region, pig production was at first the more profitable activity. With the growing demand for beef, and the enlargement of land holdings, cattle came to assume much greater importance. Today there are more than 110 million cattle in the United States – significantly more than any other livestock – some 80 percent of which are beef cattle, reared to satisfy consumer demand for hamburgers, steak and other meat products. The greatest concentrations of beef cattle are to be found in the Corn Belt.

Unlike the pigs, which are bred on farms within the Belt, the young cattle steers are mostly brought in for fattening from the rangelands to the west and in Texas. Contained within the feedlots, which may cover large areas, divided from each other by raised banks or fences, the animals are fed scientifically controlled diets of grain and processed feed,

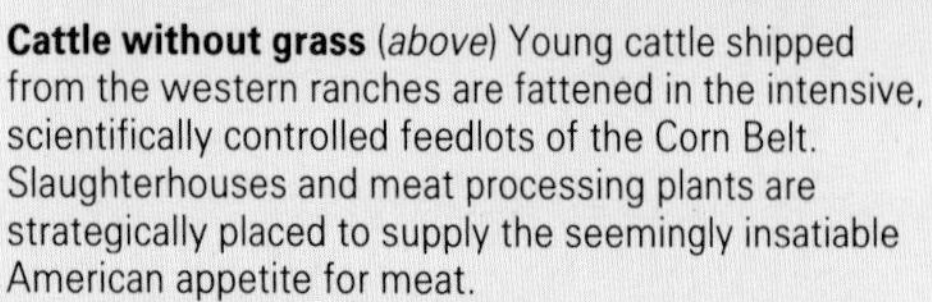

Cattle without grass (*above*) Young cattle shipped from the western ranches are fattened in the intensive, scientifically controlled feedlots of the Corn Belt. Slaughterhouses and meat processing plants are strategically placed to supply the seemingly insatiable American appetite for meat.

Plants into meat (*left*) A farmer waters his field of young corn, destined for use as cattle feed. Although corn remains an important crop, there has been an increase in the amount of soybeans grown, both for human and livestock consumption. Between 1961 and 1981, the area under soybeans more than doubled.

and injected with hormones to boost fattening, before being sent to slaughter houses and meat processors.

Traditionally most livestock was sent live ("on the hoof") by rail to Chicago, Kansas City, Omaha and St Louis, where meat packing plants were located next to the stockyards at the railheads. Since the 1960s processing plants have been established on greenfield sites in the Corn Belt itself, where the availability of cheap land has allowed huge plants to be built.

Corn and beef production are thus inextricably linked throughout the Corn Belt. Four-fifths of Iowa's farm income comes from beef cattle and pigs. However, corn occupies up to 60 percent of the cropland – a lower proportion than 20 years ago, reflecting the growing importance of soybeans as a fodder crop. Improved farming methods since 1945 have greatly increased corn yields, primarily through the use of new, improved crop varieties, more artificial fertilizers and greater mechanization. The number of man-hours required to produce 3,000 liters (660 gallons) of corn fell from 108 in 1940 to 10 by 1990. As a result, surplus production has risen, even after the feedlots have been supplied with corn.

Harvest in the sun

The preservation of food for future use is as old as settled agriculture itself. Crops harvested at the end of one season would be dried in the sun to keep them for eating throughout the rest of the year. This oldest, and simplest, method of food preservation is still cost effective even in the highly mechanized agriculture of the United States.

The movement of crops between the Old and the New Worlds, which followed the discovery of the Americas by the Spanish, worked in two directions. Not only were native American plants such as the potato, tobacco and maize introduced to Europe, but the new settlers in America brought with them many familiar crops that are now widely established there. The apricot was one of these. A native of China, its cultivation had been spread around the Mediterranean through the Middle East. It was introduced to California by the Spanish early in the 18th century, where it flourished in growing conditions similar to those of the Mediterranean.

Although the fruit can be preserved by bottling and canning, drying is still the preferred method. The dried fruit is a very rich source of iron, and can be stored for very many years.

Apricots drying on a fruit farm in Silicon Valley, California, in the United States.

FARMING FOR RICH AND POOR

THE LEGACY OF COLONIALISM · SUBSISTENCE AND SURPLUS · BREAKING WITH THE PAST

Central America and the Caribbean incorporate a range of tropical farming systems based mainly on savanna and rainforest environments. Root crops were probably being grown here 6,000 years ago; and the ancient civilizations of the Aztecs in Mexico and the Maya on the borders of Mexico, Guatemala, Belize and Honduras had cultivated the land for at least 2,000 years before the arrival of the first Spanish colonists at the beginning of the 16th century. Although some indigenous farming has remained largely unchanged, European colonization has had a long lasting influence, particularly through the introduction of plantation farming. There is largescale commercial production of cash crops such as sugar, bananas and coffee, and cattle ranching provides meat for the United States market.

COUNTRIES IN THE REGION

Antigua and Barbuda, Bahamas, Barbados, Belize, Costa Rica, Cuba, Dominica, Dominican Republic, El Salvador, Grenada, Guatemala, Haiti, Honduras, Jamaica, Mexico, Nicaragua, Panama, St Kitts-Nevis, St Lucia, St Vincent and the Grenadines, Trinidad and Tobago

Land (million hectares)

Total	Agricultural	Arable	Forest/woodland
265 (100%)	132 (50%)	33 (13%)	67 (25%)

Farmers

16.8 million employed in agriculture (33% of work force)
2 hectares of arable land per person employed in agriculture

Major crops
Numbers in brackets are percentages of world average yield and total world production

	Area mill ha	Yield 100kg/ha	Production mill tonnes	Change since 1963
Maize	8.8	16.7 (46)	14.6 (3)	+59%
Sugar cane	2.5	600.6 (101)	149.9 (16)	+41%
Dry beans	2.4	5.8 (102)	1.4 (10)	+33%
Sorghum	2.3	29.2 (198)	6.8 (11)	+547%
Bananas	—	—	8.2 (12)	+47%

Major livestock

	Number mill	Production mill tonnes	Change since 1963
Cattle	52.5 (4)	—	+45%
Pigs	26.1 (3)	—	+70%
Sheep/goats	20.6 (1)	—	+12%
Milk	—	11.0 (2)	+166%
Fish catch	—	2.0 (2)	—

Food security (cereal exports minus imports)

mill tonnes	% domestic production	% world trade
−8.9	30	4

THE LEGACY OF COLONIALISM

The traditional communal agriculture of the indigenous peoples of the Central American isthmus was completely transformed by the first Spanish settlers who established largescale production on haciendas – large family-owned estates based on the Spanish system of aristocratic landownership. The hacienda used cheap landless labor to produce crops for the new urban markets that colonization created in Central America. Although it is an inefficient system of production, relying on large amounts of land and labor, it still survives as the basic form of land organization in much of the region.

By contrast, the northern Europeans – chiefly the British and French – who colonized the scattered islands of the Caribbean and a small area of the coastal lowlands of Central America, introduced the plantation as the basic unit of organization; later it was extended into Cuba, Puerto Rico and the Dominican Republic following their break with Spain at the end of the 19th century. The plantation system was based on the largescale production of a single crop, usually sugar cane or bananas, for export, making use of the capital, technology and managerial skills of the colonial power. Like the hacienda, the plantation requires extensive land, but production tends to be more intensive and therefore more efficient. Although most plantations were originally family-owned, many have been taken over by large multinational corporations such as the United Fruit Company, Fyffes and Del Monte. This reduces the financial returns received by the producing countries.

New kinds of livestock

The indigenous people of the region had few domesticated animals until the Spanish colonists introduced horses and pigs. Horses were to become valuable draft animals, and pigs joined chickens as the staple livestock of subsistence farmers. Cattle, sheep and goats were all introduced early on, and the new livestock transformed the rural economy as well as the Indians' diet. Commercial stock raising was first established in the highland basins that are found in much of Central America, in lowland savanna areas, interior valleys and the dry steppelands of northern Mexico.

The close proximity of the huge United States market has encouraged American investment in ranching, even in the tropical rainforest areas. Rich, diverse forest has been replaced by poor grassland, but ranching is nevertheless profitable because of the high price of beef in the United States. Similarly, American and other foreign capital has been invested in a range of other agricultural and food processing ventures based on cheap land and cheap, plentiful labor. Fruit – especially bananas – coffee, sugar and tobacco are the main crops produced in this way.

The region encompasses a wide range of tropical environments and weather systems, from the Caribbean islands and the Central American isthmus, where there is rainfall all year round, to the semidesert of northern Mexico. This means virtually all the tropical crops are

grown. Many of the colonial plantations were established in areas where rich rainforests flourished; these forests were originally exploited for their valuable hardwoods, such as mahogany. Subsequent poor management by the European colonists has degraded the tropical soils, which are rich in laterite deposits containing iron and other oxides. In normal circumstances these would be leached away downward through the soil, but the removal of the tree cover and consequent

Fishing and tourism in Martinique (*above*) The tourist boom has benefited many Caribbean islanders. Fishermen now earn a large part of their living by supplying fresh fish to the tourist trade.

Hillside farming (*left*) Intricate terracing in Guatemala helps to conserve moisture as well as making economic use of every available piece of land. Vegetation is lush after the seasonal rains.

Map of agricultural zones The moist, warm climate allows a range of tropical crops to be cultivated. Rough pastures in drier areas support cattle ranching.

erosion of the soil into deep gullies exposes the laterites, making the soil more and more infertile and unproductive. This is a major problem for agriculture in some parts of the region – notably on the island of Haiti, for example.

Traditional farming

Some groups of people still use traditional agricultural practices. The indigenous Indian populations in Central America – descendants of the Aztecs, Maya and other Indian tribes, many now of mixed blood – farm today in much the same way as their ancestors. Using hand tools, they cultivate maize, beans and squash as their

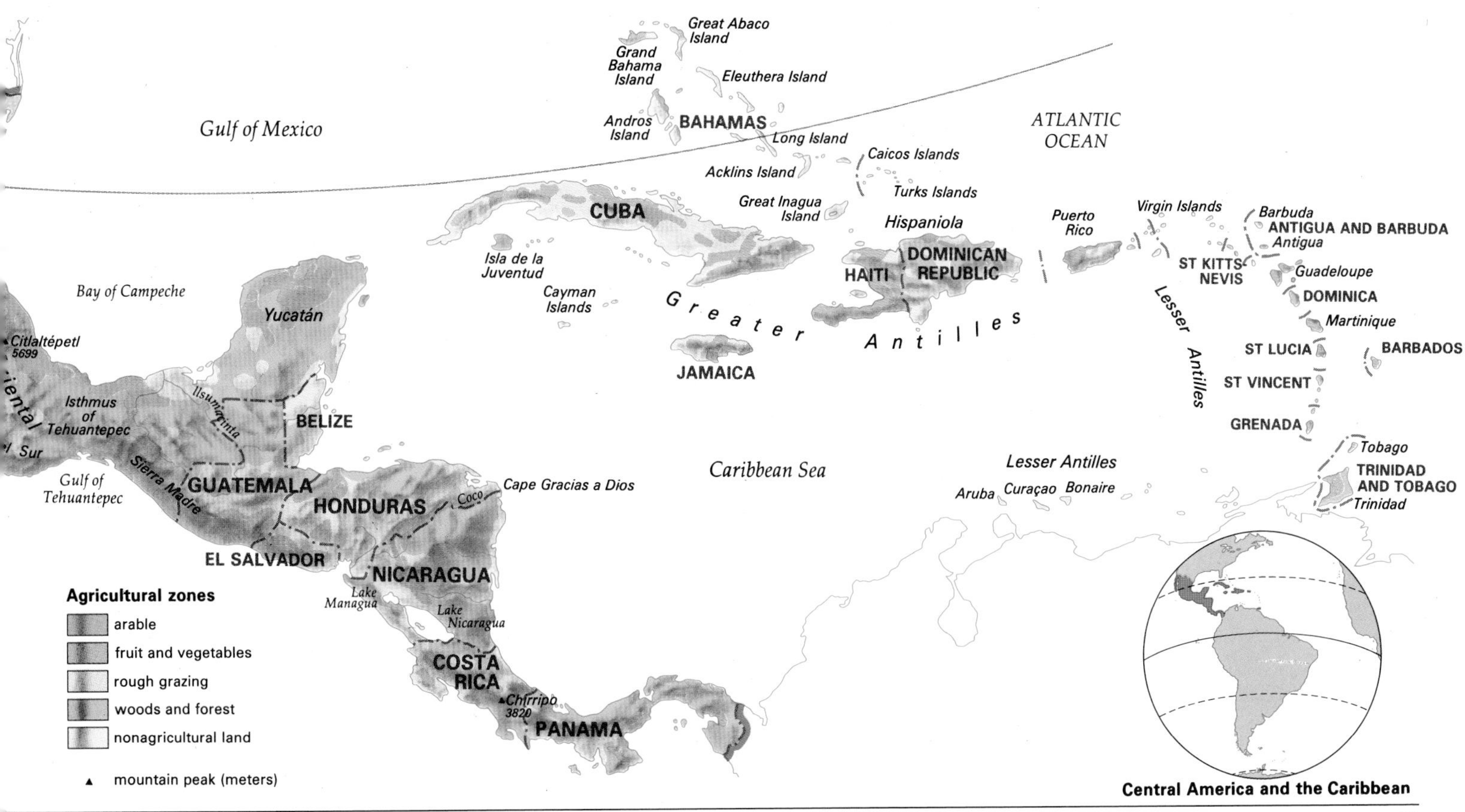

staple crops, and also rear pigs and poultry as livestock.

The Black Caribs – rare survivors of the original Indian population of the Caribbean – have retained a distinctive farming system in scattered areas on the Atlantic coastline of Central America, to which they were exiled by the British in 1797. They grow starchy root crops, chiefly cassava, sweet potato, arrowroot and peanuts, but rely heavily on fishing and hunting, with women often performing most of the husbandry.

Subsistence farming is practiced by the Afro-Americans of the Caribbean and parts of the Atlantic seaboard from Belize to Panama. These people are descendants of African slaves brought to the region by colonial powers as cheap labor for plantations, mines and forestry. By 1640 large sugar plantations that used black slave labor had been established in Barbados, and soon spread throughout the Caribbean; by 1800 Africans outnumbered Europeans by five to one. The eventual abolition of slavery from 1833 to 1865 created a new class of smallholding or subsistence farmers, as well as labor shortages on the plantations. These were filled by importing laborers from the Dutch East Indies and India.

Coffee country Young coffee plants are kept well watered in a nursery until they are transferred to the plantation. They take six to eight years to bear a full crop of berries, which grow at the same time as the plant is flowering.

SUBSISTENCE AND SURPLUS

The creation of haciendas and plantations concentrated large amounts of the best available land into the hands of relatively few owners, so preventing the development of a class of independent small farmers and a more balanced rural economy. This inequality in the distribution of land is still a severe problem in rural areas, with too many people trying to work too little land – if they possess any land at all. Today the vast majority of people working in agriculture in Central America and the Caribbean are either landless laborers or are subsistence or semisubsistence farmers eking out a living on small plots of land.

Subsistence farmers grow the traditional crops of maize, rice or beans, supplemented in different areas with other cereals such as sorghum, a variety of vegetables including squashes, potatoes and other root crops, and tropical fruits. Of all the subsistence crops, maize is the most widely grown – in Mexico it occupies over a third of agricultural land – and recent production has been increased as the result of improved seed varieties. In some parts of Cuba, Mexico and Panama, where there is high rainfall or where irrigation has been introduced, rice is the staple grain.

Subsistence farming is most prevalent in the rainforest areas of Central America where small clearings (*milpas*) are created in the jungle by felling and burning. The remaining tree trunks are removed by draft animals before the crops are sown. Without the addition of either fertilizer or manure, *milpas* are exhausted within three to four years, and in sparsely populated areas such as Belize, parts of northern Guatemala and southeast Mexico, they are frequently abandoned, being left to regain their fertility while cultivation moves to neighboring plots. Elsewhere population pressure has required a more permanent form of semicommercial cultivation using fertilizers, systematic periods of fallow, and cash sales. The continued practice of subsistence farming reveals the widespread rural poverty in the region, but there is little malnutrition as the diet is usually well balanced.

Farming for export

Commercial agriculture was initially associated with sugar production, and later extended to include other crops. Although sugar remains the principal export crop in many of the Caribbean island economies, particularly those of Cuba and Jamaica, major export production in the region now also includes bananas, cotton and coffee.

The cultivation of bananas has had a dramatic history of rapid expansion and decline due to market volatility and crop disease. The main period of expansion came after the establishment of the United Fruit Company in 1899, which developed large tracts of sparsely settled land in the hot, rainy Caribbean lowlands of Costa Rica, Panama, Honduras and Guatemala for banana plantations. However, a high incidence of disease in the years before World War II shifted production to the Pacific lowlands. Today, with the development of more disease-resistant varieties, there has been a return to the Atlantic coast, especially in Honduras and Costa Rica where bananas are a major export. They are also widely grown in the West Indies, particularly in Jamaica, Dominica and St Lucia, and the French territories of Guadeloupe and Martinique. Bananas have recently replaced sugar cane in some countries, and their production has moved from the plantation sector to smallholders who sell the fruit to corporate buyers.

Coffee farms (*fincas*) were established as early as the 1830s in the central highlands of Costa Rica, and the country's

Colonial-style cattle ranching Central American livestock haciendas are based on the Spanish model of aristocratic landownership of vast tracts of land. Despite the poor pasture, ranching is profitable because beef commands a high price.

economy is still tied very closely to the world coffee market. By the 1860s coffee growing had spread rapidly to the neighboring countries of El Salvador, Guatemala and Nicaragua. In these countries the *fincas* are situated on fertile volcanic ground and coffee is a major export crop, though its production in El Salvador and Nicaragua was drastically affected by the civil wars of the 1970s and 1980s, which had a disastrous effect on the countries' economies. Panama and Honduras also export coffee, but on a smaller scale.

Commercial production of coffee in Mexico began in the late 19th century and remained smallscale until the 1950s, when the amount of land devoted to its cultivation doubled. It is now more widely grown than sugar cane and is the country's third most valuable export.

Cotton was a boom crop in Nicaragua, El Salvador and Guatemala in the 1960s, being grown by largescale landowners on the drier Pacific lowlands and employing large numbers of landless peasants to pick the crop. The civil strife in all three countries, and poor prices on the world market, severely disrupted production of what had become their main export crop. Irrigated production in northern Mexico has now made Mexico the region's leading producer of cotton.

LIQUOR FROM AN AGAVE

In dry, rocky areas of central Mexico, fields of a spiky, gray-green plant, the maguey agave, are grown. The fermented juice, known locally as *agua miel*, or "honey water", is used to make pulque and tequila, the national drinks of Mexico. Fully grown, the maguey plant reaches a height of 3 m (10 ft), but it is harvested before maturity, when it is about six to eight years old. The juice is collected by cutting the flower bud from the plant's pineapple-like base, leaving a cavity. After several months the sap that has accumulated in the cavity is drawn off; the plant refills and the sap is drawn off again – one plant may provide up to 9 liters (15 pints) of sap before it dies. The sap, which contains up to 10 percent sugars, is fermented in vats for several days, and is widely drunk as pulque – a cloudy, whitish beer that is an important source of carbohydrates and proteins for many of Mexico's poorer people.

Tequila, a clear or sometimes golden liquor, is distilled from pulque. It takes its name from the town in the central highlands that has been the center of the drink's production since the 17th century, or even earlier, when the art of distillation was introduced to the region by the Spanish. Mexicans drink it undiluted with salt and a slice of lime; it forms the basis of the Margarita cocktail, traditionally served in a salt-rimmed glass, which has become a universally popular drink, following the growth of the tourist industry.

Commercial cattle raising

Livestock production is limited by the hot, wet, tropical climate in much of the region, which does not provide good pasture. Subsistence farmers raise a third of Mexico's livestock. Sheep farming in the highlands provides the basis for woollen mills and textile manufacturing, while cattle are much more important in the tropical lowlands and especially in the drier, semiarid areas of northern Mexico. Here the sparse pastures are only able to maintain low densities of cattle, resulting in huge livestock haciendas extending over hundreds of thousands of hectares. This is a colonial pattern that has largely been retained unchanged, with very little improvement having been made to the pastures. Commercial cattle farming in this area is frequently under American ownership, and since the early 1970s American investment has spread cattle rearing to other Central American countries, notably Costa Rica.

BREAKING WITH THE PAST

Few parts of the region are completely unaffected by modern commercial farming methods. Despite formidable physical barriers, such as the mountains of the Sierra Madre in southeast Mexico, even remote rural communities have started to farm for the market economy, abandoning ancient cultivation techniques and communal land tenure. Even so, there is a vast gulf between these semicommercial farmers and the large ranches and plantations. Although small farmers are in the majority – in Central America three-quarters of all land holdings are less than 20 ha (50 acres) – they occupy only 5 percent of the agricultural area.

Land reform programs

Unequal distribution of land, resulting in rural poverty, and low productivity due to the system of land tenure and technological backwardness, is a fundamental weakness in the agricultural economies of the area; Costa Rica, where the land is less concentrated into large estates, is exceptional in this respect. It has been a major factor in fueling the political unrest that has been endemic in the region for most of this century. A few governments, following popular revolution, have attempted to remove land from the few, and redistribute it among the landless. Most success has been achieved in Mexico and Cuba and, more recently, in Nicaragua during the 1980s.

Mexico's history of land reform goes back to the beginning of the 20th century, when over 40 percent of the land was occupied by only 8,000 haciendas; some 96 percent of rural families owned no land at all, and many people were obliged to work as virtual serfs on the large estates. After the 1910 revolution, however, the traditional communal landowning system of the Indian tribes was adopted on a large scale. This started a process of land reform that is still continuing, with land being gradually taken from the haciendas and redistributed to the former estate workers under a system of land tenure that is known as *ejido*; the *ejidores* – workers – cultivate the land either in family farms or as a collective.

Today nearly half of Mexico's cultivated land has been redistributed in this way, with the greatest reorganization in the center and south of the country. However, 30 percent of Mexico's rural families still own no land, and most Mexican governments have neglected the *ejidos*, giving them little financial or technical assistance. The quality of the land that has been redistributed is generally poor, and many *ejidores* as a consequence remain impoverished. Most of the land that has been improved in quality, particularly by irrigation, has benefited the private sector. Nevertheless, the collective *ejidos*, which are in the majority, tend to operate more commercially than the family farms.

Single-crop economies

Another major problem affecting the region is the reliance of its economies on a single crop. This is largely the result of its colonial past, because traditionally the plantations were established in order to feed the rapidly growing urban populations of the colonizing powers.

Most plantations in the Caribbean are owned by foreign corporations: British in the West Indies; United States' in Puerto Rico, the Dominican Republic and Haiti; and French in Guadeloupe and Martinique. Their control of the plantations derives from ownership either of the land or of the processing, transportation and marketing facilities.

Revenue derived from sugar exports remains the mainstay of many economies in the region. There are three main disadvantages to such dependency on a single crop: it encourages a high degree of foreign ownership; it leads to over-dependence on one main source of export revenue, as Jamaica found to its cost in 1988 when Hurricane Gilbert destroyed almost the entire sugar cane crop; and the imposition of quota agreements to protect sugar producers in the United States and

Going to market (*above*) Central America's many subsistence farmers – who are mostly of Indian origin – produce only small surpluses of their crops. They sell these at weekly village markets, which may be some distance from the farm.

A principal plantation crop (*right*) Many banana plantations are owned by American, British and French companies, but increasingly smallholders, such as this farmer on the Caribbean island of Tobago, produce the fruit and sell it to corporate buyers.

A famous export Tobacco was one of the first crops that European colonists exported from the New World in the 16th century. Each plant produces about twenty large leaves that are harvested and then cured by air, fire or flue drying. The curing process dries out the sap, and also induces chemical changes in the leaves that enhance the flavor and aroma of the tobacco.

Western Europe restricts access to these crucial markets.

Some partial solutions to these problems have been found. In Antigua and Trinidad, as well as Cuba, nationalization of the plantations has limited foreign ownership. Efforts have been made to find new markets: in Jamaica and Belize cane is converted into ethanol for use as a petrol substitute in the United States. The falling price of world sugar during the 1980s quickened the search for alternative crops, such as bananas in Dominica, and has substantially reduced the reliance on sugar in most West Indian economies. In the Leeward and Windward Islands both citrus fruits and market garden crops have become important exports. In much of the Caribbean tourism has replaced sugar as the single most important creator of revenue.

Similar diversification on the Central American mainland has encouraged greater commercial production of a wide range of fruits and vegetables. Sugar peas (snow peas or mangetout peas), for example, are grown for export in the Guatemalan highlands, and oranges for the production of frozen juice concentrate for the United States' market in Mexico's Gulf lowlands.

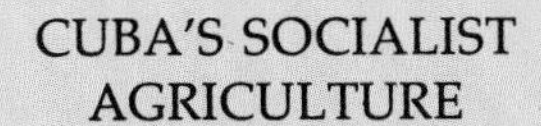

CUBA'S SOCIALIST AGRICULTURE

One of the first actions of the socialist-revolutionary government in Cuba in 1959 was to take control of the sugar cane processing industry, the island's principal source of revenue, bringing an end to foreign (chiefly American) ownership of the plantations and sugar mills. Although the government took over ownership of landholdings of 65 ha (160 acres) or more, the remainder – belonging to 80 percent of farmers – were left in private hands. There were subsequent spasmodic attempts to reduce the numbers of these small landholders through so-called "microplans", which encouraged both early retirement and land transfers to the state in order to ensure the "fulfillment of socialist principles". State control was less rigidly enforced than it was by the communist regimes of China or the Soviet Union, though state farms and collectives often held the best land.

Farming on about a quarter of the surviving private smallholdings is for subsistence purposes only, but the practice of growing small amounts of fruit and vegetables for sale is widespread. Traditionally tobacco for production of the world-famous Havana cigars was grown in these smallholdings. After the revolution output fell as a result of the loss of Western markets, particularly in the United States. More emphasis has been placed upon growing food crops and sugar cane, and little attempt has been made to extend the area under tobacco. The government has maintained the economy's dependence on sugar exports, but has replaced the United States' market with the Soviet Union and other countries of the eastern bloc.

The story of cane sugar

Sugar cane (*Saccharum officinarum*) is a giant, thick, perennial grass of the plant family Graminae. Raw sugar is derived from the sweet sap in its stem, which can grow up to 4.5 m (15 ft) tall and 2.5–5 cm (1–2 in) in diameter. The plant needs a minimum of 1,250 mm (49 in) of rain annually, and a short dry season to aid maturation. It is grown in most of the world's tropical regions; it can be planted and harvested by hand, so was suited to areas where labor was plentiful.

The cultivation of sugar cane probably originated in the South Pacific and spread from there to India and then to the Arabian countries and Europe. Christopher Columbus (1451–1506) is believed to have brought sugar cane to the New World on his second voyage west across the Atlantic in 1493, though it is possible that it had already been introduced to the Arawak and Carib peoples of the Caribbean by a route across the Pacific.

By the 17th century sugar plantations had been established throughout the West Indies. Islands such as Jamaica and Barbados were developed by British colonizers primarily for their ability to produce sugar for the home market, and were part of the notorious "Atlantic triangle" that linked Britain, West Africa and the West Indies in trading slaves and sugar. Ships sailed from British ports to collect slaves from West Africa to work on the sugar plantations. They then returned to their home port laden with cargoes of refined sugar, rum and molasses.

Barbados was the first island to experience the "sugar revolution" brought about by the use of slave labor on plantations. So many slaves were introduced that by the early 19th century the population density had reached 585 people per square kilometer (700 people per square mile), and Barbados had become known as a "city with sugar cane growing in the suburbs". Some 240 plantations, each of about 80 ha (200 acres) accounted for four-fifths of all land on the island.

The colonial plantations were adversely affected in the 19th century by slave uprisings and the eventual emancipation of the slaves, which made labor more expensive, as well as by soil exhaustion, competition from Cuba and the other Hispanic islands, and by the growth of the European sugar beet industry and loss of monopoly access to the British market. However, the modernization of plantations, made possible by more capital and land, more efficient management and more sophisticated technology, ensured the survival of a substantial sugar economy in the Caribbean.

Refining the sugar

Modern refineries serve large areas of plantation land and are easily accessible to bulk raw sugar carriers, minimizing transportation costs. Sugar is removed from the canes in one of two ways – by milling or by a diffusion process, which was originally developed to extract sugar from beet. In the first process, rapidly revolving cane knives cut the cane into chips, and the juice is then extracted by crushing and shredding the cane in large metal rollers that squeeze it from the fiber. In the diffusion process the sugar is separated from the finely cut stalks by dissolving it in hot water.

In the next stage of processing the extracted juice is clarified to remove the non-sugar components. Evaporation then removes the water, and the raw syrup is boiled until crystallization takes place and a mixture called massecuite forms. The raw sugar crystals are separated from the massecuite by centrifugal machines; molasses, which is distilled to make rum or used as feed for farm animals, is crystallized out of the remaining juice. Nothing is wasted: the cane residue, left over in the processing, is used to fire the boilers in the processing plants.

A natural sweetener All green plants make sugar. Sugar cane is one of two plant species grown commercially for their naturally high sugar (sucrose) content; the other is sugar beet, grown in temperate regions of the world.

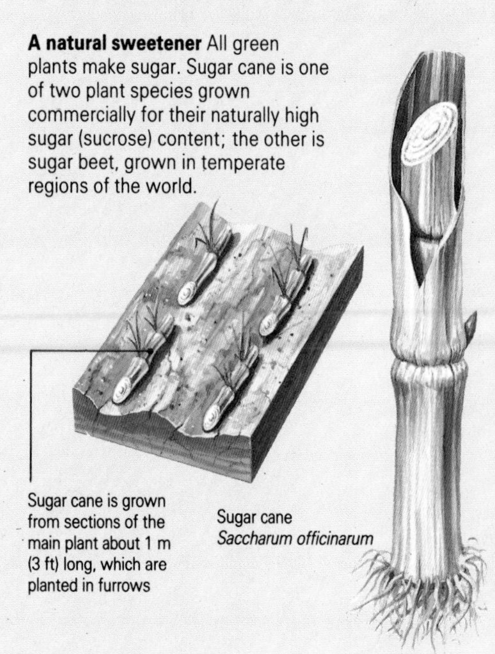

The harvest Sugar cane is harvested – mostly still by hand – when sugar content and moisture are greatest. The stems are stripped of leaves before being sent for immediate processing.

FOREST PLOTS TO VAST HACIENDAS

THE OLD AND THE NEW · A CONTINENT OF CONTRASTS · UNREALIZED POTENTIAL

South America has tremendous diversity of farming – the result of the continent's geographical location, which extends from the tropics to the subantarctic. There are equatorial, tropical and temperate areas with plentiful all-year rainfall, but the Atacama Desert on the Pacific coast is as arid as anywhere on Earth. Cultural differences influence patterns of agriculture, which embrace shifting cultivation of temporary clearings in the Amazon rainforest and extensive commercial cereal farming of the Pampas plains of Argentina. Traditional cultivation of indigenous crops was augmented by the introduction of new crops, animals and farming techniques by European colonizers after the middle of the 16th century. Despite these natural riches, few South American countries have profitable agricultural economies.

COUNTRIES IN THE REGION

Argentina, Bolivia, Brazil, Chile, Colombia, Ecuador, Guyana, Paraguay, Peru, Surinam, Uruguay, Venezuela

Land (million hectares)

Total	Agricultural	Arable	Forest/woodland
1,753 (100%)	617 (35%)	116 (7%)	900 (51%)

Farmers

24.2 million employed in agriculture (24% of work force)
5 hectares of arable land per person employed in agriculture

Major crops
Numbers in brackets are percentages of world average yield and total world production

	Area mill ha	Yield 100kg/ha	Production mill tonnes	Change since 1963
Maize	19.7	21.2 (58)	41.6 (9)	+133%
Soybeans	13.6	18.9 (99)	25.6 (26)	+6,178%
Wheat	9.6	18.6 (80)	17.9 (3)	+78%
Paddy rice	7.4	21.5 (65)	15.9 (3)	+98%
Sugar cane	5.4	627.7 (105)	337.5 (35)	+197%
Coffee	4.7	6.7 (120)	3.2 (51)	+26%
Bananas	—	—	15.2 (23)	+38%

Major livestock

	Number mill	Production mill tonnes	Change since 1963
Cattle	257.8 (20)	—	+73%
Sheep/goats	129.9 (8)	—	−12%
Pigs	53.9 (6)	—	+37%
Milk	—	29.3 (6)	+93%
Fish catch	—	12.0 (13)	—

Food security (cereal exports minus imports)

mill tonnes	% domestic production	% world trade
+4.2	5	2

Plowing in the Andes Among the changes that European colonists made to South American farming was the introduction of draft animals to pull plows. Farming at high altitude is arduous work. Subsistence farmers like this grow wheat and barley as well as indigenous crops such as potatoes, maize and beans.

THE OLD AND THE NEW

A variety of agricultural economies existed in South America long before European conquest and settlement began in about 1520. Primitive shifting agriculture predominated in many areas, but in other parts well-organized systems of permanent cultivation had developed: the wealth of the Inca empire, which by the 15th century stretched from southwest Colombia to central Chile, was based on agriculture. Food stores were held in common for lean years, cultivation terraces were constructed on steep Andean hillsides, and stone aqueducts were built to carry irrigation water.

The Incas used hand tools such as digging sticks and hoes, but the wheel had not been invented so travel was either on foot or by canoe. Foodstuffs were transported by llama over an intricate network of trackways to towns or to areas of food shortage. The llama can carry a heavy pack, but Inca farming was hampered by not having animals that could pull loads or be ridden.

South America has been called the "kitchen garden of the world" because so many of its indigenous crops, unknown elsewhere until they were exported by European colonists, are now of widespread importance. Maize, manioc (cassava), sweet potatoes, potatoes and many types of bean – major world crops today – had been cultivated for centuries by Amerindian peoples. Indigenous crops flourishing in other continents include cocoa, cotton, groundnuts, pineapple, tomato, and various types of pepper and squash. Coca (cocaine) and tobacco also originate from South America.

European innovations

The European settlers, with their new crops, livestock and agricultural techniques, had an immediate impact on South American agriculture. Wheat, barley, alfalfa, sugar cane, grapes, citrus fruit and garlic were some of the many crops they introduced; cattle, horses, sheep, goats, pigs and poultry were added to the limited range of indigenous domestic animals such as llamas, alpacas, guinea pigs and muscovy ducks.

From early settlements on the Caribbean coast of Colombia, Spanish colonists

Map of agricultural zones Subsistence farming is practiced in a range of environments. In lowland areas there is some largescale commercial farming and, where conditions are drier, cattle ranching.

followed the Andes down the west side of the continent, attracted by the more temperate high-altitude climates and the more developed agricultural systems. Continuing southward, they reached central Chile where they found a familiar Mediterranean climate to which they could adapt many crops. Portuguese settlers colonized the moist tropical coast of Brazil, establishing plantation agriculture there as early as 1532, when sugar cane was introduced from Madeira. They also grew cotton and fruits for European markets. Sugar cane and banana plantations were established by Spanish, British, French and Dutch colonists in the tropical coastlands from northern Peru round to Guyana and Surinam.

Important developments in agriculture took place in the 19th century. The rich temperate grasslands around the Plate river in Argentina and Uruguay were opened up for cereal and beef production. After the introduction of refrigeration ships in 1877, cattle raising expanded dramatically in response to the new accessibility of the European market. Farther south in Argentina, cooler conditions were better for sheep farming, while in Brazil there was a steady movement inland to use the tropical grasslands of the Brazilian plateau for cattle ranching. Coffee, which is still South America's main agricultural export, was cultivated especially in the São Paulo area of Brazil and in Colombia.

Agriculture today

About half of South America's work force is engaged in farming, which nevertheless contributes only one-fifth of the total economic product. Yields are generally low; subsistence farming is still widespread; agricultural exports remain small.

The rapidly expanding population and the drive for economic development, sometimes of a speculative nature, are the causes of the continued clearance of forest for agriculture, a source of great controversy, particularly in the Amazon rainforest. Nevertheless, half the continent remains sparsely populated. The cold deserts of Patagonia, the high Andean plateaus, the Atacama Desert and the vast Amazon Basin have failed to support significant productive agriculture.

Fishing and forestry are both of great economic importance in some countries of the region, and account for 16 percent and 10 percent respectively of total South American exports. Commercial fishing is concentrated in the rich, cool waters off the Pacific coasts of Ecuador, Peru and Chile. Forestry is most important in cool temperate southern Chile, in Brazil's tropical rainforests and in Paraguay. Together they provide 85 percent of South America's forestry exports.

A mountain of maize The only cereal that is native to South America, maize is now grown all over the world. It is the staple crop of the region; it contains up to 15 percent protein, though this is of lower nutritional value than that of other cereals.

A CONTINENT OF CONTRASTS

The diversity of natural conditions, compounded by cultural differences, has produced many forms of agricultural organization. Farming systems are subject to infinite regional variation and often merge with one another. Nevertheless, several major types of farming are broadly characteristic of the region.

Commercial farming systems

Haciendas (which are known as fazendas in Portuguese-speaking Brazil) are large estates under single – and traditionally European – ownership, and range in size from a few hundred to several thousand hectares. They are based on the *latifundia* system of landholding transferred to South America from Europe, and exist throughout the Andean countries from Venezuela to Chile. In mainly arable areas the estates are divided between the home farm, where the owner's crops are tended by tenants as labor rent, and very small plots (*minifundia*) tilled by the tenants for their own food supply. In the higher arable areas of Peru and Bolivia wheat, barley and potatoes are grown commercially on the home farm, while the *minifundia* typically produce potatoes, quinoa grain and beans for family subsistence.

In poorer, drier areas the haciendas are cattle ranches or, at high altitudes, sheep and llama ranges. They are frequently run by waged cattle hands. Cattle ranching predominates around the peripheral zones of the Pampas in Argentina, the Chaco district of northern Argentina and western Paraguay, and across the llanos grasslands of Venezuela. In spite of regional variations, haciendas do have features in common: they frequently have absentee owners, they are extensive, and productivity is low.

Plantations are found mainly in the tropical coastlands, and derive from the colonial system of using imported slave labor for specialist production of a single crop. Plantations are still monocultures, frequently owned or financed by overseas ventures, but unlike haciendas they are associated with reasonably high levels of mechanization and productivity. Sugar cane, the original plantation crop, is still widely produced, along with cocoa, palm oil and coconuts in Brazil, bananas along the humid coasts from Ecuador to Brazil,

The ubiquitous pig (*below*) Grubbing for scraps wherever they can find them, pigs are efficient converters of vegetable and other organic wastes into meat, and are kept as a valuable source of protein in countless backyards.

A valuable crop (*above*) Brazil and Colombia are two of the principal producers of coffee, the beans of which – roasted, ground and brewed in hot water – make a drink that is consumed by about one-third of the world's population.

and rubber in the lower Amazon. There are also large vineyards in central Chile and western Argentina, yerba maté (also known as Paraguay tea) plantations in northern Argentina, and large coffee fazendas in southeast Brazil.

Modern commercial farming for the sale and export of produce has developed in many areas previously used for subsistence farming, and there are both large, labor-efficient estates and much smaller intensive farms. Modern farms in the Argentinian Pampas and in Uruguay grow wheat and maize, and raise cattle on pastures that have been improved with sown grasses and alfalfa. Cattle from the drier ranches are fattened before being sent on to meat processing establishments. There are important dairy farms close to the population centers of Buenos Aires in Argentina and Montevideo in Uruguay, and sheep farms in southern Argentina and the Falkland Islands.

South-central Chile has well-developed commercial mixed farms with dairy cattle, wheat, barley and temperate fruits such as apples. Just to the north, farm products include beef cattle, maize, wheat, citrus fruits and small vineyards. Coffee is grown in the smaller, more specialist farms, as well as in large plantations in Brazil and Colombia. Southern Brazil has many mixed farms with cereals, tobacco, pigs and dairy produce. These commercial farms are typically worked by the owner's family and local laborers, or by tenants who pay cash rents.

Subsistence farming

Farming for purely family or community food needs is usual in remote and less advanced areas. In the northern and central Andes settled subsistence farming is commonly found at high altitudes, above the better land that is occupied by haciendas. At altitudes of about 3,500-4,000 m (11,500-13,000 ft) farmers graze sheep, llamas and alpacas, all for their wool, on mountain pastures (*páramos*). Below, at about 3,000 m (10,000 ft), they grow hardy grains, potatoes and vegetables, and keep poultry and guinea pigs for eating. These isolated communities, which are mostly of Amerindian origin, carry on the ancient traditions of communal farming, sharing the key activities of building, harvesting and storage.

Primitive shifting cultivation is practiced mainly in tropical rainforest clearings in Brazil, and in similar lowland environments in Colombia, Venezuela and eastern Peru; in the high Andes squatter farmers cultivate temporary plots on unoccupied land. The crops raised vary according to the conditions, but the essential feature of all shifting cultivation in the region is the abandonment of plots after three to five years, when soil fertility and crop yields have diminished.

SLASH-AND-BURN FARMING IN THE RAINFOREST

Vast equatorial rainforests are the natural vegetation of the Amazon Basin. They cover just under half of Brazil, and extend into adjacent countries. High rainfall and a very warm environment produce luxuriant vegetation, though the underlying forest soils are generally poor. Indigenous Indian peoples have practiced shifting slash-and-burn farming techniques here for at least 3,000 years.

A small patch of forest is cleared during a drier period of the year so that the wood can be burned before the onset of heavier rains. The plot, now covered in fertilizing ash, is then planted. The most important crop is manioc, which can be harvested all year and so provides a constant food supply. This small, bushy plant produces tubers similar to potatoes in appearance; these have to be boiled before they can be eaten in order to extract the toxins they contain. Maize and beans are also staple crops, and the diet is supplemented with wild fruit, nuts and game.

The plot has to be abandoned after three to five years, when the soil has become less fertile and crop yields have fallen. A new clearing is then prepared, if possible within walking distance of the existing settlement of round timber and thatch dwellings. If not, the whole community builds a new settlement beside the new clearing.

Because it is practiced on a small scale, and is transient in nature, slash-and-burn farming has only had a temporary effect on the ecosystems of the rainforest. Regrowth soon begins on the abandoned plots, and the forest quickly regenerates. By contrast, the modern commercial exploitation of the Amazon causes irreversible damage by largescale bulldozing, destruction of the vegetation cover and rapid depletion of soil nutrients because of inappropriate land use.

Farming on the Equator In Ecuador the Andes rise from sea level to over 6,000 m (20,000 ft), so conditions for farming vary from fertile equatorial lowlands to treeless pasture below the snowline at 4,600 m (15,000 ft).

Altitude (meters)	Subsistence farming			Cash crops and animals	Temperature (°C)
6,000					−6
4,600			Ice and snowfields		+1
3,200	potatoes		Heath and scrub	sheep llamas	+10
2,000	beans maize wheat	potatoes barley seeds	Cold zone	dairy cattle	+17
1,000	maize rice	beans	Temperate zone	coffee cotton tobacco sugar	+23
Sea level	manioc rice	plantains	Hot zone	bananas sugar cocoa beef cattle	+27

UNREALIZED POTENTIAL

Low productivity is the critical problem besetting South American agriculture. Although the region is still not densely populated – in 1990 there were roughly 300 million people, compared with 225 million people in under half the area in the United States – the population is nevertheless on the increase. The recent growth in population has not been matched by a comparable improvement in agricultural output. By 1980, South America had become a net importer of staple foods, with only the southern countries of Argentina, Chile and Uruguay managing to produce a food surplus. The export of cereals and meat products is still relatively insignificant, accounting for only 13 percent of the value of agricultural exports, mostly from Argentina, Brazil and Uruguay. By far the most valuable exports are coffee and cocoa, totaling 44 percent. Once again, this is almost all from only three countries – Brazil, Colombia and Ecuador – indicating few surpluses in the other countries.

Causes of low productivity

Technological backwardness and the hostility of the terrain in some areas partly account for poor productivity, but the fundamental cause of agricultural inefficiency is the inequitable nature of South American land ownership. A minority of very large landholdings occupy the better land at the expense of smaller units, which are scarcely viable. Surveys made in the 1950s of the pattern of land ownership showed that the percentage of small farms in South America varied from 22.5 percent in Brazil to 90 percent in Ecuador, but the farmland they occupied ranged from only 0.5 percent to 16.6 percent. Conversely, the number of large farms varied from 0.4 percent in Ecuador to 6.9 percent in Chile, but occupied between 45 percent and 81 percent of the total farmland.

The predominance of the haciendas means that ownership and power has rested with a disproportionately small class of landlords, leaving tenant farmers and paid laborers economically dependent on them. There has been little incentive for owners to invest in machinery and modern techniques, as they have always had access to a permanent source of extremely cheap labor.

Other factors weaken the agricultural economy still further. Access to markets is inhibited by poor transportation, storage and marketing facilities. Technical, advisory and support services – which are esential if agriculture is to develop – are also frequently inadequate.

The extent of change

Agrarian reform has been a central issue over the last few decades. All South American countries have attempted to solve the problems of farm size and land tenure, with differing degrees of commitment and success. Some hacienda lands have been redistributed under programs that compensate the previous owners. The land reforms undertaken by Chile's socialist governments (1964–73) allowed owners to retain 40 ha (99 acres) of their choice; the rest, for which they received compensation, was redistributed among peasant workers. By 1973, when the government was replaced by a military junta, nearly half Chile's farmlands had been redistributed in this way. Although some land was subsequently returned to haciendas, medium-sized commercial farms with improved fruit and vegetable production have now become established in the country.

Even where land reform has been seriously attempted, difficulties remain. The possession of land does not necessarily help the new farmers if they lack the necessary backup services such as legal, financial and technical assistance, as well as transportation and marketing facilities. Some rural workers, displaced by agricultural reforms, have been forced to migrate to urban shanty towns.

A number of new crops and crop varieties, and new methods, have been introduced, particularly in the commercial farming areas. The production of rice on the northern Caribbean coastlands, and of soybeans in Brazil, have been notable successes. The latter are grown for export as cattle feed in countries that produce food surpluses of their own, at the expense of Brazil's domestic food production. Crop yields in general have not increased since the 1950s, and rises in production have largely been achieved by the extension of agricultural land.

Fields of soybeans *(right)* surround the headquarters of a commercial farm in Brazil. Savanna soils in the south have been improved to expand production of this valuable export crop, but the profits generally go to multinational corporations outside the country.

Llamas in the Andes *(below)* This herd of llamas is being driven across a salt lake high in the Peruvian Andes. Llamas are a species of the camel family that can live and work at very high altitudes. They are the Andes' principal livestock: they provide meat, wool and leather and can also carry heavy loads.

A truckload of papaya makes the hazardous journey to market. Many kinds of tropical fruit, both of native origin (as these are) and introduced, are grown in the region. Pineapples and bananas are both important plantation crops.

NARCOTICS: A PROBLEM EXPORT

In several countries in South America, where agricultural performance is poor, the production and sale of illegal drugs is the most lucrative export earner. This agricultural subculture has flourished since the 1960s, supplying the markets of North America and Western Europe with cocaine and heroin.

The Amerindian people of the Andes have long been aware of the narcotic properties of the leaves of coca plants, which are native to the region. Since the last century they have been cultivated for supply to pharmaceutical companies. The cocaine extracted from the leaves is used in medicine as a valued anesthetic. But in recent years there has been a tremendous expansion of coca cultivation on the mountain slopes of Bolivia and Peru to supply the illegal drug trade. The high sale value of cocaine means that even small coca plots are extremely profitable. Illegal cocaine is thought to be Bolivia's most valuable export, while Peru produces about half the world's supply.

Much of the coca paste is converted into powder in Colombia. The notorious power of the Colombian drug barons is also based on heroin. This drug is extracted from the white juice of the opium poppy, which is grown in small, isolated clearings in Colombia and to a lesser extent in other South American countries. The cocaine and heroin are illicitly exported via certain Caribbean islands to Florida in the United States. Some drugs travel to Brazil and North Africa to enter Europe through Spain.

The highly addictive nature of these illegal drugs assures the continuance of a profitable market for them, despite the efforts of authorities worldwide to stem the flow. Although the Bolivian and Peruvian governments have both expressed the intention of offering farmers subsidies to grow other crops, profits from coca production are so large that they are unlikely to be able to afford sufficient inducement without outside financial aid.

Cattle ranching

Cattle ranching – the rearing of cattle on extensive rangelands – provides the most evocative image of South American farming. Horse-riding cowboys, the *gauchos* of Spanish-speaking countries or *vaqueiros* of Brazil, are still associated romantically with individual independence, wide-open spaces and untamed frontiers. It is a use of the land and a way of life that still prevails in many areas.

Cattle were introduced from Europe in the 16th century. Large hacienda ranches were soon established on the higher pastures of the Andes from Venezuela to Chile, and became the dominant landholding system on lowland tropical grassland areas (*llanos*) such as those of Venezuela and the Gran Chaco area of Paraguay and northern Argentina. Ranching is the main farming activity both in the drought-prone thorn scrub (*caatinga*) country of northeast Brazil and in the tropical grassland expanses (*campo cerrado*) of the Plateau of Brazil. More advanced methods of ranching have evolved on the temperate Pampas grasslands and the *campanha* of southern Brazil.

Extensive areas of land are devoted to ranching. South America has 21 percent of Third World cattle; Brazil alone has over 100 million head, the fourth largest herd in the world. Exports are limited, however, with cattle products representing only 6 percent of South American farming exports by value. Practically all these (94 percent in 1985) come from just three countries – Argentina, Brazil and Uruguay.

Cattle ranching illustrates the pervading characteristics of agriculture in South America. Generally there is low productivity, unrealized potential, and persistent contrasts between traditional and modernized farming methods. In many areas free range cattle roam across unimproved natural grasslands that have low stock capacity. For estates to be viable they have to be large.

The Plateau of Mato Grosso in western Brazil typifies traditional cattle ranching in a contemporary frontier zone where the population has doubled since 1960. Improved zebu types of cattle, which are better able to withstand periods of drought and are able to resist disease, are bred for their hides and for salt or dried beef. Animals may take six years to achieve slaughter weight in the natural *cerrado* ranges. Ranching operates with a small labor force of *vaqueiros*, each responsible for as many as 2,000 steers. The cattle are driven eastward into São Paulo and Minas Gerais states to be fattened on improved pastures.

Farther north the rapid clearance of Amazon rainforest for cattle grazing lands is the controversial "last frontier". There

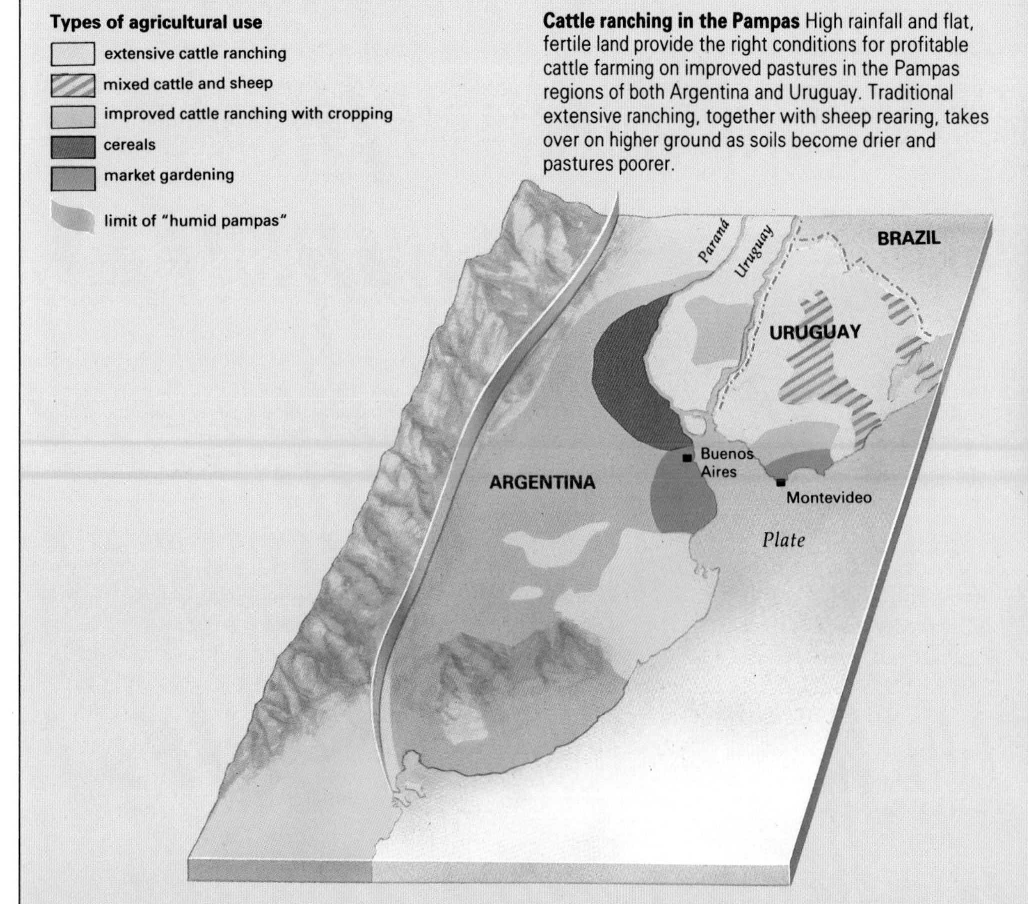

Cattle ranching in the Pampas High rainfall and flat, fertile land provide the right conditions for profitable cattle farming on improved pastures in the Pampas regions of both Argentina and Uruguay. Traditional extensive ranching, together with sheep rearing, takes over on higher ground as soils become drier and pastures poorer.

Contemporary cowboys Cattle ranching in the Mato Grosso in Brazil. Zebu strains of cattle, originally from India, with the characteristic hump and heavy dewlap, are well suited to dry, tropical conditions, and have largely replaced European-descended breeds.

are over 5 million cattle and 7.5 million ha (18.5 million acres) of cattle pastures; these are often in the hands of powerful overseas interests. Serious environmental misgivings may lead to further policies to restrict any extension of cattle ranching in the future at the expense of the Earth's richest ecosystems.

By contrast with most ranching in these tropical pioneer zones, much more productive cattle ranges have developed on the Pampas of Argentina and Uruguay. Gentle gradients and fertile soils are combined with warm temperate conditions. From the mid-19th century new settlers established both large cattle estancias and smaller farms for the production of hides and salt beef. From the late 1870s refrigeration made the European markets accessible for beef exports. Quality and productivity have been improved by introducing British breeds, especially Herefords, and growing rye grasses, clovers and alfalfa in the richer humid Pampas areas, where more than 500 mm (20 in) of rain falls each year, around Buenos Aires and west of Montevideo. The peripheral drier areas have more traditional extensive ranching. Excellent road and rail networks provide links with the ports, with their tanneries, corned beef canneries and large chilling and freezing plants (*frigerificos*). Some 65 percent of South American meat exports are derived from this exceptional cattle farming zone.

FARMING IN NORTHERN LATITUDES

FARMING NATIONS · AGRICULTURE BASED ON LIVESTOCK · COOPERATION AND SUPPORT

Agriculture in the Nordic countries is highly efficient and profitable, despite their northerly situation. Denmark, the smallest of the five states, has one of the most successful agricultural systems in the world. Here, and in southern Sweden, there are extensive areas of good agricultural land. There is much less cultivable land in northern Sweden, Finland, Iceland and Norway, but farming is well adapted to the prevailing conditions. Mixed farming predominates throughout the region. Livestock – cattle, pigs, sheep and poultry – are found everywhere; dairy farming is particularly important in Denmark and in southern Sweden. The cool waters of the North Sea and the North Atlantic Ocean are rich fishing grounds; the coniferous woodlands of Finland, Norway and Sweden are also an important natural resource.

COUNTRIES IN THE REGION

Denmark, Finland, Iceland, Norway, Sweden

Land (million hectares)

Total	Agricultural	Arable	Forest/woodland
117 (100%)	12 (10%)	9 (8%)	60 (52%)

Farmers

677,000 employed in agriculture (6% of work force)
13 hectares of arable land per person employed in agriculture

Major crops
Numbers in brackets are percentages of world average yield and total world production

	Area mill ha	Yield 100kg/ha	Production mill tonnes	Change since 1963
Barley	2.2	36.1 (155)	8.1 (4)	+46%
Wheat	0.9	47.5 (204)	4.4 (1)	+129%
Oats	0.9	31.2 (169)	2.8 (7)	–5%
Rapeseed	0.5	18.9 (132)	0.9 (4)	+360%

Major livestock

	Number mill	Production mill tonnes	Change since 1963
Pigs	13.4 (2)	—	+25%
Cattle	6.5 (1)	—	–29%
Milk	—	13.4 (3)	–8%
Fish catch	—	6.0 (6)	—

Food security (cereal exports minus imports)

mill tonnes	% domestic production	% world trade
+2.6	15	1

FARMING NATIONS

The natural environment imposes significant restraints on Nordic agriculture. Snow and ice cover most of Finland and Sweden, as well as much of the tideless Baltic Sea, during the winter. The warming influence of the North Atlantic Drift modifies winters in Denmark, and the coastal areas of Iceland and Norway. The growing season is generally short, and in over a third of the region late frosts – even in summer – pose a threat to farmers.

Agricultural activity responds to a pronounced seasonal rhythm of long hours of daylight during the summer and of darkness during the winter. At higher altitudes and latitudes the corresponding climatic extremes make conditions very difficult for agriculture. Iceland's agriculture suffers from a lack of sunshine; western Norway's from a surfeit of rain. Only the brown forest soils of Denmark and southern Sweden can support intensive agriculture; elsewhere soils are often poor, with widespread peatlands. Crop rotation, drainage and fertilization have improved the soil in many areas.

The farming landscape

Bronze Age finds by archaeologists suggest settled farming began in Denmark some 3,000 years ago. Rock paintings show that there were hunting and fishing settlements along the Norwegian coast very much earlier than this. Place-names and other evidence have charted the northward advance of farming up the coasts and into the wooded interiors. The mixed forests that once covered Denmark and southern Sweden were cleared to make way for agriculture; Nordic farming has always been combined with fishing, forestry, hunting and mining.

Until the 18th and 19th centuries most cultivated land was generally held in an open-field system. Its reorganization into unitary holdings, which meant that it could be farmed more efficiently, was often accompanied by the breakup of village settlements and the establishment of isolated farmhouses. In Iceland, farms had always been widely separated from each other.

Legacies from the past can be found in the farming landscape throughout the region. Mounds of boulders and stone-built walls, many of them centuries old, bear witness to the backbreaking toil of clearing stones from cultivated fields. In Norway, where timber has traditionally been the principal building material, the log barns and storehouses cluster around farmhouses, each district having its own distinctive style.

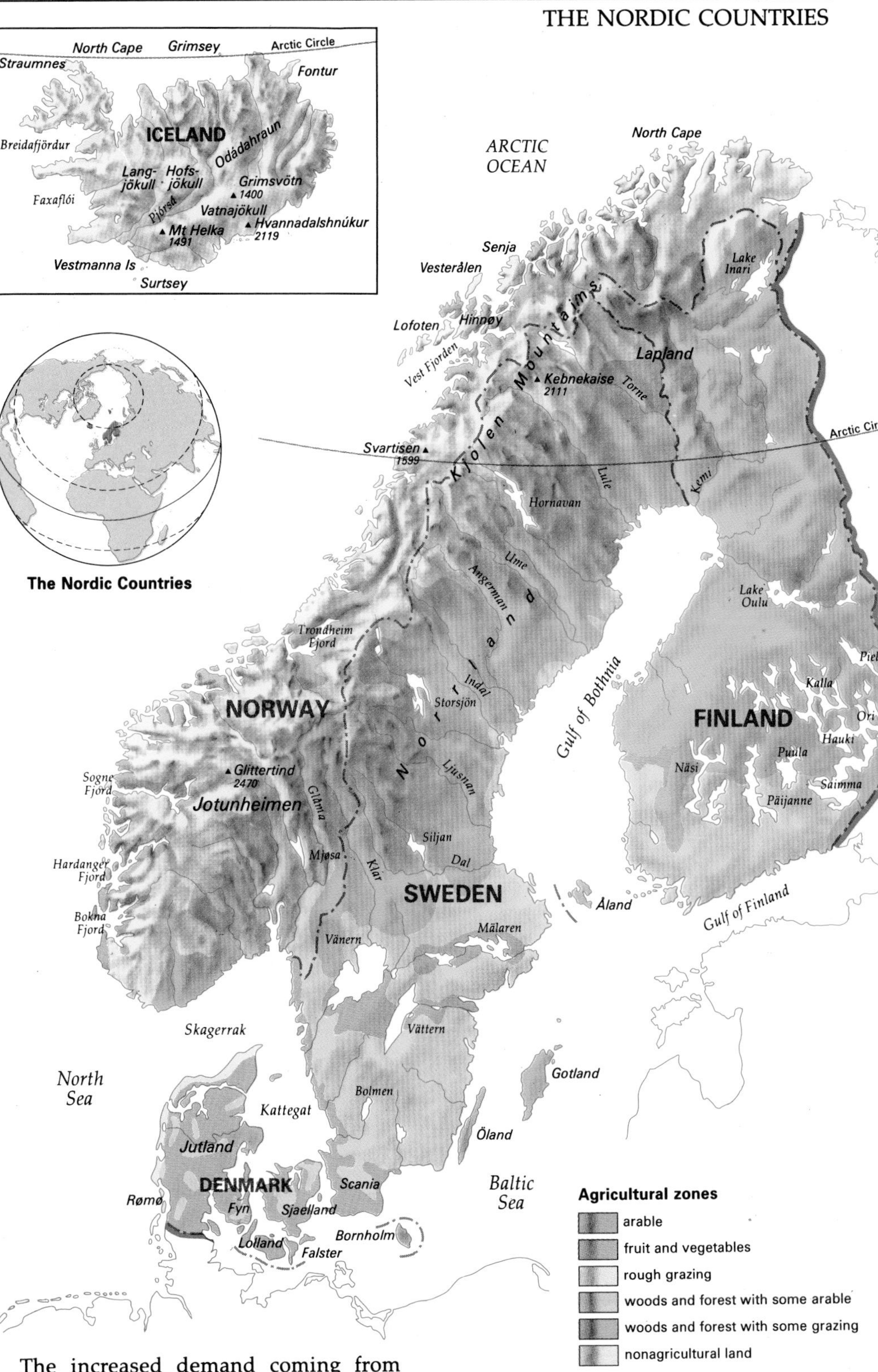

Fragile growing season (*above*) The growing season is short, with long hours of daylight. The fields are plowed after the thaw, but frosts can strike even in summer and destroy crops.

Map of agricultural zones In this region of mixed farming the most productive arable land is in lowland Denmark and southern Sweden. Farther north cultivation is limited by the mountainous terrain and the short growing season, and farming is complemented by forestry and fishing.

Industrialization came relatively late to the Nordic countries, whose economies were largely agricultural until the early 20th century. A hundred years ago Nordic agriculture was so poor that rural poverty and emigration were common. Today, thanks to plant and animal breeding, investment in new technology, and high standards of quality control, there are overall production surpluses.

Denmark led the way in commercializing its agriculture. Until the 1870s, when the European market began to receive grain from North America, Danish farmers had concentrated on grain production. Unable to compete with the low cost of imports, they changed to dairy farming, with complementary pig and poultry rearing. Denmark had established itself as the principal supplier of bacon, butter and eggs to the breakfast tables of Western Europe by World War I, and is still important as such today.

The increased demand coming from Western Europe's rapidly expanding urban populations also benefited the fishing industry of Scandinavia's Atlantic coast at the start of 20th century. Simultaneous advances in canning and refrigeration techniques meant that it was able to extend its marketing operations. At the same time, Baltic Scandinavia was able to profit from a fall in shipping rates to export increasing quantities of timber, paper and woodpulp to Western Europe.

Most of the region's 23 million inhabitants come from farming, fishing or forest-working backgrounds, and even though they are today mostly employed in service industries, the people still have a remarkable affinity with the land. Farming is a part of Nordic culture, with town dwellers regularly returning to the rural areas, where many of them own second houses. Access to the countryside is unhindered – except in more densely populated Denmark – and the popularity of outdoor recreations brings the farmer and the city dweller into close and frequent contact. So, too, does hunting. Hunters from all walks of life in Norway, Sweden and Finland join in annual culls of hundreds of thousands of moose (elk) to keep numbers under control.

Lush farmland Danish farmers turned to dairying in the 19th century, when their markets for wheat were undercut by cheap North American grain. Cattle and pig farming dominate Denmark's highly successful agriculture – the farms are small but efficient.

AGRICULTURE BASED ON LIVESTOCK

In the mixed farming that is characteristic of the region, animal husbandry is the most important element; in Denmark 80 percent of farm revenue comes from livestock. Dairy farming and pig production are often interrelated: the byproducts of dairying – such as skimmed milk – are used to feed the pigs. The number of beef cattle has been slowly increasing over recent years, in response to growing customer demand for meat products. Iceland has very little cultivated land, but there is widespread grazing for sheep, which here and in the Faeroe Islands take precedence over cattle as the principal livestock. Commercial poultry and egg production are most profitable in the southern parts of the region, where fodder crops are easily grown and there is greater access to markets.

Outdoor grazing is possible for only five or six months of the year at the most. During the long winters livestock are kept in the large brick, concrete or timber farm buildings that characterize Nordic farms. In Norway, the cattle were traditionally moved from winter accommodation to high mountain pastures in the summer, but improved fodder production has made this arduous – though picturesque – practice less common. Young cattle and goats are still moved to better upland pastures, and many sheep are transported to them by truck from southwestern Norway. The summer migration of sheep is also still a regular practice in Iceland. Forest grazing has virtually disappeared in Sweden and Finland.

Horses and ponies still play an active part in Nordic agriculture. In Iceland small native ponies are numerous; they are used for mustering sheep. Riding horses are bred for export in Denmark, as well as for domestic agricultural use, and the sturdy native horses of Norway and Finland are still widely used to complement the tractor in winter forestry.

Cultivating the land

Nordic farmers are almost entirely small-scale owner–operators. The arable area of an average holding is less than 30 ha (74 acres); in Norway it is often so small that it is calculated in decares rather than hectares. The size of the farms is not only the result of the system of dividing inherited land; much farmland is also physically fragmented by intrusive bedrock, boulders, lakes and swamps.

Fodder crops, which are essential to the region's livestock-based agriculture, occupy the greatest area of arable land. Only limited areas of land are permanently under grass; longterm leys – where grass is alternated with crops every five years or so – are common. The grass crop is mostly converted to silage, replacing hay, which is liable to rot in wet summers. Ley farming is crucial in northern Scandinavia, where the grass crop takes three-quarters of the field area.

Barley is the chief grain crop, accompanied by wheat in more favored areas. Cultivation of oats and rye has declined, but oilseed rape adds its vivid yellow to the springtime scene in many areas. On the Danish islands and in southern Sweden, fodder beet and, to a lesser extent, sugar beet are common crops.

Potatoes are grown for the farmer's own use as well as for the market. Commercial fruit production is concentrated in the Danish islands and along the margins of the Hardanger and Sogne fjords of western Norway, where apples, cherries and pears thrive, and raspberries are unrivaled in size and flavor.

Iceland's fishing economy *(above)* Fish are Iceland's only abundant food resource, making the country susceptible to market prices and fluctuations in the size of the catches. Agriculture is limited by the rocky terrain and the harsh climate.

Specialist fruit production *(below)* Raspberries are an important crop in the commercial fruit-growing areas of western Norway and Denmark. Most small farms also grow some fruit and vegetables.

SEA RANCHING IN NORWAY

For a number of reasons, including overfishing, annual catches from the North Atlantic have been declining in recent years. The Norwegian fishing industry has responded to this by turning increasingly to sea ranching – so called because it follows the same principles of stock management as commercial livestock farming: the "ranches" are specially constructed systems of tanks and cages for the intensive breeding of saltwater fish. The first fish to be farmed in this way were salmon and sea trout, and crustaceans such as oysters, lobsters and mussels; more recently sea chard, bass, bream, turbot, halibut and even cod have been added to the stock.

Sea ranches are mostly family run; by 1990 they employed 4,000 people directly, with roughly the same number working in related industries. There are ambitious plans for expansion: target production for 1992 is 150,000 tonnes of fish as opposed to the 95,000 tonnes produced in 1989. Most of the fish are exported in freezer trucks that supply the European market, while 14,000 tonnes of farmed fish are airfreighted annually to the United States.

Sea ranching is a very sophisticated business that requires high levels of scientific research, technical expertise and longterm capital investment. Five experimental stations have been set up to investigate the special problems of sea ranching, such as water pollution, genetic disease and the vulnerability of stock to natural predators. The cages, tanks, walkways and other ranch structures are expensive to install and to maintain; they can easily be damaged by wind, waves, ice or sea traffic. For this reason, landbased saltwater tank systems are also used.

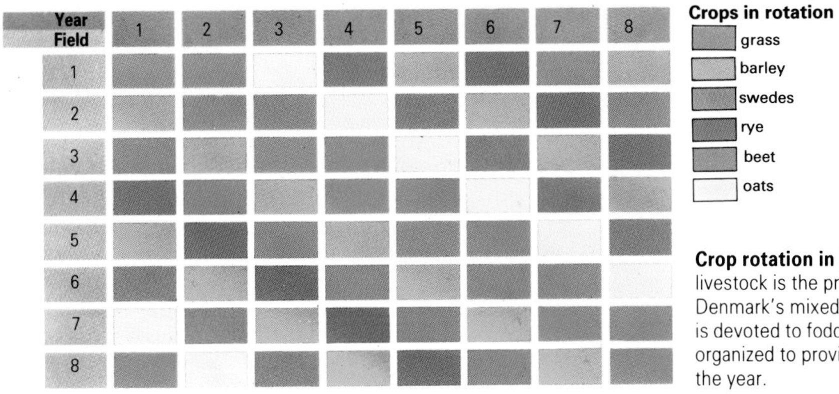

Crop rotation in Denmark Raising livestock is the principal activity on Denmark's mixed farms. Arable land is devoted to fodder crops, which are organized to provide feed throughout the year.

Alternative rural economies

Away from the better agricultural lands of the south, farming is often combined with fishing or forestry. Fish from lakes and rivers provide a modest supplement to farm incomes. Along the Atlantic coast of Norway, where cod and herring are seasonally abundant, fishing and farming have always been closely linked; farming statistics still differentiate between farmer–fishers and fisher–farmers, though the two activities have in fact become more independent in recent years as fishing enterprises have grown in scale. Iceland's economy is highly dependent on fishing: the Icelandic government declared a 320 km (200 mi) exclusion zone around its coast in 1975 to keep foreign fishing fleets from plundering its most valuable resource.

For many smallholders in Sweden, Finland and interior Norway, farming and forestry are inseparable, with timber cropping taking place side by side with crop farming. Both the privately owned timber lots and the commercially owned forests are an important source of employment during the winter months. Softwood timber and timber products are an important source of export revenue; timber is also used extensively for building, fencing and fuel. Forestry is particularly successful in Sweden and Finland. This is explained by good forest management as well as by the relatively rapid growth of the trees.

The timber industry Forestry has been revolutionized over the last few decades. Recognized as a valuable renewable resource, forests are now scientifically managed using a high degree of mechanization. The timber business generates rural employment as well as significant export revenue.

COOPERATION AND SUPPORT

All Nordic farms require high investment of capital. Buildings, necessary to protect livestock against harsh weather conditions, account for the principal outlay. Equipment comes second – mechanical harvesters for grain, seeds and potatoes, tractors and dairying equipment. Underground drainage of arable land is also expensive: in Finland and Iceland open ditches are still common.

Support for farmers is available, through both subsidy and advice. Agricultural cooperatives were established in Denmark in the late 19th century to coordinate the marketing of dairy products. They then spread across the whole region and now include cooperative research and advisory institutions, meat and dairy processing factories, purchasing societies, banks, and mortgage and insurance societies. Management of the cooperatives is mostly in the hands of elected farmers, and collectively they wield considerable political influence.

Financial support is forthcoming from national governments, except in the case of Denmark, which as a member of the European Community (EC) conforms to the Common Agricultural Policy (CAP). The amount of subsidy is determined by the distance of a farm from its sources of supply and principal markets, as well as by the degree of climatic hardship that it experiences. Norway provides the most financial assistance to farmers: in 1988 the average subsidy was $10,000 per farm.

The services provided to rural communities are of a high standard. Roads are generally good, and even in the most remote areas nearly all farms have electricity and telephones. Some local services such as hospitals and cooperative stores have disappeared, but these have usually been replaced with mobile dental and medical clinics, shops and libraries.

Greater efficiency

It is a measure of the efficiency of farming in the Nordic countries that, except in Iceland, there are surpluses of farm products. The situation is more complicated in Denmark, where the imposition of EC milk quotas has caused dairy production to be cut.

At the same time, the increasing world demand for wood products calls for increasingly scientific programs of felling, maintenance, drainage and fertilization in the woodlands of Finland, Norway and Sweden. Some Swedish farmers have recognized the limited new demand for farm products and are returning land to forest plantation. This move has been encouraged by government policy, as the widely diversified products of the softwood industries account for about 20 percent of the country's exports. The forests provide rural employment, and their management is linked to both nature conservation and recreational facilities.

The modernization of methods of forestry has also brought about many changes. Whereas axes and saws were once the main tools of a mostly seasonal workforce, there are now highly skilled forest workers, using a range of mechanical equipment, who operate throughout the year. Timber is managed as a renewable resource that is grown, harvested and replanted like any other crop.

Greenhouse cultivation Since the 1930s farmers in Iceland have built their greenhouses near hot springs, of which there are many in this young volcanic island. The warm water is used to heat the greenhouses, allowing cultivation of crops such as tomatoes, grapes, vegetables and even bananas.

Finding a balance

Although it is generally profitable and successful, Nordic farming has its problems. For example, the fragmentation of ownership can reduce effective land management. Again, in the marginal agricultural lands of the north, commercial farming – though technically feasible – is uneconomic. In order to survive, many small farmers have to rely heavily on elaborate systems of government subsidy, as well as off-farm incomes. By contrast, because of overproduction, in some areas farmers have been paid to withdraw land from cultivation and to dispose of surplus dairy cattle.

Laws have also been introduced to encourage farm amalgamation, increasing the average farm to a viable size. Since the 1950s Nordic farms have fallen in number, with a corresponding enlargement in their average size. In Denmark there are restrictions on the maximum size of farm holdings to prevent smaller operators from being squeezed out of business by larger competitors.

Given the problem of surplus agricultural production, it should theoretically be possible to supply all essential needs from only the best farmland, without resorting to farming in the high-cost, climatically sensitive areas. The strong agricultural representation in all five national parliaments, which is generally opposed to radical changes in agriculture, often constitutes a problem in its own right. In any case, farming is so much a way of life in the Nordic countries that social arguments in its favor tend to outweigh the logic of economics.

FUR FARMING

Fur farming – breeding animals purely for the value of their pelts – is an important agricultural specialty in Finland, particularly in the north where other agricultural options are limited by the climate. The most commonly bred animals are mink, fox and racoon. Hybrid breeding has increased the size of the animals, and so of the pelts, as well as the range of their natural colors: there are, for example, eight different shades of mink. A quarter of all foxes are bred through artificial insemination. Many of the highly bred animals, especially the "super" foxes that yield an exceptionally large pelt, are very susceptible to disease, making rigorous care necessary.

Feeding is carefully controlled. The animals need twice as much food in the winter as in summer, and their diets of fish, cereals, vegetables and slaughterhouse by-products are seasonally adjusted. Much research has been done on the best methods of caring for the animals in winter, though the cold, dry prolonged winters of the far north naturally produce the finest pelts.

Almost 5,000 fur farms employ more than 25,000 people – many of them women – in what has become a highly professional industry, with its own training establishments, two research stations and a central storage warehouse. Production increased rapidly in the postwar years: by 1990 4 million mink pelts were being produced every year, as opposed to only 70,000 in 1950, and there have been corresponding increases in the production of other types of skins. The development of fur farming has favored the rise of an exclusive trade in the manufacture and sale of fur garments. These are handled by some forty wholesalers, some of whom are direct descendants of the great Russian furriers of the tsarist era. Most fur products are exported, the principal markets being those of Hong Kong, Japan, South Korea, Switzerland and the United States.

The Finnish fur farming industry has been threatened in recent years by international pressure groups opposed to fur farming and the wearing of fur garments. Their campaigns have had a marked influence on public opinion, particularly in Western Europe where fur prices have fallen sharply. Decreasing demand has already taken its toll: after 1986 the number of fur farmers was in decline.

Reindeer herding: the Sami specialty

Reindeer husbandry is the traditional occupation of the Lapps, or Sami people, who live in northern Norway, Sweden and Finland. There are now an estimated three-quarters of a million semidomesticated reindeer in these areas, probably more than at any time in the past. They are managed by only a very small number of Sami – about 2,000 in each of the three countries. This makes Scandinavian reindeer husbandry extremely intensive.

Sami families generally live on smallholdings on the fringe of permanently settled areas, and supplement their incomes with fishing, hunting and craft work. In Norway and Sweden, the Sami have an exclusive right to engage in reindeer husbandry, and receive limited state subsidies. In Finland, Finns as well as Sami who live in reindeer-herding areas may also own reindeer. There are no state subsidies. Reindeer owners, who are taxed according to the estimated size of their herds, are organized into herding associations. Representatives of these associations meet every year at a "reindeer parliament".

The herds, which vary greatly in size, are not nomadic but migrate seasonally along fairly well-defined routes, occupying lower wooded ground in winter and upland grazing land in summer. There has always been freedom of movement across the open frontiers of Finland, Sweden and Norway, though herds tend to remain within the national territories. There is no movement of reindeer from Norway and Finland across into the Soviet Union.

The seasonal rhythm of activity for the herders reaches peaks in late winter, when calving takes place, and in the fall, when the reindeer are rounded up, sorted according to their markings, and selected for culling. About one-quarter to one-third of any herd is culled each year. Meat from the carcasses is then smoked, salted, refrigerated or canned.

It has been calculated that a herd of at least 500 reindeer is necessary to provide its owner with an income equivalent to that of the average industrial worker. If the herd is small, then losses due to predators, traffic accidents and starvation (thousands of which occur every year) can be very costly to its owner. The labor required to tend reindeer is roughly one day per animal per year, so that a herd of 500 reindeer would need 500 days of labor (including extra help required for the roundup). Herding has been brought up to date with the use of snowmobiles and mobile telephones, and particularly large herders even resort to helicopters and light aircraft.

A cause of conflict

Reindeer husbandry is undoubtedly the most efficient way of using the northern tundra lands, where conditions prohibit most other forms of agriculture – but it causes problems where it impinges on other activities. There is continual conflict between reindeer herders and farmers, foresters, road and rail authorities, and hydroelectricity operators. While reindeer can damage farmland and woods, chemical fertilizers and insecticides can harm reindeer. The construction of highways, railroads and hydroelectric dams may interfere with established grazing rights and migration routes.

The Sami are a vocal ethnic minority who campaign strenuously to preserve their traditional rights. For this reason, reindeer husbandry – the symbol of the Sami way of life – has become a national issue in the three countries, even though it is of only limited economic importance. While some experts consider that the number of reindeer now exceeds the carrying capacity of the grazing lands, the herders point out that modern methods of forestry encourage grass and herb growth at the expense of the lichens that are the reindeer's principal source of winter fodder. In any case, many reindeer die of starvation every year. A totally unforeseen environmental disaster struck in 1986 when fallout from the explosion at the Chernobyl nuclear plant in the Soviet Union contaminated the lichen pastures, necessitating the slaughter of thousands of reindeer.

Herd management (*above*) Sami herders tag their reindeer for identification purposes. The fall is one of the busiest times of the year, when the reindeer are rounded up and corralled. Some of them are culled for meat and leather products.

Reindeer by the thousand (*right*) There are more semidomesticated reindeer in northern Scandinavia than at any time in the past. Some critics believe that the number of reindeer now exceeds the carrying capacity of the grazing lands.

AN AGRICULTURAL PATCHWORK

THE GROWING PACE OF CHANGE · FARMING AS A BUSINESS · ADAPTING TO CHANGING DEMAND

Agriculture in the British Isles is still extremely diverse – the result of the varied climate and soils, market forces and government policies. On the British mainland (England, Scotland and Wales) there is a broad division between the drier eastern lowlands, where arable farming of cereals predominates, and the wetter uplands in the west, where dairy and sheep farming are more common. Since the end of World War II changes in farming techniques and practice have led to fewer, larger and more mechanized farms – a trend that has been encouraged by government policies, particularly since Britain and Ireland became members of the European Community (EC) in 1973. Farming in both countries has become increasingly specialized, with intensive livestock production replacing the traditional mixed farming.

COUNTRIES IN THE REGION

Ireland, United Kingdom

Land (million hectares)

Total	Agricultural	Arable	Forest/woodland
31 (100%)	24 (78%)	8 (25%)	3 (9%)

Farmers

793,000 employed in agriculture (3% of work force)
10 hectares of arable land per person employed in agriculture

Major crops

Numbers in brackets are percentages of world average yield and total world production

	Area mill ha	Yield 100kg/ha	Production mill tonnes	Change since 1963
Barley	2.1	51.4 (221)	10.8 (6)	+49%
Wheat	2.0	60.2 (258)	12.3 (2)	+220%
Rapeseed	0.4	34.8 (244)	1.4 (6)	+45,533%
Sugar beet	0.2	400.6 (116)	9.6 (3)	+40%
Potatoes	0.2	356.8 (227)	7.5 (3)	−19%
Oats	0.1	46.8 (254)	0.6 (1)	−71%
Vegetables and dry peas	—	—	4.3 (1)	+29%

Major livestock

	Number mill	Production mill tonnes	Change since 1963
Sheep	29.6 (3)	—	−14%
Cattle	18.1 (1)	—	+8%
Pigs	8.9 (1)	—	+10%
Milk	—	21.1 (5)	+41%
Fish catch	—	1.2 (1)	—

Food security (cereal exports minus imports)

mill tonnes	% domestic production	% world trade
+3.0	12	1

Harvest time This apparently timeless scene shows signs of the recent changes in Britain's agriculture. Hedgerows have been removed to allow access for large harvesting machines. These separate the grain from the straw, which is bound into cylindrical bales.

THE GROWING PACE OF CHANGE

The agricultural innovations that revolutionized farming in much of Europe during the 18th century led in Britain to the creation of a landscape of small farms, each containing a patchwork system of small fields. Agriculture generally was dominated by livestock farming on permanent grass pastures, with some grain production, mainly wheat, in the east. Increased efficiency and mechanization had made some headway before 1914, when the outbreak of World War I disrupted imported food supplies, particularly those that came from the colonies. These had previously provided Britain with cheap foodstuffs.

The government consequently encouraged home production, particularly of arable crops, by guaranteeing that the prices farmers received for certain crops would not fall below a specified level. This was achieved through a system of deficiency payments – the difference

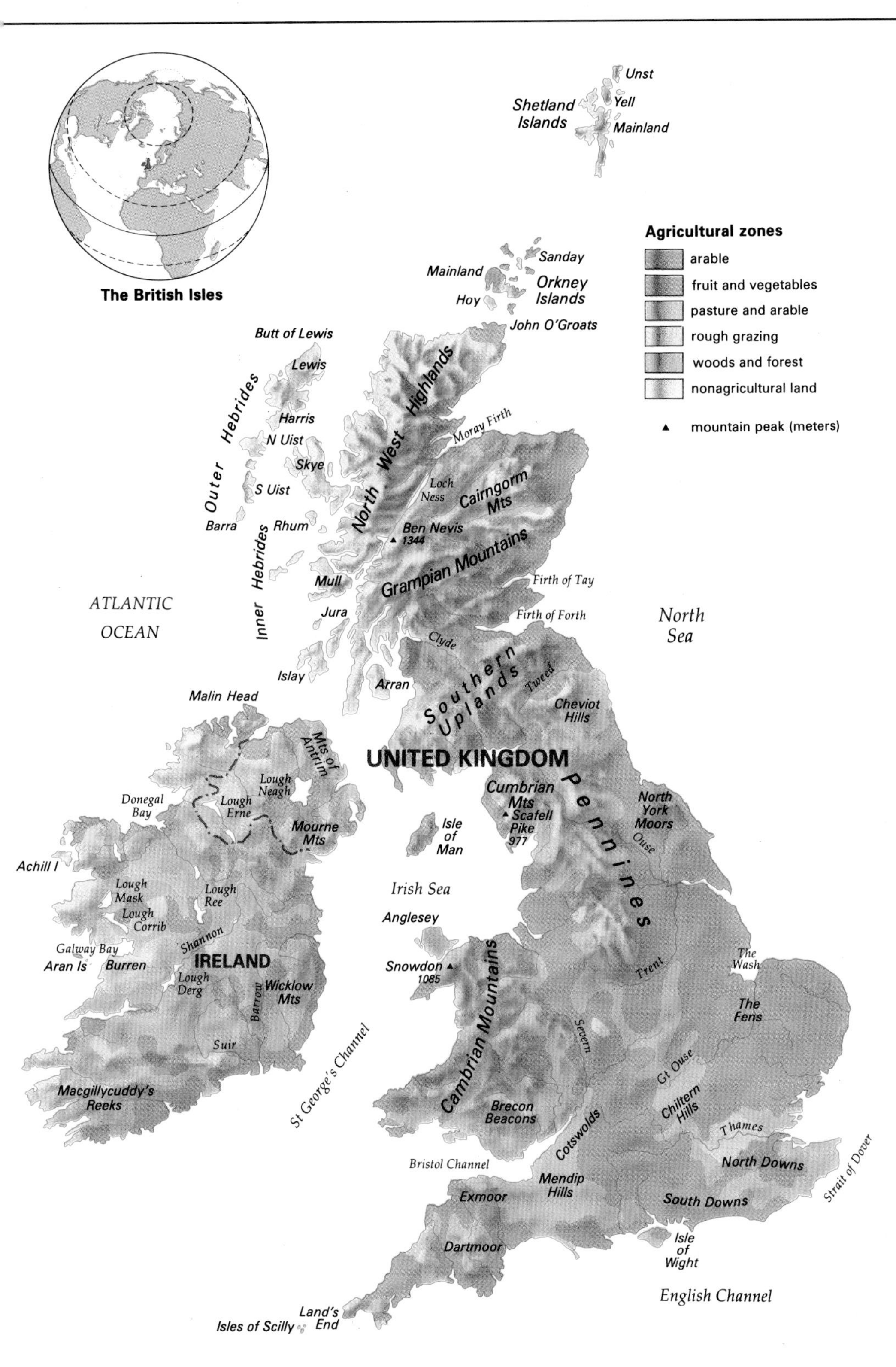

Map of agricultural zones The rapid growth of commercialized methods of production has led to greater specialization, with arable farming becoming concentrated in the drier eastern lowlands and dairy farming in the wetter uplands of the west and north.

between the market and the guaranteed price. These helped maintain a reasonable income for assisted farmers, and the emergency measures continued after the war. The severe depression of the world economy in the interwar years increased the need to protect British agriculture, and so legislation was introduced to stabilize farming.

Grants and subsidies

World War II marked the start of a profound change in British agriculture. Arable crops were again encouraged by the government, through grants and subsidies. As a result the area of arable land increased by 2.3 million ha (5.6 million acres) between 1939 and 1944, a phenomenal rise of 63 percent. After the war the aim was to ensure that there were adequate national food supplies, so many wartime policies were continued. Prices were guaranteed and farmers received deficiency payments for certain crops (cereals) and livestock (beef cattle and sheep). A dramatic exaggeration of the east–west production pattern resulted, with production in each area becoming more specialized.

Producer-controlled marketing boards, such as the Milk Marketing Board, introduced with limited success in the 1930s, were strengthened in 1947 – an action that had a particularly far-reaching impact on the distribution of dairy production. The Milk Marketing Board had been set up to reduce the cost of transporting milk to dairies and ensure year-round supply of milk. Prices and transportation costs were now standardized so that proximity to major urban centers was no longer essential. As a result dairying became more important in the isolated western areas of the British Isles.

Deficiency payments continued to be made until Britain and the Republic of Ireland both joined the EC. Under the EC's Common Agricultural Policy (CAP), which came into full effect in Britain in 1977, farmers receive guaranteed prices for most products – but not horticulture, potatoes, pigs and poultry – through intervention buying. The EC sets guaranteed prices for goods; if market prices fall below this level, it will intervene to purchase them, thus maintaining the farmer's profits. Together with import controls, intervention buying has established a relatively risk-free economic environment for farmers, who have responded by intensifying production.

The aim of the EC's policies was to ensure a fair standard of living for farmers and guarantee regular supplies of food at reasonable prices to consumers. Under the original CAP the more a farmer grew, the greater was the subsidy he received. Larger farms therefore got the lion's share of CAP money, allowing them to flourish at the expense of smaller ones; agricultural specialization increased even further, and production in northern parts of the region of goods such as cereals, beef and milk rose. The CAP was so successful in increasing production that surpluses were created in certain products.

The growth of grain and beef "mountains" and milk "lakes", together with increasing concern over the environmental consequences of a more intensified and specialized system of farming, led after 1984 to changes in policy. In order to reduce, rather than continue to expand, agricultural production, the EC introduced milk quotas, farm diversification and subsidies to encourage farmers to take land out of production (set aside).

Cattle in the morning mist The hills of the west, where conditions are wetter, provides pasture for dairying and the fattening of beef cattle. Once a part of traditional mixed farms, dairy farming has become increasingly specialized. The introduction of milk quotas in 1984 to curb overproduction reduced the profitability of some farms, and the number of dairy herds dropped sharply as a result. Beef production has become concentrated in upland areas of poorer quality pasture, or in specialized intensive units.

FARMING AS A BUSINESS

Farming systems in the British Isles have been transformed since World War II as part of a general process of modernization or, in some cases, industrialization. As farms have become larger, with a greater proportion being owner-occupied rather than rented, they have become more capital intensive. They have also become more dependent on other sectors of the economy such as food processing companies and the suppliers of fertilizer, pesticides and machinery. Farms closely involved with these sectors have become akin to businesses, typifed by contract farming of peas and intensive techniques of livestock production, which have strictly controlled "factory-like" regimes. Some of these trends have especially characterized the arable farming of the eastern part of Britain. The seed suppliers, grain merchants and specialist machinery contractors who support arable farming came to be located in this area, reinforcing the specialization.

Arable farming

Wheat production in these areas has seen a dramatic increase since World War II, at the expense of barley and oats, owing to the higher prices offered under the CAP and the introduction of improved strains. In 1983 the total wheat-growing area exceeded that of barley for the first time. Developments in winter wheat varieties have also enhanced yields, and 95 percent of all wheat is now sown in the fall. Barley has increased in importance in western and northern areas, as new varieties can withstand the cool, wet conditions.

The growing of vegetables, fruit, flowers and bulbs, which all come under the heading of horticulture, remains concentrated in the relatively small areas of Britain that have good soil and a high number of frost-free days, though recent technological developments in greenhouse production have reduced dependence on a favorable climate. New methods of transporting and refrigerating perishable goods have removed the need to be close to major urban centers.

THE EGG AND POULTRY INDUSTRY

Most poultry in the British Isles are commercially raised. As the chickens are kept indoors where the environment is closely monitored, all that is required to set up a poultry farm is a small area of concrete, a suitable building and large amounts of capital. Large companies, able to supply the capital required, are increasingly becoming closely involved in poultry production – an example of an industrially organized type of agriculture.

There are two kinds of indoor poultry production: broilers for meat, and battery hens for eggs. The numbers of broilers – chickens up to seven weeks old – kept in Britain have increased, mainly as a result of the move in consumer tastes away from eating red meat, for health reasons. There are similar consumer concerns about the high levels of cholesterol found in chicken eggs, and demand for these has steadily declined since 1970.

The market for battery-hen eggs has been further depressed by consumer concern for the welfare of hens raised in artificial indoor conditions. There are worries, too, over diseases such as salmonella food poisoning, caused by bacteria that are widespread in hens and may be passed on in food to humans. Consequently free range eggs, produced from hens allowed to roam freely outdoors, command a premium price in the marketplace, and many people are prepared to pay higher prices for free range birds as well.

Factory farms Overcrowding is common in these units, and sickness or injuries through pecking may go undetected.

As a result, traditional horticultural areas in the southeast and in west-central England have suffered a relative decline. Horticulture survives here because of the expertise of local growers, many of whom have formed cooperatives to coordinate the sale of produce and to share the costs of pesticides and specialized machinery. Such farms are small and fragmented, contrasting with the field-scale cultivation of vegetables such as potatoes and peas, which has expanded greatly in the eastern fenland areas of England where farms are large and production more efficient; these large farms are often under contract to supply supermarket chains and food processing companies.

Dairying and livestock production

Dairying is one of the most widespread types of farming in the British Isles, and it remains popular because it is the only farming activity that provides a regular monthly income. It is particularly concentrated in the permanent pasturelands of the west, where it was traditionally the most profitable enterprise of numerous small, mixed farms. Government policy over the years encouraged such farmers to increase their landholdings, mechanize milk production and expand the size of their herds. When milk quotas were introduced in 1984 to reduce overproduction, many farmers were forced to give up dairying (except in Ireland, where production was allowed to increase): in 1980, 27 percent of farms in Britain had dairy herds; in 1987 the numbers had dropped to 17 percent. The farmers most severely affected by quotas were those in areas where there were few profitable alternatives – just the areas where the government had previously encouraged farmers into dairy specialization.

Livestock production of beef and sheep has become increasingly concentrated in the marginal western and northern areas of Britain and Ireland on the poorer quality upland pasture. In areas where land quality is higher, livestock farming can still be profitable in specialized intensive livestock units. This has led to increased links with feed companies and in turn to difficulties over diseases such as bovine spongiform encephalopathy (BSE). The disease was introduced into cattle from sheep by feeding them high protein compounds, including waste animal products, to boost growth.

Pig and poultry factories

Pig production traditionally took place in smaller, mainly dairy, farms, but as it has become more intensive both specialization and pig numbers per farm have increased: in some respects the production of both pigs and poultry is now closer to that of factories than other types of farming. As the animals are housed permanently indoors, the units can be located anywhere in the region regardless of environmental conditions. However, pig production has become particularly important in the major arable areas of the east, where supplies of large quantities of grain for feed are readily available and transportation is no problem.

ADAPTING TO CHANGING DEMAND

While the modernization of British agriculture has led to greater efficiency and productivity, it has had major environmental and social effects. The removal of hedgerows to increase field sizes and allow easier access for tractors and combine harvesters, as well as specialist machinery for cultivating crops such as potatoes and sugar beet, has destroyed wildlife habitats and caused loss of soil through wind or water runoff. Increased use of fertilizers has raised nitrate levels in water, particularly in eastern Britain. Pesticide spraying and stubble burning close to centers of population causes increasing concern. Mechanization has reduced the need for a large labor force, making many farm workers redundant and leading to rural depopulation in some parts of the region.

Technical innovation has been developed and promoted by the government. The EC encourages farmers to grow certain new crops by providing a relatively risk-free economic environment under the CAP. Self-sufficiency within the EC is a major goal, and crops that the EC is deficient in are encouraged. In the early 1970s, for example, the EC Agricultural Commission decided to offer farmers high prices for oilseed rape in order to make good a large Community deficiency in vegetable oil and protein animal feed, which can be extracted from rape seed. This inducement was so successful that the area in England and Wales under oilseed rape cultivation increased from 13,670 ha (33,780 acres) in 1972 to 276,360 ha (682,900 acres) in 1986.

In terms of land area, oilseed rape is now more important than potatoes, oats and sugar beet, being the third largest crop in cultivation after wheat and barley. Initial cultivation of oilseed rape was in the south and has spread northeastward with the introduction of varieties more suited to the colder climatic conditions. However, in 1987, as its policies to expand production had been so successful, the EC then began to curtail oilseed rape cultivation by first cutting, and then removing, levels of price support.

A technological treadmill

For many farmers the introduction of technological innovations has not been entirely beneficial. They have been placed on a "technological treadmill". In order to maintain their incomes they have been forced to adopt new methods to increase production and to remain competitive. Often these have caused them to go heavily into debt. Not only are innovations expensive, but the cost of inputs such as fertilizer has also increased at a faster rate than the prices received for the end product. As a result, many farmers have become part-time farmers, taking jobs outside agriculture in order to supplement their incomes, or they have abandoned farming altogether.

The use of new technology and government incentives has been so successful in raising levels of food production in the

City meets country (*above*) Many members of Britain's predominantly urban population have little opportunity to experience the ways of rural life. In some cities, farms have been set up on derelict land. These allow children to care for livestock and grow a few crops.

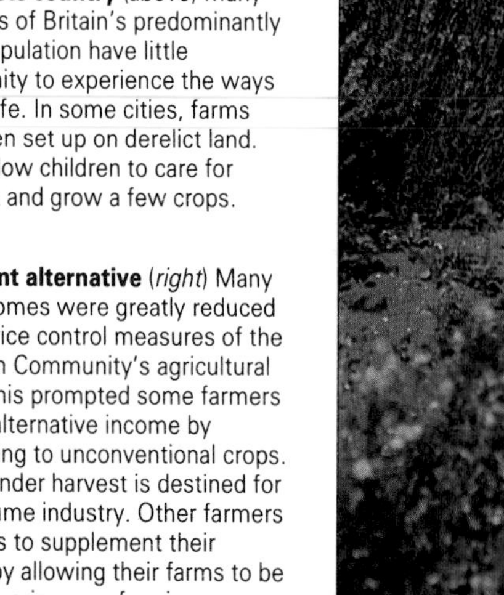

A fragrant alternative (*right*) Many farm incomes were greatly reduced by the price control measures of the European Community's agricultural policy. This prompted some farmers to seek alternative income by diversifying to unconventional crops. This lavender harvest is destined for the perfume industry. Other farmers find ways to supplement their income by allowing their farms to be used for various nonfarming activities, or by finding outside jobs.

FARMING CLOSE TO THE CITY

Farmers with land on the edge of towns and cities have to take account of a number of factors that do not affect farming in general. The most obvious of these is that when land on these urban fringes is taken for residential, industrial or road development, it can either provide a farmer with the capital he or she needs to survive, or it fragments the farm, making it small and unviable. Additionally, the prospect that agricultural land might be subject to future urban development affects not only its current use and value, but also what system of farming to adopt in the meantime – intensive systems to maximize land use before development takes place, or more extensive ones to minimize financial risk. The farmer's choice is further complicated by land ownership patterns, past experiences of land loss and his or her own attitude toward farming.

Proximity to the city brings both problems and opportunities. Although incidents of trespass, vandalism and theft become more numerous closer to centers of population, a wider range of employment openings nearer to hand means that many farmers are able to supplement their agricultural incomes with outside jobs. The converse of this is that people who work in the city sometimes purchase smallholdings on the urban fringe and take up farming as a hobby.

Farmers may also benefit from increased market opportunities. They can supply local shops and restaurants or diversify into activities such as farm shops, commercial letting or recreation. Many suburban farms, for example, work closely with education authorities in providing facilities for school educational visits, giving children the chance to see a working farm.

British Isles that a different problem now exists. After 1984 attention switched from finding ways to increase food production toward methods of reducing it. However, farmers are reluctant to produce less, even with government payments to prevent their incomes falling. One solution might be a return to more mixed systems of production, and to organic farming in particular. The latter has been viewed as a way of overcoming the linked problems of overproduction, declining farm incomes, environmental degradation and unhealthy food. Another possibility is to increase income from farm diversification into other activities.

Alternative enterprises

Farm diversification can take a number of forms. Farmers can produce unusual foods that are not in surplus and are not subject to government control. Examples of farming for such specialized markets are snail production, or the cultivation of evening primrose for the medical market. Another way is to process conventional agricultural products on the farm and sell them for a higher price than unprocessed ones. Making jam on the farm itself, for example, will generate more profit than selling fruit to a large jam-processing company. Similarly, selling direct to the customer through a farm shop will also be more profitable. Farmers can also diversify into tourism and offer accommodation, recreational facilities, farm museums and visitor centers. Ventures that are entirely unrelated to agriculture are also possible money-makers: for example, rented units for craft workshops, offices or residential accommodation can be set up in disused farm buildings.

Such diversification may seem to be the answer to agricultural overproduction or declining farm incomes, but it is not a solution open to everyone. Alternative enterprises often require large amounts of initial capital as well as specialized production and marketing skills. A further drawback is that 40 percent of farms in Britain are either wholly or partly rented from landlords, who are often reluctant to allow tenant farmers to indulge in unconventional farming activities.

Jersey's early crops The Channel Islands take advantage of their better climate to supply early crops of vegetables, such as these cauliflowers, to the mainland. Most Jersey farmers rely heavily on immigrant labor. Farmland on the island reaches the town limits, as there is intense pressure on land.

Hill farming

Hill land – normally defined in the British Isles as that above 240 m (800 ft) – is found mainly in western and northern areas. Compared with their lowland colleagues, hill farmers suffer from various disadvantages. With increasing height above sea level and the more rugged countryside, the range of crops that they can grow diminishes, and the quality of grassland for dairying and livestock fattening declines. As a result the rearing of sheep and beef cattle to be sold on for fattening elsewhere is the main farming activity. The harsh weather conditions, poor soils, isolation from the major markets and the often fragmented farm units also make life difficult for hill farmers.

In Scotland, in the far north of the region, crofting is important in the uplands, with sheep breeding and rearing accounting for 90 percent of farm income. Crofts are typically small – half of them are less than 5 ha (12 acres) – though there is access to large areas of common grazing land. As 85 percent of holdings provide less than two full days of work each week, crofters rely on additional occupations and government support.

The problems of hill farming were recognized in the early 1940s by the government, which introduced ewe and cow subsidies as well as farm improvement grants, and increased expenditure on roads, electricity and water supplies. This scheme was discontinued after 1963, but a new package was introduced in 1967 that attempted to encourage the reseeding and fencing of hill pastures.

Types of sheep Domesticated sheep provide people with wool, milk and meat. Although Britain was noted for its fine wool-producing sheep in medieval times, the first two functions declined in importance after the Industrial Revolution, and the types of longwool and Down sheep most typically found on British farms are the result of selecting breeds for meat production; the wool has secondary importance only. The major wool-producing regions are in the southern hemisphere, from Merino sheep.

"Less Favored Areas"

In 1975, some 41 percent of land in Britain was designated by the EC as falling within the category of Less Favored Areas (LFAs). This means they are recognized as having limited potential for agricultural improvement, low levels of farm income and a declining population dependent

The upland in winter Large areas of Britain are so hilly that they are unsuitable for cultivation and are left to grass. Breeds of hill sheep need to be tough to survive the bleak conditions. Lambing usually begins in April when the snows are over.

mainly on agriculture. This designation was expanded in 1985 to cover 53 percent of the country. Payments are made to compensate farmers for the disadvantages of hill farming, to encourage farm modernization and to dissuade them from leaving the countryside.

As under previous schemes, farmers are paid annually for each head of livestock kept and are offered improvement grants. However, in common with most CAP payments, this system has benefited the larger farms rather than the small ones, which are most in need of aid. It has also provided an incentive for farmers to improve their rough grazing land, leading to a reduction in the diversity of upland habitats and wildlife.

Increasing concern over the pressures on the environment and the need to take measures to protect it led in 1986 to the designation of Environmentally Sensitive Areas (ESAs). Five of the nine original ESAs fell within the hill farming LFAs. Farmers within these areas are given financial inducements to practice traditional farming methods that help to preserve the characteristic landscape rather than altering it.

Competition in the hills

Farming in the hills is subject to growing competition from other activities. A policy of planting coniferous trees in upland areas was begun earlier this century to reduce Britain's heavy reliance on imported timber. Forestry, which tends to be viewed by the government as a more economic use of uplands than hill farming, continues to expand.

Furthermore, increasing numbers of people are visiting hill areas for holidays and weekend breaks. Some 70 percent of national parks are in the uplands of England and Wales; their aim is to preserve and enhance the natural beauty of the landscape and promote public enjoyment of the countryside. Three-quarters of the national park land is privately owned, mainly by farmers, who must notify the park authorities when farm improvements are necessary. Because any changes to their farm must not adversely affect the appearance of the landscape, it becomes difficult to maintain a viable farm business, and many farmers have had little alternative but to abandon agriculture altogether.

The yellowing of Britain

The recent rapid increase in the cultivation of oil seed rape has had a marked impact on the British landscape: in spring its vivid yellow flowers paint great swathes of color across the countryside that contrast with the greens of the young wheat crop.

Oil seed rape (sometimes known as colza) is one of a large number of crops grown mainly for their oil content. A native of Europe and Asia, it is an annual plant that grows to a height of about 1.5 m (5 ft). The pods that form after the flowers have fallen contain hundreds of tiny seeds.

The oil that is pressed from the seeds is used for cooking, for fuel and as a lubricant, as well as in soaps and in the manufacture of synthetic rubber. The oil cake – the residue after the oil has been pressed out – makes a protein-rich and mineral-rich animal feed. It may be fed directly to larger animals or ground up as oil meal.

As the climate becomes warmer – a result of increased carbon dioxide and other greenhouse gases in the atmosphere – Britain's yellow revolution is likely to be followed by the introduction of other unfamiliar crops. Such crops, including sunflowers and lupins, will have an equally striking effect on the landscape.

The yellow revolution: a rape field in flower in southern Britain.

A FARMING SUCCESS STORY

A NORTH–SOUTH DIVIDE · FRANCE'S "RURAL REVOLUTION" · PROBLEMS OF SUCCESS

France is the "green giant" of Western Europe: in area it occupies almost a quarter of the European Community (EC), and it commands a vast range of land resources. Some 60 percent of the country is devoted to agriculture and 30 percent to woodland – yet only 7 percent of the labor force works the land. The volume of its production is second in the world for wine, third for milk and butter, fourth for meat, fifth for eggs, sixth for wheat and barley, and seventh for sugar, and its agricultural exports are surpassed only by those of the United States. France's farms, forests and fisheries account directly for 8 percent of the value of all exports and indirectly for a further 8 percent, from goods processed from them. Despite its obvious successes, however, French agriculture today is not without its problems.

COUNTRIES IN THE REGION

Andorra, France, Monaco

Land (million hectares)

Total	Agricultural	Arable	Forest/woodland
55 (100%)	31 (57%)	18 (33%)	15 (27%)

Farmers

1.4 million employed in agriculture (6% of work force)
12 hectares of arable land per person employed in agriculture

Major crops
Numbers in brackets are percentages of world average yield and total world production

	Area mill ha	Yield 100kg/ha	Production mill tonnes	Change since 1963
Wheat	4.9	55.6 (238)	27.4 (5)	+119%
Barley	2.0	52.7 (226)	10.5 (6)	+59%
Maize	1.7	71.5 (197)	12.5 (3)	+352%
Sunflower seed	1.0	25.4 (178)	2.7 (13)	+10,979%
Grapes	1.0	88.7 (118)	9.2 (14)	–4%
Rapeseed	0.7	36.0 (253)	2.7 (12)	+1,255%

Major livestock

	Number mill	Production mill tonnes	Change since 1963
Cattle	22.8 (2)	—	+13%
Pigs	12.4 (1)	—	+38%
Sheep	10.6 (1)	—	+19%
Milk	—	28.6 (6)	+14%
Fish catch	—	0.8 (1)	—

Food security (cereal exports minus imports)

mill tonnes	% domestic production	% world trade
+26.0	49	12

A NORTH–SOUTH DIVIDE

France contains three major lowland areas (the Paris basin in the north, and Aquitaine and the Mediterranean coastal belt, both in the south). These are separated and fringed by uplands (the Massif Central, Pyrenees, Alps, Jura Mountains and , Vosges). Soil fertility is excellent in much of the lowlands, but in many upland areas it is poorer.

There are three main climatic regimes. The Mediterranean south has dry summers and mild winters, while the Atlantic west, whose coasts are warmed by the North Atlantic Drift, boasts moderate winter temperatures and abundant rainfall. The third major climatic regime covers northeastern France. Influenced by the climate patterns of the continental landmass of central Europe, winters here are harsh and the growing season is limited, especially in the uplands.

The first traces of farming date from Neolithic times (between 8,000 and 7,000 years ago), when hunter–gatherers in southern France started to domesticate sheep and pigs. During the Bronze Age, about 2000 BC, areas of woodland and scrub were cleared, using simple plows, and cereals, hemp and flax cultivated. Agricultural production expanded under Roman rule (from 57 BC) and important trade in wine, wheat and other foodstuffs developed. As the Roman empire began to break up, from the 3rd century AD on, conditions declined until the medieval period, when trade in farm products flourished once more.

Local agricultural systems varied considerably but there was a major contrast in practices between the north and south of the country. These reflected not only differences in climate and terrain but also variations in customs, laws and social organization. In northern France, for example, agriculture was organized around the village, or manor, but in the south the rights of the family and the individual were more prominent.

Under the northern system land was cultivated to produce cereals for two successive years and then left fallow for grazing and manuring by communal

The mighty Charolais Prized for its great size and rapid growth, this French breed has been successfully exported to other countries for beef production, notably to the United States. French beef production is concentrated in the center of the region; dairy farming is important in the northern lowlands and central and eastern uplands.

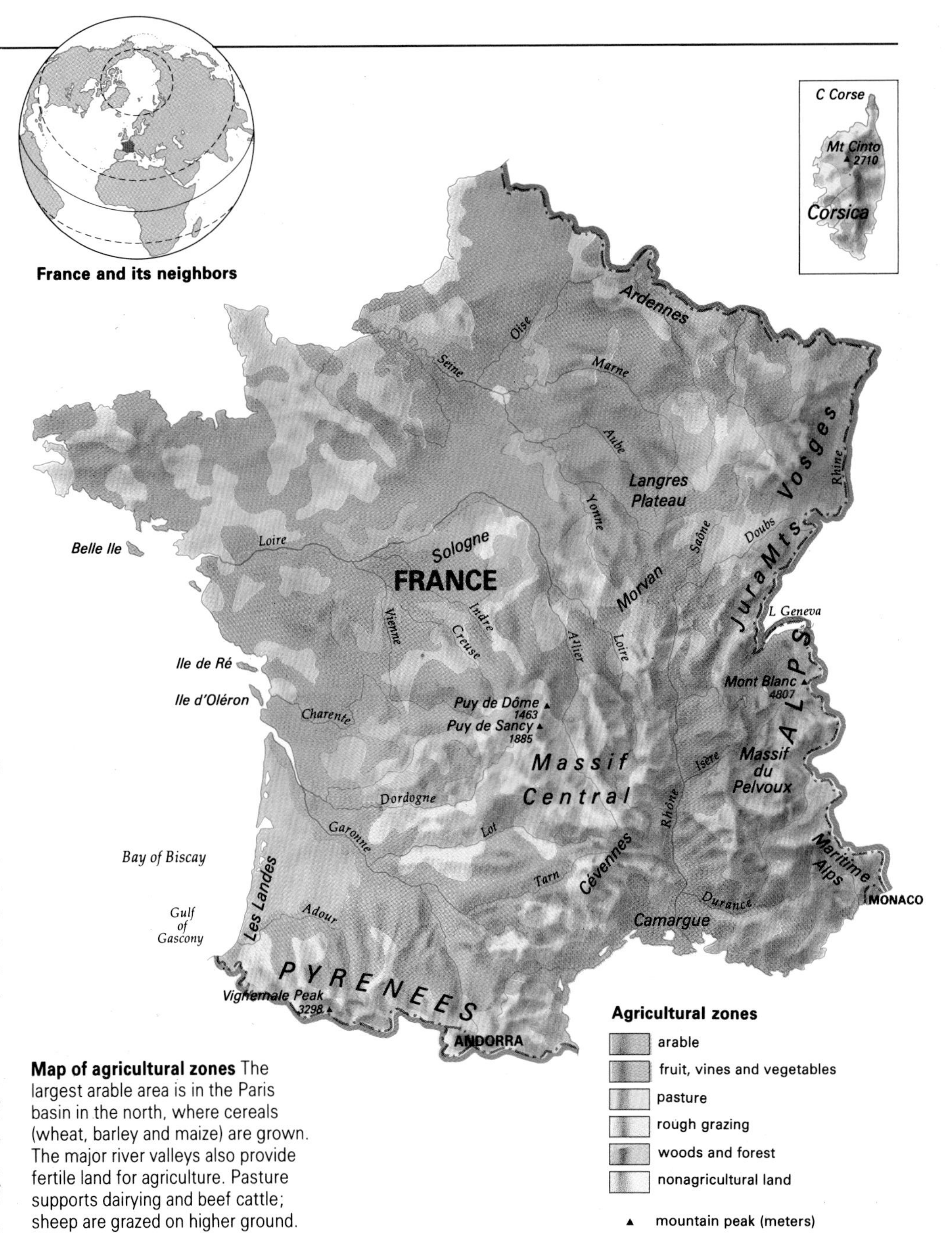

Map of agricultural zones The largest arable area is in the Paris basin in the north, where cereals (wheat, barley and maize) are grown. The major river valleys also provide fertile land for agriculture. Pasture supports dairying and beef cattle; sheep are grazed on higher ground.

flocks in the third year. In the drier conditions of southern France cereal production alternated with a fallow year, and sheep were shifted to summer pastures in nearby uplands (transhumance). Both systems generated a range of cereal and animal products but neither was very efficient in restoring soil fertility. The normal way of increasing food output was by bringing more land into cultivation.

Entrenched farming methods

French farming improved very slowly from medieval times until the 1850s. Potatoes, maize, tobacco and various fodder crops were introduced to the region. However, most French peasants did not specialize but concentrated instead on subsistence farming, producing a range of cereals, vegetables and animal products (polyculture). Occasional surpluses would be marketed, but commercial farming was restricted to the immediate surroundings of towns and cities and to places linked to them by rivers or easy land routes. France differed from England, the Low Countries and northern Italy because it never experienced an "agricultural" revolution.

During the French Revolution (1789–99) feudalism – the medieval form of landholding based on a complex system of rights between landlord and tenants – was abolished and new laws for land inheritance were passed. The effect of these changes was to strengthen the hold of peasant-family farming, which in turn blocked radical innovations in the countryside. Therefore, in order to increase food supplies, moors, marshes and hillsides continued to be cleared or drained so that they could be cropped.

The building of the railroads in the mid 19th century precipitated the change to agricultural specialization. Many farmers responded to these new opportunities but thoroughgoing change was hindered by the traditional peasantry and by protective trade legislation, which sheltered farming from the sharp shocks of free trade and modernization. In contrast to farming in Britain, Denmark and the United States, low crop yields, modest mechanization, only limited applications of fertilizer and a surplus of labor were prevalent. In 1901 over 40 percent of the French labor force worked the land; by 1945 36 percent still remained in the farming sector. The prospects for French agriculture looked bleak so long as farming methods remained unchanged.

FRANCE'S "RURAL REVOLUTION"

After the end of World War II in 1945 France entered a phase of rapid urbanization and industrialization. Overmanned and underproductive agriculture advanced far more slowly, and poverty on family farms was widespread. During the late 1950s farmers rioted in many areas, demanding greater price support and government investment in farm modernization. Laws passed in 1960 and 1962 launched a "rural revolution" by helping to reduce the labor force, by modifying the size and layout of farms, and transforming many aspects of agricultural practice. Pension supplements encouraged elderly farmers to amalgamate their land with that of neighboring farms or to release it to younger farmers, who were eligible for special training grants and start-up awards. The combination of government schemes, retirement and natural wastage had trimmed the farming workforce from just over 5 million in 1955 to 1.5 million in 1990.

More than thirty regional organizations encourage farm amalgamation, creating less fragmented, more manageable holdings, and assist young farmers to acquire property. They deal with two-fifths of all farmland that comes onto the market, and helped to reduce farm numbers from 2.3 million in 1955 to just over 1 million in 1990 – but the average farm is still only 28 ha (70 acres) and half of all holdings are less than 20 ha (50 acres) apiece. At the other extreme the Paris basin, where largescale, market-oriented farming developed early on to serve the capital's demand for cereals and other foodstuffs, contains some giant farms of 750 ha (1,850 acres) or even more. Many kinds of cooperative have been set up, ranging from neighbors who share equipment, through organizations for buying and using machinery on a communal basis, to larger enterprises that commission, purchase, process and eventually market the farmers' crops and livestock.

One vital aspect of this "rural revolution" has been that many French farmers have realized they can no longer concentrate on subsistence farming. Instead they must respond to commercial pressures and raise loans to modernize and compete, rather than rely on savings as their parents and grandparents had done. They have had to learn how to think and behave as businessmen.

Dramatic productivity gains

Arable crops occupy over half of all farmland in France, but are particularly dominant in the Paris basin. Cereal production doubled in the 1950s and then tripled in the quarter century after 1960, even though progressively less land was used. This transformation was effected by the use of mechanization, new fertilizers and new plant strains in response to a soaring demand for cereals to be used as animal feed. The greatest changes of all, however, were brought about by the establishment in 1957 of the European Economic Community, later known as the European Community (EC). The introduction of price guarantees under the Community's Common Agricultural Policy (CAP) offered very considerable incentives for seemingly endless increases in production.

Wheat and barley production have both doubled since the "rural revolution", whereas maize output rose tenfold between 1955 and 1980. Over the same period the amount of land under sugar beet doubled, and output rose from 1.5 million tonnes to 4 million tonnes. Despite the imposition of EC quotas since then, France remains a leading producer and exporter of beet sugar.

Until the 1850s commercial farming of fruit, vegetables and vines was restricted to areas close to major towns and alongside navigable rivers. Thereafter rail transportation enabled many farmers in southern and western France to deliver their crops farther afield, to Paris and other cities. The commercial production of fruit and early vegetables gradually became concentrated in the Mediterranean south and in the valleys of the Rhône and Garonne, while farmers from along the northern coastal fringe of Brittany in the northwest of the region specialized in producing potatoes, cauliflowers, artichokes and onions. New plant strains and a more extensive use of irrigation and modern fertilizers have boosted output even further.

Cattle are raised in most parts of France except the Mediterranean south. There are, however, major concentrations in lowland areas of northern France (from Flanders, through Normandy and Brittany, to the Loire) and in the uplands of central and eastern France. Large, high-yielding dairy herds in northern France contrast with much smaller herds (10–20 head of stock) on many upland farms. Dairying accounts for a sixth of the total value of French agricultural output and yields a wide range of products including butter, yogurt and cheese as well as fresh milk. Skimmed milk and various other by-products are fed to the large numbers of pigs that are raised in these major

The cereal harvest The commercialization of farming has encouraged the consolidation of strip fields into larger blocks of land. Many farms are now run cooperatively. Greater field size increases profitability, but also the problem of soil erosion.

dairying areas. Beef cattle are more characteristic of central France, which is the home of the famous Charolais and Limousin breeds, whereas sheep are most numerous in the Causses – the limestone uplands of the Massif Central.

More than a quarter of France is forested, and the extent is increasing as marginal farmland is abandoned. The ancient oak and beech forests of northern France have been carefully managed for centuries for their valuable hardwood timber. By contrast the evergreen woodlands of the Mediterranean south have only modest economic value, though they are important for controlling soil erosion on hill slopes.

CHEESE

President Charles de Gaulle (1890–1970) once remarked that it was impossible to govern a nation that contained more than 365 cheeses. Cheese is an important element in the French diet; on average each Frenchman consumes as much as 19 kg (42 lb) a year.

Hard cheese was made as far back as Roman times but soft varieties are more recent: for example, Camembert – the popular soft cheese of Normandy – was invented in 1790. In the 1930s over 400 types of cheese were being produced; by 1990 the total had fallen to about 200, most of which were made from cow's milk. Some, such as Roquefort, use goat's or ewe's milk.

Before rapid transportation and chilling techniques became available there was a simple logic to the geography of cheese making. Fresh cream cheese (*fromage frais*) and soft cheeses, which did not keep long, had to be made in dairying areas within easy access of centers of demand, notably Paris. By contrast, the pressed, hard and often cooked cheeses such as Cantal or Gruyère would keep for long periods and could be produced in distant uplands with high summer pastures, such as the Massif Central, Jura and the Alps. Most French cheese is now factory produced, but old-established techniques also flourish. Since 1966 strict regulations have controlled production of more than two dozen traditional cheeses to guarantee their high quality.

King of cheeses Roquefort is one of the world's finest blue cheeses. It is made during the lambing season from ewe's milk in creameries all over France, but is always matured in the humid limestone caves near the town of Roquefort-sur-Soulzon in southern France.

PROBLEMS OF SUCCESS

France's massive postwar increase in farm productivity has not been achieved without some sacrifices. Mechanization has changed the scale of operations and has required farms and fields to be enlarged and remodeled, with the removal of surplus hedgerows, trees and ponds to accommodate the recent proliferation of tractors and other highly specialized equipment. Accelerated runoff and wind erosion have resulted, leading to serious ecological damage in Brittany and several other areas, and the loss of protective habitats for wildlife. The annual application of chemical fertilizers rose fourfold from 1955, to 6 million tonnes in 1990, creating problems of water pollution and longterm soil management.

New plant strains have had a major impact on the location and productivity of many crops. Until the 1960s maize was grown almost exclusively in southwest France, where the combination of moisture and heat was ideal. Since then new hybrid varieties have conquered the deep, fertile plowlands of the Paris basin, where portable irrigation systems provide water to boost growth during the summer months. Since the 1960s the widespread cultivation of the American Golden Delicious variety has led to major changes in apple growing; it is no longer concentrated in parts of the cool north, but has expanded massively into Languedoc and the southwest, where irrigation enables maximum benefit to be drawn from summer heat. Irrigation has also permitted a marked increase in the production of salad crops, leading to an intermittent "salad war" between producers in southern France and EC competitors in Spain and Italy.

Big business

Even more controversial has been the rise of factory-farming techniques in livestock raising, and the implantation of hormones and other growth-inducing substances into veal calves, pigs and poultry. This kind of intensive production is particularly concentrated in Brittany and the north; it provides a good illustration of the way in which many farmers have increasingly become part of a complex business chain, whereby cooperatives, producers' associations and multinational organizations specify both the quality and the quantity of livestock and crops to be delivered, and allow the farmers very little freedom to maneuver.

"Lakes" and "mountains"

Since the spectacular transformation of its agriculture France has been the major contributor to the EC's "lakes" of surplus milk and table wine and "mountains" of grain and butter. In recent years the CAP has been drastically revised in an attempt to avoid further surpluses. Quotas have been imposed for a range of products; many farmers are faced with falling incomes and the challenge of having to devise yet more ways of organizing their enterprises in order to make a profit. Farm-based tourism and production of smaller quantities of high-quality goods, which command a high price, are just two possible ways. Set aside – the process that allows farmers to receive compensation for leaving a proportion of their holdings fallow – has proved unpopular in the more productive farming areas.

A RELUCTANCE TO CHANGE

Deep in the south of France, the rural area of Bas-Languedoc was experiencing serious depopulation, excessive reliance on viticulture and very modest industrial development during the first half of the 20th century. To remedy this, a seminationalized development corporation – the Compagnie Nationale d'Aménagement de la Région du Bas-Rhône et du Languedoc (CNABRL) – was founded in 1955. Using water from the Rhône and other rivers, irrigation was to be the key to economic diversification. In order to cope, the layout of many farms was reorganized, farmers were encouraged to abandon vines in favor of fruit and vegetables, marketing was improved and food-processing industries introduced.

Despite considerable achievements in eastern Languedoc, results have not really lived up to expectations. Many farmers have proved very reluctant to make use of irrigation water to grow different crops, because of sentimental attachment to viticulture, the relatively high cost of irrigation water itself, and the existing overproduction within the EC of many of the crops, such as apples, peaches and salad crops, that they were being encouraged to grow.

By 1990, 8,500 farmers – barely a quarter of the regional total – received irrigation water from the CNABRL. This was well below capacity, so some of the supply was diverted to nearby towns, and the water was even used for irrigating areas of vines.

Changes in irrigation *(left)* There has been a marked increase in the use of portable irrigation systems in recent years. This has not only boosted the productivity of traditional crops, but has also led to an increase in fruit and vegetable growing in the south, where the crops can benefit from high summer temperatures.

Traditional farming *(right)* Many small farmers are being driven out of business by their larger competitors. However, the more enterprising among them may benefit from the upsurge in consumer demand for the healthier free range and organic farm produce.

A second major problem is the widening "development gap" that has emerged between modernized, highly productive holdings in the Paris basin and some irrigated areas in southern France, and low-income farms in upland areas. In an attempt to halt depopulation and prevent piecemeal desertion of the land, special policies have been implemented by the French government and the EC to support upland farming, create jobs in light industry and rural tourism, and sustain local services. Even so some stretches of upland France are being abandoned to scrub or planted with timber.

Modernization has given rise to casualties as well as winners in recent decades. Hundreds of thousands of "marginal" farmers in France have simply retired or gone out of business. Active modernizers entered heavily into debt as they borrowed large sums for new equipment and buildings and for improved livestock. Their position became even more precarious after 1973, when costs of farming inputs (labor, machinery, fertilizers) started to rise more steeply than commercial returns from farm products.

Established capitalist farmers who were in close touch with the demands of the market initially profited from the new situation, but they too soon experienced static or declining incomes. In the light of the revised CAP, which aims to reduce overcapacity, their future profitability is far from guaranteed. One fact is certain: there will be far fewer full-time farms and farmers at the start of the next century than there were at the end of the 1980s. Two-fifths of French farmers are over 55 years old, many sons are reluctant to follow their fathers on to the land, and the number of young people entering farming from outside is modest.

The wines of France

Age-old skills High quality wines are allowed to mature in wooden barrels. The size of the barrel, the type of wood, the temperature, humidity and length of storage all influence the character of the wine. The effervescence in good sparkling wines comes from natural secondary fermentation in the bottle.

The major vine-growing areas in France today – Alsace, Bordeaux, Burgundy, Champagne, the Loire and the Rhône – were in existence in Roman times, and by the Middle Ages even larger amounts of land were devoted to viticulture; small vineyards then surrounded Paris and other northern cities beyond the present-day limit of commercial vine growing.

Experimentation over more than two thousand years by landowners, wine makers and merchants has produced a great variety of wines, which are now subject to rigorous classification. Cheap and modest *vins de table* ("table wines") must satisfy minimum standards but are often blended with imported wines. They account for two-thirds of French production. Next come *vins de pays* ("country wines"), first given official recognition in the 1960s in order to confer status on good but little-known local wines. They must be produced within specified areas from designated grape varieties and have a minimum alcohol content. More than sixty *vins délimités de qualité supérieure* (VDQS; "wines of superior quality") are subject to even tighter controls, and regulations are exceptionally demanding for wines crowned with *appellation d'origine contrôlée* (AOC; "controlled name of origin"), to merit which a wine has to be grown within a specific area. Adding the names of the estate (château or domaine) on which the wine was made and of the wholesaler gives a further indication of the excellence of a wine.

To yield good grapes for wine making, vines require well-drained but not rich soil, and early rain followed by strong and continuous sunshine during late summer and early fall – conditions that are best satisfied on southeast-facing river terraces, valley slopes and areas of chalk and limestone in the southern half of France.

The challenge to modify microclimates and soils and to experiment with grape varieties and methods of making and storing wine has, therefore, always been stronger in more northerly locations. Many of the best French wines come from places where environmental conditions require the greatest ingenuity and skill, such as Champagne to the east of Paris, and Burgundy and Alsace in the center and east of the country. The most renowned wines of all come from the area in the west around Bordeaux at the mouth of the Garonne.

Wine production in crisis

Although Italy now produces a greater volume of wine, the vineyards of France remain preeminent for the variety and excellence of their wines. Demand for high-quality wines remains strong at home and abroad, with West Germany, Belgium, Britain and the United States being leading purchasers. By contrast, the production of cheap table wine is in crisis, with supply greatly exceeding demand. Competition from spirits and soft drinks (sales of which have risen in the market partly as a result of government

A family affair The grape harvest is a great local event that takes place as soon as the grapes are fully ripe. This is tested by measuring the amount of sugar in the grapes. If the vines do not receive sufficient heat during the growing season, the grapes will be low in natural sugars and the resultant wines will have too much acidity. On small vineyards, or on those where the terrain is steep and awkward, grape-picking is still done by hand and everyone – young and old – gathers to help. The larger vineyards may use mechanical harvesters that shake the fruit from the vines.

Relying on its roots A vine is able to survive on poor soils by developing a root system that is able to penetrate layers of moisture-retaining sands and gravels deep in the soil. The deeper the roots, the less likely the vine is to be affected by adverse conditions on the surface, such as drought or floods.

Top surface is pebbly and sandy with few roots

Roots spread horizontally along a layer of crumbly marl dug in as a fertilizer when the vines were planted

Compacted sandy layer

Roots descend to layer of gravel and sand with some rich organic matter

Roots fill out pockets of sand and gravel within layers of sands, through which water drains

campaigns to reduce alcoholism) has depressed annual wine consumption in France from 130 liters (28.6 gallons) per person in 1959 to 85 liters (18.7 gallons) per person in 1990.

As a result, while the vineyards that produce quality wines are being expanded, grants are being allocated by the EC to encourage vine growers to grub up the vines producing humble table wines. Many of these are in the Languedoc area to the west of the Rhône, where wine production expanded rapidly with the growth of the railroads after 1850, and where thousands of small farms deliver grapes to wine cooperatives for manufacture into cheap red wine. Today the number of vine growers is falling rapidly, and the total area under vines in 1990 (1.1 million ha/2.7 million acres) was less than half the peak recorded in 1875 (2.5 million ha/6.2 million acres).

Scented fields

Provence, in southern France, has long been noted for its perfume industry, based on the cultivation of several aromatic plant species that are native to the Mediterranean area. Lavender, a member of the mint family, is one of these. The plant is grown for its oil and for its purple flowers, which are dried and sold in bunches or in sachets.

The distinctive fragrance of lavender is derived from an oil that is found in small glands embedded in hairs covering the flowers, stems and leaves of the plant. It is extracted by distillation and used in the production of fine perfumes and cosmetics. An inferior grade of oil, known as spike oil, is used in a number of industrial processes, including soap manufacture and to aid the application of paint to fine porcelain.

The commercial cultivation of plants for perfume is an example of a highly specialized form of agriculture with close ties between the grower and the industry that processes the crop. There is obviously a limited market for such crops, and so only a small number of farmers are needed as suppliers. Similar specialization centers on the production of many other plants for culinary or medicinal purposes, among them several species that are closely related to lavender, such as mint, marjoram (oregano) and thyme.

A field of lavender in flower in Provence, in the south of France.

HIGH INTENSITY FARMING

IN THE STEPS OF THE PAST · USING THE LAND TO THE FULL · LEADING THE WORLD IN YIELDS

Human intervention in the Low Countries, where people have changed the very shape of the land, together with favorable natural physical features, has led to intense agricultural production. Throughout the region there is a rich variety of farming systems in the valleys and deltas of the Schelde, Meuse and Rhine rivers. Behind the coastal dunes of northern Belgium and the Netherlands lie the fertile reclaimed lands of the polders. To the east arable, livestock and mixed farming is carried out on the undulating sandy plains that rise up to the Ardennes plateau in the south. Some of the highest yields in the world come from the Low Countries; although as little as 3 percent of the workforce is employed directly in farming, agricultural produce accounts for a substantial proportion of exports.

IN THE STEPS OF THE PAST

The tradition of mixed farming that characterized much of the Western Europe in the medieval period is still evident in the present-day pattern of agriculture of the Low Countries. The balance between arable farming, with its cycle of crop rotation, and livestock rearing was established very early. Then, too, began the age-long struggle to drain the coastal marshlands of the Netherlands, reclaim and improve the heathland of central Belgium by manuring the sandy soils, and clear the forest of the Ardennes plateau in northern Luxembourg. Nearly a third of Luxembourg remains forested and a fifth of Belgium; in the Netherlands only 10 percent of the land is wooded.

Agricultural land use is remarkably diverse – the result of different soil types and drainage conditions. Over the years farming has consequently evolved into many different specialized areas. Since medieval times, when coastal marshlands were first drained and the land was protected against invasion from the sea, the fertile peat and clay soils have been used for fresh vegetables, cut flowers, flower bulbs or green salad crops to supply the nearby urban markets. Farther inland, a relatively short winter and a mild maritime climate encouraged the growth of grasses, especially on the heavier clay soils, and pasture was developed to support improved breeds of dairy cows, especially Friesians. Milk was sold either fresh in the urban market or processed into butter and cheese. On

COUNTRIES IN THE REGION

Belgium, Luxembourg, Netherlands

Land (million hectares)

Total	Agricultural	Arable	Forest/woodland
7 (100%)	4 (53%)	2 (25%)	1 (15%)

Farmers

326,000 employed in agriculture (3% of work force)
5 hectares of arable land per person employed in agriculture

Major crops
Numbers in brackets are percentages of world average yield and total world production

	Area mill ha	Yield 100kg/ha	Production mill tonnes	Change since 1963
Wheat	0.3	59.8 (256)	1.8 (–)	+25%
Sugar beet	0.2	534.2 (154)	12.5 (4)	+105%
Potatoes	0.2	445.1 (283)	9.4 (3)	+70%
Barley	0.2	52.8 (227)	1.0 (1)	+11%
Vegetables	—	—	4.3 (1)	+62%
Fruit	—	—	0.9 (–)	+8%

Major livestock

	Number mill	Production mill tonnes	Change since 1963
Pigs	20.1 (2)	—	+305%
Cattle	7.9 (1)	—	+22%
Milk	—	15.7 (3)	+42%
Fish catch	—	0,5 (1)	—

Food security (cereal exports minus imports)

mill tonnes	% domestic production	% world trade
−5.9	176	3

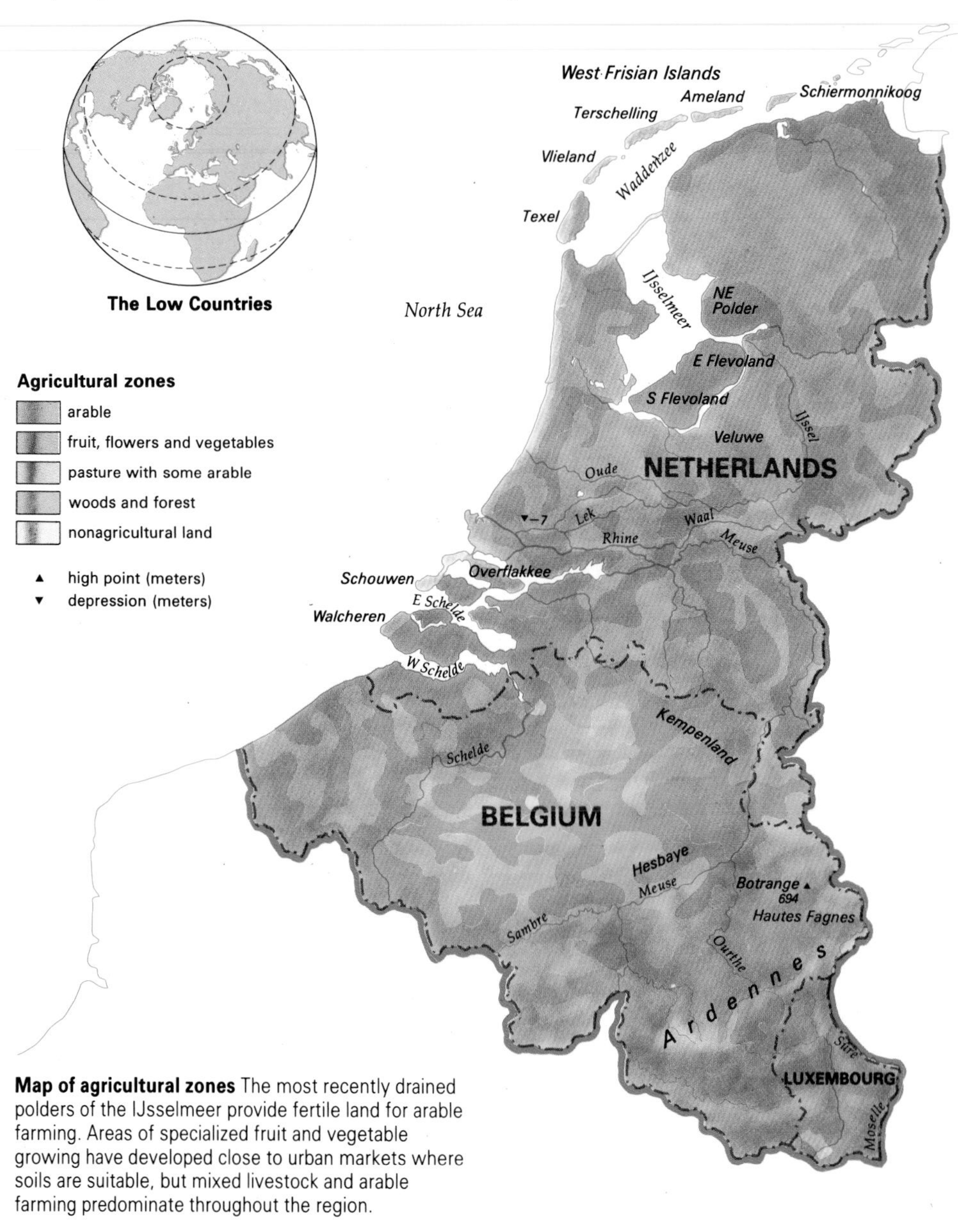

Map of agricultural zones The most recently drained polders of the IJsselmeer provide fertile land for arable farming. Areas of specialized fruit and vegetable growing have developed close to urban markets where soils are suitable, but mixed livestock and arable farming predominate throughout the region.

Sheep on the polders This peaceful rural landscape has been created by human engineering. Much farmland in the Low Countries has been reclaimed from the sea, marshes or lakes to form polders like this one in Belgium. Windmills, used for pumping out water, are characteristic of the scenery.

higher slopes, pastures were grazed by beef rather than dairy cattle.

On the lighter soils farmers developed cropping systems in order to produce field vegetables such as potatoes, cereals and, especially in Belgium, sugar beet. This last crop was introduced in the 19th century, and its cultivation was deliberately supported by governments anxious to be independent of seaborne supplies of sugar cane from tropical regions in time of war. Its production is sustained by arrangements under the Common Agricultural Policy (CAP) of the European Community (EC), to which all three Low Countries have belonged since its inception in 1957.

The urban impact

The major influence on the evolution of farming has been the emergence of the large, prosperous, urbanized population of the Low Countries, giving farmers a continually expanding market for food with relatively low transportation costs. Agriculture lost much of its labor to the towns and cities during the period of industrialization in the 19th century. Today less than 3 percent of the Low Countries' workforce is directly employed in farming, one of the lowest proportions anywhere in the world.

To maintain productivity levels with a small labor force, farming has had to become both intensive and highly mechanized, whether it is based on horticulture, dairying, crop or animal farming. The high protein diet demanded by the growing urban population has boosted intensive methods of rearing livestock, such as pigs, so that domestic supplies of animal feed have had to be supplemented by imports – approximately 53 percent of animal feed used in the Netherlands today is imported.

The intensive farming systems of the Low Countries have been supported by the parallel development of a large, complex food industry. Industrial corporations supply the animal feed, seeds, fertilizers, pesticides and machinery that characterize modern intensive agriculture, and they now process most of the farm produce, including butchering meat and packaging fruit and vegetables. This situation is reflected in the fact that agriculture accounts for only 4.5 percent of the gross national product of the Netherlands, yet agricultural produce makes up 24 percent of the value of all its exports. In Belgium the figures are 2.6 percent and 11 percent, and in Luxembourg 3.5 percent and 9 percent.

Reaping the benefits Land is limited, but farmers use high levels of inputs – chemical fertilizers, herbicides and pesticides – to maintain productivity and increase yields. The generally flat landscape is ideally suited to heavy machinery.

USING THE LAND TO THE FULL

One of the most important types of farming in the Low Countries is horticulture: the cultivation of fruits, vegetables and flowers in the open or under glass. It is very intensive, and takes place where dune sand has mixed with the fertile peat and heavy clay soils of the flat polders (areas of reclaimed land) that lie below sea level along the inner fringe of the coastal sand dunes. This horticultural production comprises 15 percent of the Netherlands' total output. It is less important in Belgium, accounting for only 6 percent. The coastal polder land in the northwest of the country extends over some 500 sq km (190 sq mi), while market gardening is also a feature of the Brussels area, especially in the greenhouses at Hoeilaart to the southeast of the city.

Dairy cows are grazed intensively on the grasslands that cover the lower-lying clay soils of the Netherlands and form either permanent pasture or are grown in rotation with other crops. The cheeses that are made – particularly the red-rinded Edam and the yellow-rinded Gouda – find their way to many corners of the globe through markets such as the one at Alkmaar in the district of North Holland. In both the Netherlands and Belgium, milk accounts for some 29 percent of agricultural output and is the single most important farm product.

Rich soils and poor

Much of the farmland of the Low Countries is used for arable crops – over half of Belgium, slightly less in the Netherlands and Luxembourg. There is specialized crop farming on the newer polders of the IJsselmeer and in central Belgium. On recently reclaimed polders over a third of the land will be given over to cereals, especially to wheat and barley, up to a fifth to sugar beet, and the rest to oilseeds, potatoes, flax, alfalfa and hay. In central Belgium, where the land is higher, there are fertile loam soils. Farms are larger, growing wheat and sugar beet as the main crops; there is some specialization in other crops such as flax, tobacco, fruits, potatoes and other vegetables.

Where the land is less fertile, farming often includes a mixture of crops and livestock. This becomes quite noticeable

in the undulating country in the east of the Netherlands, where the land rises to as much as 90 m (300 ft), with broken areas of heath and woodland covering the less fertile sandy soils. On land that has been drained and manured cereals, potatoes and fodder maize are cultivated, interspersed with pasture grazed by dairy cows, beef cattle and some sheep.

In northern Belgium, including the lowlying Kempenland (Campine region) that straddles the border with the Netherlands, coarse sands and gravels overlie an impermeable subsoil, giving areas of poor drainage. Where drainage allows, forest and grassland is mixed with crop growing. In southern Luxembourg, cereals and fodder crops are farmed together with dairy and beef cattle, and vines are grown on the valley sides of the Sûre and Moselle rivers.

In contrast to the fertility of the polders, the Ardennes plateau, with its steep-sided river valleys, moors and forests, has thin, poor soils. It is hardly surprising that rural depopulation is common in this unprofitable farming area. Those who stay have small farms mainly under pasture for raising beef cattle and sheep, mixed with cereals such as oats, rye and barley, potatoes and fodder crops.

A patchwork of small farms

There are quite marked differences in the size of farm enterprises in the different countries of the region; for example, the average dairy herd in the Netherlands is 40 cows, twice the size of Belgian and Luxembourg herds. There is still greater disparity in the numbers of pigs, ranging between average herds of 51 animals in Luxembourg, 152 in Belgium and 284 in the Netherlands. However, the average size of farms in both Belgium and the Netherlands is 15 ha (37 acres), whereas in Luxembourg they are considerably larger at an average of 28 ha (69 acres). These differences clearly reflect the more intensive development of livestock farming in the Netherlands compared with the rest of the region. Nonetheless, small, intensively farmed holdings dominate the farming landscape throughout the region.

A sweet crop Sugar beet – a temperate crop – is widely grown throughout the Low Countries, and particularly in Belgium. The natural sugar stored in the roots is extracted by shredding them. The sugar is then separated in heated water, and the remaining pulp residue is used as animal fodder.

TRANSFORMING THE PHYSICAL ENVIRONMENT

Some of the most productive farmland is to be found in the polders. These are areas of land reclaimed by drainage from the sea, or from marshes and lakes. The first such lands to be used for agriculture were created as early as the 10th and 11th centuries, when sea walls were built out from the coastline to retain and protect accumulated sand and silt deposits. The creation of inland polders started in the 14th century using techniques developed in Flanders (now in Belgium): windmills, used to pump water out of the bogs, became a dominant feature in the landscape over the next centuries. Dutch engineers developed great expertise in drainage methods, later exporting their skills to lowlying areas around the world, particularly to Britain in the 17th century, and currently to Bangladesh.

Autonomous syndicates, and later the public water authorities, organized and coordinated the drainage works, including building and maintaining field drains, canals, sluice gates, sea dikes and pumping mechanisms. Drainage of the vast inland lakes such as the Haarlemmermeer (which covers 18,000 ha/44,480 acres) took place in the 19th century. It was made possible by the development of steam operated, and later electric, pumps. This created dry land lying 6–7 m (20–30 ft) below sea level that could be used for farming.

By the 1920s it was possible to construct a dike 30 km (18 mi) long to isolate the Zuider Zee from the North Sea; the lake is now called the IJsselmeer. Although the polders formed by this grand scheme have not, as originally planned, been used entirely for agriculture (industry, transportation networks and new cities taking their not inconsiderable share), the farmlands of the Netherlands have been enlarged by approximately 10 percent through these reclamation schemes.

LEADING THE WORLD IN YIELDS

Farmers in the Low Countries produce among the highest yields of crops and livestock in the world. By applying the very latest agricultural technology as it becomes available, they have led the way in breeding higher yielding dairy cows, and developing artificial insemination of cattle, intensive methods of pig breeding and fattening, and new higher-yielding varieties of potatoes. The close attention that has been paid to research, educational and advisory services has ensured that new farm technologies are widely understood and used.

The Netherlands has the highest wheat yields in the world, averaging more than 7,000 kg per hectare (6,230 lb per acre), and they are high in the rest of the region too. Barley yields for both the Netherlands and Belgium are 5,000 kg per hectare (4,450 lb per acre), and 3,500 kg per hectare (3,120 lb per acre) in Luxembourg. Milk yields have risen steadily to over 5,000 liters a year per cow in the Netherlands, and to 4,000 liters a year in Belgium and Luxembourg. However, the actual number of dairy cows has fallen since 1984, when milk quotas were imposed on all dairy farmers in the European Community.

These high yields of crops and animals have their cost. They can be sustained only by inputs of chemical fertilizers, purchased animal feed, pesticides and herbicides, farm plant and machinery; together they account for up to 58 percent of the value of the final agricultural output. With farm prices falling in real terms in recent years, farmers have had to borrow increasingly large sums of money to sustain their high input–high output farming systems; absolute debts in the Netherlands are the highest per hectare in Western Europe. Loans rose from 19 percent to over 25 percent of the capital employed in agriculture between 1963 and 1988.

While agriculture in the Low Countries still generates the highest farm incomes in Western Europe, despite the small farm sizes, incomes outside the farming sector have been rising more rapidly. To keep abreast of current earnings more and more farmers and their families have been taking off-farm jobs. In northern and central Belgium farmers have traditionally worked part-time in local industry and other urban jobs; in the Ardennes tourism and foresty are now providing alternative or supplementary employment.

Labor-saving methods Milking parlors are fully automated for maximum efficiency. The yield from each cow is measured before it is piped to the dairy, and the quantity of feed the animal requires is automatically adjusted while the milking takes place.

Working together

In the face of their financial difficulties, farmers have tried to maintain their income in a number of ways. Many have grouped together to market their produce cooperatively with the aim of driving a harder bargain. Cooperative marketing is strongest in the Netherlands, where as much as 90 percent of milk production is sold in this way.

Another strategy has been for individual farmers or cooperatives to enter into forward contracts with food processing companies. Under this arrangement the farmer receives a guaranteed market and price for a farm product, but in return must meet exacting standards of quality, quantity and delivery times. Sugar beet, peas, potatoes and calves are frequently marketed in this fashion.

Farmers have also sought to increase incomes by expanding production despite falling real prices. The intervention of the CAP has encouraged this by preventing prices falling farther and faster than they would have done in a free market. As a result the Low Countries are not merely self-sufficient in many of the main farm products – they also achieve annual farm

surpluses. The Netherlands now has a clear role in supplying farm produce to neighboring Eruopean countries: pigmeat, poultrymeat, eggs, fresh vegetables and potatoes are all exported. Other products such as milk and sugar beet merely contribute to the Community's food mountains – as the stockpile of surplus farm products is known.

Counting the environmental cost

The impact of the intensive farming practices of the Low Countries on the environment is only now being realized. A number of problems have arisen. Both nutrients from fertilizers and animal manures are washed by the rain into rivers, lakes and coastal marshes, destroying and modifying natural and seminatural ecosystems. Intensively farmed soils are being increasingly compacted by heavy farm machinery and eroded by wind and by surface water runoff. The traditional landscape is being remodeled by the removal of field boundaries and woodlands to maximize the area of cultivated land, and to create the large fields needed for the efficient operation of large modern farm machinery.

A problem of excess *(above)* Animal manure is spread over a field. This efficient system solves two problems at once: it disposes of farmyard muck and returns nutrients – nitrogen, phosphorous and potassium – to the soil. But intensive livestock farming means that there is now too much manure.

Processing cheese *(below)* Surplus milk is processed into butter and cheese where it joins the European Community surpluses. The famous Dutch cheeses – Edam and Gouda – are now mass-produced in modern factories.

TOO MANY ANIMALS

Agricultural pollution is one of the most pressing environmental problems in the Low Countries; it is caused in the main by the development of intensive dairy, pig and poultry production. The large numbers of animals and birds, kept at high densities, produce considerable volumes of manure – in the Netherlands alone, with its 114 million head of livestock, nearly 95 million tonnes of animal manure are produced every year.

Spreading the manure on farmland as fertilizer – the traditional method of disposal – creates problems of its own. Potassium, phosphates and nitrogen are leached from the manure, especially by winter rain. These nutrients then enter the watercourses, and eventually pollute inland lakes and groundwater drinking supplies, encouraging the buildup of algal growth.

There are also limits to how much manure can be disposed of in this way. Each year there is a surplus of some 23 percent, created by calf-fattening and dairy units (40 percent), pig breeding and fattening units (42 percent) and poultry units (18 percent). In the Netherlands strict laws now control the volume and even the time of year when animal manure can be spread on the land. Alternative methods of disposing of the surplus manure, such as refining it to produce an acceptable garden fertilizer, remain uneconomic at present. Nevertheless, the growing problem of balancing the production of needed food supplies against the environmental damage of modern agriculture is one that will have to be resolved during the 1990s.

A city of glass

Although the Netherlands are internationally renowned for their fields of bulbs that color the countryside in the spring, Dutch flowers, particularly tulips grown for export, are increasingly likely to have been raised in the vast "glass city" of Westland, a triangle of land on the coast to the north of the Rhine. A second area farther to the north, where the bulbfields lie, is also dominated by horticulture under glass. Together they cover some 9,300 ha (22,980 acres) of land, mainly within the "Green Heart" of the Randstad conurbation.

Greenhouses were originally established by growers who were involved in intensive horticultural production in the open, supplying nearby urban markets. They depended on the rich silts and clay soils of the reclaimed polders. But today most greenhouse producers bring in new soil to refresh the growing medium and to keep plant diseases at bay. Greenhouse production is commonly carried out alongside cultivation in the open, and the farmland surrounding the greenhouses is still given over to horticulture.

Today's greenhouses depend on advanced agricultural and computer technology. Within the artificial, protected environment of the greenhouse, the computer determines the amounts of water, fertilizer, temperature and light to be applied according to instructions keyed in by the grower. Soil and air temperatures are manipulated by a heating system using local supplies of subsidized natural gas. Internal watering systems include the use of overhead sprinklers, ventilation prevents overheating, pests and diseases are controlled, and plant nutrients are provided through fertilizers. This high input–high output system has been so profitable that successive subdivisions of the land have produced hundreds of small greenhouse businesses. Today's typical greenhouse grower has a production unit that is measured in terms of square meters rather than hectares.

The vegetables, fruits and flowers that are grown vary in different localities. Vegetable growers, whose activities make up 47 percent of the greenhouse area, concentrate on crops such as tomatoes, cucumbers, lettuces, leeks, carrots and spinach. Fruit production is now giving way to tree nurseries.

Industry in bloom

Forming 50 percent of the greenhouse industry, the production of flowers is the most important sector of greenhouse farming – and it is still expanding. The greenhouses produce many popular cut flowers such as roses, chrysanthemums, freesias and carnations, as well as pot plants – mainly ficus, dradaena, kalanchoes, begonias and azaleas.

Many of the flowers cut by hand in the afternoon will be on their way to the United States, Asia and other parts of Europe the following day, airfreighted through Amsterdam's Schiphol airport in refrigerated containers or pre-chilled boxes. West Germany receives 45 percent of the Netherlands' exports of flowers, followed by France (14 percent) and Britain (10 percent). As distant a market as Japan accounts for 1.3 percent.

Some ten auction markets have grown up to control the flower trade. The world's largest is at Aalsmeer, serving the northern greenhouse area, Naaldwijk serves the south, with others at Rijnsburg and Bleiswijk. To guarantee the all year round capability of the flower trade, about 12 percent of the throughput of the auction markets is imported. This compensates for the slight seasonal bias of production toward the summer months.

The Netherlands' "Green Heart" (*right*) This aerial view, which shows only a tiny part of the intensive horticultural area in the southwest, gives an idea of the extent of the greenhouse industry. Many of them are actually covered with plastic, rather than glass.

Computerized cultivation These tomatoes are being grown with the aid of the most up-to-date technology. The cultivator programs the greenhouse computer to control all aspects of the environment, such as water, heat and light.

A mania for bulbs

Plants have long been cultivated for pleasure and as objects of beauty, and ornamental gardens are almost as old as agriculture itself. Many plants now considered to be garden plants were originally bred for their medicinal and healing qualities. The introduction into Europe of new plants from the New World in the 16th century acted as a great stimulus to botanical curiosity, and encouraged experiments in plant breeding as well as the collecting of rare and unusual plant specimens from all around the world.

The horticultural industry of the Netherlands dates back to the 17th century, when a trade became established between European and Turkish nurserymen in bulb plants such as anemones, hyacinths and tulips – all natives of southern Europe and the Middle East – that the Turks had developed from wild species. As the fashion grew for tulips with ever more varied shapes, colors and markings, a new profession of plant growers came into being.

The craze reached its height in the Netherlands in the 1630s, when homes and estates were mortgaged so that bulbs could be bought and resold at higher prices without ever leaving the ground. Many Dutch families were ruined when the market crashed in 1637, but the bulb industry survived. It remains an important part of the Netherlands' horticultural output, with millions of bulbs being grown for export every year.

Carpets of color The tulip fields of the Netherlands are world famous, visited and photographed by thousands of visitors every spring.

A HIGH, DRY LAND

THE INFLUENCE OF THE PAST · AGRICULTURAL DIVERSITY · CATCHING UP

Agriculture in the Iberian Peninsula has been shaped by two fundamental influences. Deep-rooted traditions of land ownership have interacted with environmental factors to produce extreme variation in farming systems and methods. There are intensively cultivated small farms in the lush green northwest; vast estates cover the parched brown landscapes of much of the center and south. The smallholdings produce a wide variety of subsistence crops. The large estates are more commercial, concentrating on the few crops best suited to the poor environment. Pockets of good, irrigated land, mainly on the east and south coasts, produce several crops of fruit or vegetables each year. Generally, agriculture is characterized by wheat, olives and vines, with livestock also playing a part. Commercial forestry is of growing importance.

COUNTRIES IN THE REGION

Portugal, Spain

Land (million hectares)

Total	Agricultural	Arable	Forest/woodland
59 (100%)	34 (58%)	18 (30%)	19 (33%)

Farmers

2.5 million employed in agriculture (13% of workforce)
7 hectares of arable land per person employed in agriculture

Major crops

Numbers in brackets are percentages of world average yield and total world production

	Area mill ha	Yield 100kg/ha	Production mill tonnes	Change since 1963
Barley	4.5	22.1 (95)	9.9 (5)	+391%
Wheat	2.5	24.9 (106)	6.3 (1)	+28%
Maize	0.8	52.4 (144)	4.2 (1)	+145%
Grapes	1.8	44.1 (59)	7.8 (12)	+29%
Oranges	—	—	2.5 (6)	+46%
Vegetables	—	—	11.4 (3)	+55%
Other fruit	—	—	5.9 (3)	+144%

Major livestock

	Number mill	Production mill tonnes	Change since 1963
Sheep/goats	26.4 (2)	—	−10%
Pigs	18.7 (2)	—	+156%
Milk	—	7.0 (2)	+93%
Fish catch	—	1.8 (2)	—

Food security (cereal exports minus imports)

mill tonnes	% domestic production	% world trade
−3.5	17	2

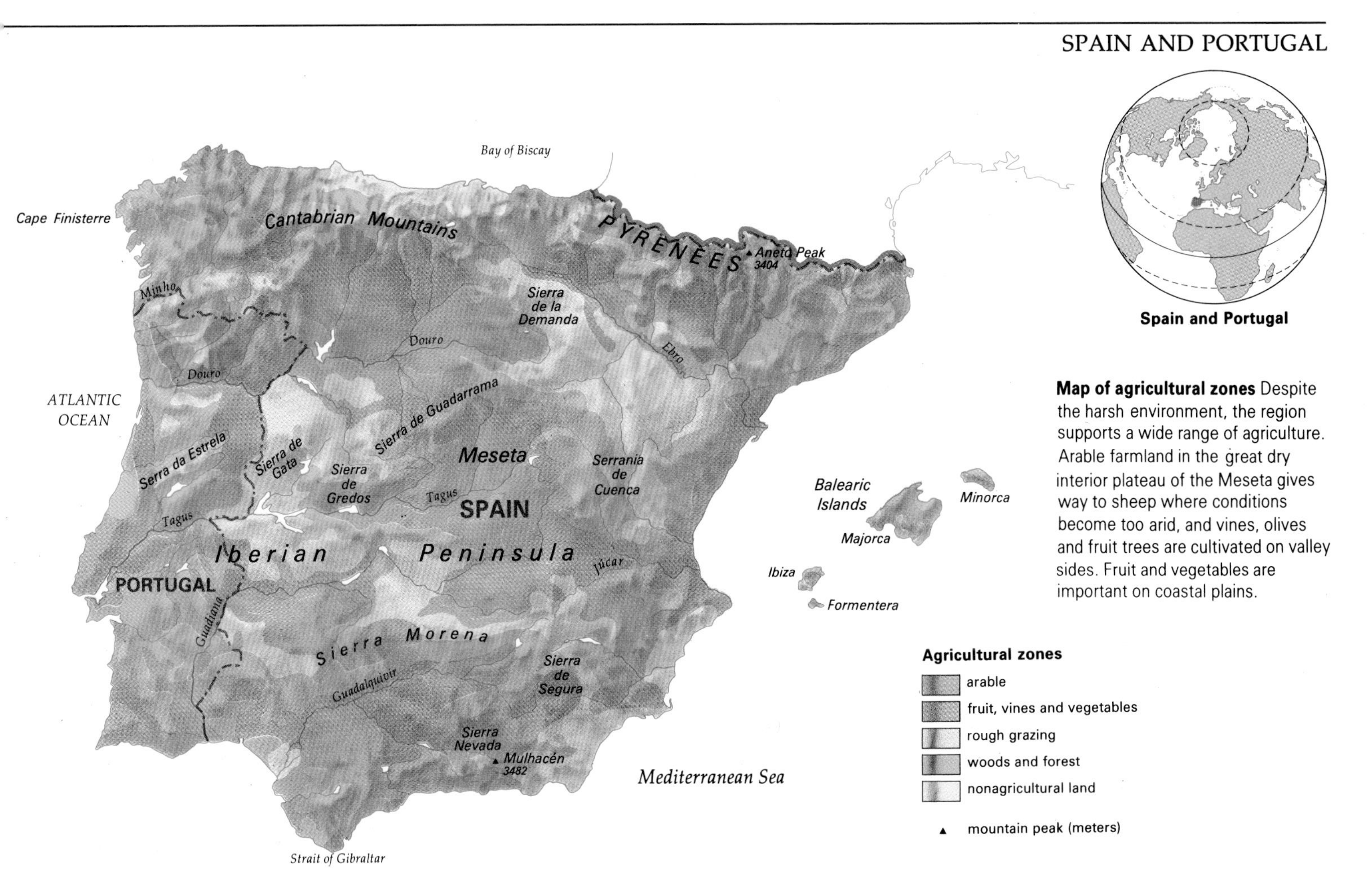

Map of agricultural zones Despite the harsh environment, the region supports a wide range of agriculture. Arable farmland in the great dry interior plateau of the Meseta gives way to sheep where conditions become too arid, and vines, olives and fruit trees are cultivated on valley sides. Fruit and vegetables are important on coastal plains.

THE INFLUENCE OF THE PAST

Most of the Iberian Peninsula is affected by drought. Rainfall in the Pyrenees and in the Cantabrian Mountains in the north can be over 1,000 mm (40 in) a year, but the encircling mountain ranges prevent humid airmasses from the Atlantic Ocean reaching the central plateau, or Meseta. In some places the average rainfall is under 200 mm (8 in) a year. It is so distributed that the poorest soils often receive the most rain, while the best land – on the east coast and in the Guadalquivir valley in the southwest of Spain – can be rainless for long periods. The availability of water is the most critical factor in Iberian agriculture.

Marked regional differences both in the farming landscape and in the organization of farming are the result not only of environmental factors, but also of strong cultural influences. The countryside in the north is dotted with small family farms, many of them still run in much the same way as they were hundreds of years ago, reflecting a long history of relative independence from invasion. Most of these farms are reasonably productive, but those in the extreme northwest are adversely affected by poor soil and ancient inheritance laws. These permit excessive subdivision of the land, resulting in tiny, uneconomic plots of land called *minifundia*. In Portugal in the early 1970s, for example, 33 percent of northern landholdings were under 4 ha (10 acres). *Minifundia* also exist in some northern areas of the central Meseta.

Andalucian harvest The great estates of the south, once symbols of social division and agricultural inefficiency, are now commercially run using modern methods and machinery. They produce the bulk of the region's arable crops, notably wheat and barley. The harsh, dry climate of the area restricts yields, which are low by European standards.

Roman and Arab influences

The center and south of the peninsula was successfully colonized in the distant past by invaders from the eastern Mediterranean, including the Phoenicians, the Romans and, from the early 8th century AD, the Arabs (Moors). Under the Roman empire irrigation methods and an agricultural economy based on large estates, or *latifundia*, were introduced to Spain and Portugal. The Arabs made further great improvements to agriculture, particularly in the south and southeast where their presence was strongest; they applied the skills they had learnt in North Africa to construct intricate irrigation systems. These supplied water to small, intensively cultivated plots of land – *huertas* – on which a wide variety of crops were grown, including many they introduced to the region, such as rice, sugar cane, cotton, oranges, apricots and almonds.

Resistance to Arab rule, beginning in the 11th century, led to the reconquest of all of Spain and Portugal by 1492. In the heart of the Spanish crownlands, in the center and south of Spain, vast tracts of land were granted to the Church and aristocracy in return for their crusading services in the reconquest. The establishment of this class of large landowners reinforced the *latifundia* system.

With the exception of the provinces of Murcia and Valencia in the southeast, the intensive agricultural system introduced by the Arabs was abandoned. Once fertile arable land in southern Spain was left to revert to pasture. Irrigation channels were filled in and the land used for sheep and cattle. For four centuries sheep farming dominated Spanish agriculture. Rural communities were often close to starvation as they tried to scratch a living from the remaining poor land. By the 19th century the balance had again tipped in favor of arable farming. Wheat, olives and vines were the principal crops.

The basic organization of the *latifundia* survived all the vicissitudes of Iberian agriculture. They were a source of wealth and social status for their largely absentee owners, but they brought misery to the rural poor who worked on them. They were generally inefficiently run, relying on the exploitation of cheap seasonal labor. Throughout the 19th century and well into the 20th century the poor productivity of the *latifundia*, and the social injustice they exemplified, was one of the great issues of Iberian politics.

Northern smallholdings Different traditions of land tenure have led to predominantly small family farms in the north of the peninsula. On this Basque farm in the western Pyrenees, all members of the family, including the elderly women, help with threshing wheat.

AGRICULTURAL DIVERSITY

At the beginning of the 20th century the labor force in Spain and Portugal was predominantly employed in agriculture, mostly as peasant farmers or landless laborers. Since then, the region has seen a steady fall in the agricultural population, more so in Spain than in Portugal. There have been large increases in irrigation and in commercial fruit production and forestry. Both countries have had difficulty in responding adequately to the increased demand for meat and dairy products from their growing, relatively affluent urban populations. This is due to poor pasture, and to the fact that until recently cereal crops – particularly wheat – took precedence, even on unsuitable land. Wheat production has now fallen in favor of fodder crops and meat and dairy production.

Arable crops are grown on a small scale everywhere. The *latifundia* provide the bulk of commercial production. Wheat and barley dominate the great dry interior known as the Meseta, from the Alentejo in southern Portugal into central Spain, where the low rainfall restricts the cultivation of other cereal crops. The Meseta accounts for 42 percent of Spain's arable land. Maize and rye are more important in the northern and western coastal areas. Rice is grown on the flat irrigated lowlands of Catalonia, Valencia and Andalucia in eastern and southern Spain. Industrial crops are also important and are concentrated in southern Spain. Tobacco plantations cover some 24,000 ha (53,900 acres). During the 1980s the area devoted to sunflowers increased dramatically, and by 1990 was well over 1 million ha (2.5 million acres). Cotton production has also boomed.

Olive trees grow throughout most of the southern half of Iberia, but commercial olive oil production is centered in Andalucia and in southeastern Portugal. Together, Spain and Portugal produce approximately a third of Europe's olive oil. Average annual production in the 1980s was 2 million tonnes in Spain and 200,000 tonnes in Portugal. Spain is one of the world's largest producers of citrus fruits; while the bulk is grown in the southeast, citrus groves are also found throughout the region. Spanish almonds – another traditional southern crop – provide a quarter of the world's supply.

Nearly all the region's quality wines come from the north. The only exception is sherry, which comes from the district around Jerez de la Frontera in the southwest of Spain. Lower quality wines are produced in bulk in La Mancha, just north of the main olive-growing area.

Dairy and pig farming are particularly important in the small farms of the north, where the animals are traditionally stall-fed. There is also a limited amount of dairy farming on irrigated pastures farther south. Much of the pasture is poor in quality, best suited to grazing sheep or goats. Sheep roam extensively over the central Meseta, while goats thrive in the harsher conditions to the south.

Agricultural success

In the eastern coastal lowlands of Spain the climate allows year-round cultivation of a wide range of fruit and vegetables under irrigation. Early crops can be sold for high prices in the winter, while high summer temperatures are exploited to produce four to six crops per year of some

Wines for all tastes *(above)* Spain and Portugal produce large amounts of wine, ranging from mass-produced cheap table wines to the fine oaky reds and fortified wines – sherry, port and madeira – for which the countries are famous.

Almond groves *(left)* Almonds were introduced to the peninsula by Arab settlers in the 8th century, along with oranges, and they are still major crops in southern areas. They form an important part of the local cuisine, almonds providing a valuable source of protein.

vegetables and salad crops. Some new crops, such as watercress, have been introduced, destined for sale in the supermarkets of northwest Europe.

The three-tier system of cultivation here makes efficient use of every available piece of land. Higher hillsides are used for grazing sheep or goats, lower slopes are terraced for olives or vines and the intensive cultivation of fruit trees, such as apricot or orange. Vegetables such as potatoes or lettuces are grown on the coastal plain. Although landholdings are generally small, yields are high. The irrigation systems introduced by the Arabs have been maintained and improved over the centuries; the farmers of the area are hardworking and businesslike, and they market their produce through well-organized cooperative societies to the rest of the peninsula and northern Europe.

A new agricultural success story has been unfolding in Almeria, on the coast south of the Sierra Nevada. Once Spain's poorest province, it is now a boom area thanks to a "greenhouse revolution" over the past 30 years. By 1985, 11,600 ha (28,650 acres) were covered by greenhouses, usually made of plastic. They are owned mostly by peasant farmers, organized into cooperatives, who grow tomatoes, salad vegetables, eggplant (aubergines), zucchini (courgettes), green beans and melons. The crops, grown mainly for export, are harvested twice yearly when

PORT WINE IN THE DOURO VALLEY

Port is one of the world's great wines. It is Portugal's second most important agricultural export after woodpulp. The grapes from which it is produced are grown on coarse schist soils in the upper Douro valley in northern Portugal. The growing area covers 24,000 ha (59,300 acres) divided into roughly 80,000 individual vineyards owned by 30,000 growers. Port takes its name from the town of Oporto, from which it was traditionally exported.

The vines are cultivated on an extensive system of terraces that were traditionally supported by a rambling network of stone walls. The high cost of maintaining these walls and the introduction of mechanization has led to a new system of larger, sloping terraces. The harvested grapes were formerly trodden in large granite tanks – *lagars* – now replaced by modern vinification tanks, permitting much greater control over the fermentation process.

Port is made by adding *aguardente* – a clear spirit distilled from wine – during the fermentation of the grapes in a proportion of one part *aguardente* to every four parts of wine. The *aguardente* halts fermentation and raises the alcoholic strength of the final product. Port is greatly improved by aging, first in the cask and then in the bottle. Until 1986 all port wine had to be transported to the coastal town of Vila Nova de Gaia, where it was matured in great warehouses, or lodges, belonging to family-owned firms of shippers who dominated the trade. A change in the law now allows producers to mature wines in, and export them from, the upper Douro valley itself.

they are out of season elsewhere in Europe. The sunny climate and huge deposits of underground water explain the success of the greenhouse revolution, though expansion has more recently been halted due to dangerously low levels in the underground wells.

Commercial forestry

Large programs of reforestation have made commercial forestry an important new element in the agricultural economy. Woodpulp accounted for 32 percent of Portugal's agricultural exports in the mid 1980s, with prepared timber adding another 8 percent. Most of the woodpulp is produced from conifers, particularly maritime pine, cultivated in the uplands of the northwest where the amount of rainfall is sufficiently high. Eucalypts are also widely grown, mostly for the production of high-grade cellulose, replacing the indigenous deciduous trees, such as oak, sweet chestnut and walnut. Cork oaks thrive in the dry southern areas of the peninsula. Iberia is the world's largest supplier of cork.

Manufacturing cork at Alentejo in Portugal. Cork is obtained by stripping the bark off cork oak trees that are at least 20 years old. The sections of bark are then softened by boiling, and flattened. Cork is used for insulation and in making linoleum flooring, as well as for bottle corks.

CATCHING UP

The coexistence of *latifundia* and *minifundia* has long served as a stumbling block to agricultural success in much of the Iberian Peninsula. During the 20th century various attempts have been made to restructure landownership through settlement schemes. Spain's socialist government passed laws in 1983 allowing it to redistribute land from inefficient large estates to landless peasants. Despite their traditional image of inefficiency, however, most large landowners now run their estates on a commercial basis, with high levels of mechanization and much improved yields. As a result, the *latifundia* are now largely successful in purely economic terms. However, no complete solution has yet been found to the social problem of the landless poor in rural southern Spain.

There was dramatic social and political upheaval in Portugal in 1974, when the right-wing dictatorship was overthrown. Under the communist–socialist alliance that followed, vast amounts of land were expropriated from the estates of the rural elite. Just under 1 million ha (2.5 million acres) were redistributed in the cereal-growing lands of the Alentejo to the south of the country. Much of the land was seized spontaneously by peasants who established workers' cooperatives, but the swing to the right in Portuguese politics during the 1980s produced legislation that returned considerable amounts of land to the original owners.

The governments of both countries are taking steps to rationalize the *minifundia* of the north. These include incentives to consolidate farms and reduce the number of fields, and improvements to the rural infrastructure of roads, electrification and buildings. These measures have so far had only limited success.

Lack of water is the single greatest constraint on Iberian agriculture. In 1900 approximately 1 million ha (2.5 million acres) of Spanish agricultural land had guaranteed irrigation. This figure has improved significantly: between 1955 and 1978 the irrigated area doubled to reach 3 million ha (7.5 million acres), or 10 percent of agricultural land.

Portugal has a slightly higher proportion of irrigated land – 680,000 ha (1.7 million acres) or 15 percent – but most of this is supplied by outdated dams and canals. More and improved irrigation is a priority, especially as pressure on water supplies has been increased by the emphasis that has been put on water-hungry crops, such as citrus fruits and peaches,

New face of farming Accession to the European Community put increased pressure on farmers to improve productivity. Modern, intensive farms like this one at Consuegra in central Spain are likely to become more common.

instead of traditional "dry" ones, such as figs, olives and vines.

Average crop yields are generally lower than in the rest of Europe. Spain's yields are higher than those of Portugal. Wheat yields in the early 1980s were 2,200 kg per hectare (1,958 lb per acre) in Spain and 1,200 kg per hectare (1,068 lb per acre) in Portugal, compared with a European average of 4,200 kg per hectare (3,740 lb per acre). Low yields are partly due to the harsh physical environment, but they also reveal insufficient capital investment.

The difference in productivity between the two countries is reflected in their ability to supply domestic demand and the success of their exports. Spain produced just under 90 percent of its cereal requirements in the late 1970s, whereas Portugal produced less than 45 percent. Both produce surpluses of Mediterranean crops such as citrus fruits and olives, but Spain's surpluses are proportionally much higher. Likewise, the deficit in livestock and animal products is lower in Spain than in Portugal. Portuguese agricultural imports are three times higher than exports in value, whereas Spain's substantial exports of fruit and vegetables almost balances its imports of animal feed and livestock products.

Moving into the European Community

Iberian agriculture was put under the spotlight when both countries joined the European Community (EC) in 1986. Their membership not only exaggerated some of the underlying problems of the EC's Common Agricultural Policy (CAP), but also increased the incentive for agrarian reform. Spain and Portugal's surpluses of citrus fruits, olive oil, wine and some vegetables led to a considerable increased in the supply of these products in the Community as a whole, endangering the profitability of other EC producers. At the same time, Spain and Portugal were forced to pay much higher prices for animal feed imported from the EC than they had when they obtained it at lower world prices. Although a series of complex transitional agreements was set up to minimize the short-term effects of these price differences, they are likely to lead to a sifting out of less profitable farm enterprises, with many small farmers being forced to abandon farming.

Arrangements were made by the EC to give Iberian farmers considerable financial support during the first 10 years after accession. Portugal was due to receive a total of 70 million European Currency Units (ECUs) for structural improvements to its agriculture. These included schemes for rural electrification, road building and irrigation. The grant was designed to encourage technically more advanced agriculture in the north, and greater productivity in the south. However, one effect of integration into the EC will be the reduction of the farming populations of both countries. In 1985, 23 percent of Portugal's and 13.7 percent of Spain's working populations were employed in farming. This contrasts with an average of 8.7 percent in the other 10 EC countries. Many farmers in Spain and Portugal therefore stood to lose from entry into the Community.

TRADITIONAL VALUES AND TECHNICAL CHANGE

On many of the small farms in northern Portugal the ox cart is still a far more familiar sight than the tractor, and traditional methods are used to work the land. The farmers have adopted very few of the innovations of modern agriculture: manure is still used instead of chemical fertilizers, and indigenous seed varieties are preferred to the new hybrid ones. This resistance to change has been regarded as innate conservatism; however, recent research has shown that their traditional practices are in fact founded on highly rational land-use decisions, and that farmers are willing to adopt new techniques and equipment if they are certain that they will derive benefit from them.

The real reasons for the lack of change are that the farmers have little access to information, and they lack the capital to buy costly modern agricultural supplies. Many small farmers are prepared to learn new ways if this will enable them to farm successfully. However, they also prize the dignity of their way of life and the traditional social values of their communities.

The pressures from the EC for more technically efficient agriculture and a reduced farming population place these farmers in a dilemma. They must either freely absorb the capitalist values of the Community, or have these thrust upon them. In any event, it seems inevitable that their rural way of life must change.

Tourism and agriculture

The "economic miracle" that took place in Spain in the 1960s was due in large part to the opening up of the country to tourists. The holidaymakers who flocked to the Mediterranean beaches of the Costa Brava and the Costa del Sol for cheap breaks in the sun brought much-needed revenue to the country. Ever since then, tourism has been of growing importance to the economies of both Spain and Portugal.

Between 1960 and 1980 the number of tourists visiting Spain rose from 5 million a year to over 30 million. In the mid 1980s receipts from tourism accounted for more than 4 percent of Spain and Portugal's domestic revenues. The areas most affected by tourism are the east and south coasts of Spain, and the Algarve in southern Portugal; the great cities in the interior of the peninsula also attract large numbers of visitors. Recently there have been attempts to encourage tourism elsewhere in the region – in the less developed, as yet unspoiled rural areas such as the Minho in northern Portugal and Galicia in northwest Spain.

Losses and gains

The rapid growth of tourism has had important effects on agriculture in the tourist areas, though it is not always easy to separate changes due directly to tourism from those that are the result of broader changes in the economy.

The development of tourist resorts has often been uncontrolled and has resulted in ribbons of high rise hotels or apartment blocks along the coasts. Prices for land have soared, making it uneconomic to continue using it for agriculture. In addition, the wages to be obtained from the tourist trade are usually much higher than income from farming. Consequently many farm workers have left the land to serve tourists in the hotels and bars, and local agriculture has declined as a result. Coastal resorts have frequently grown up around old fishing villages, and although tourism generates a demand for fresh fish, fishermen have also found tourism a more profitable way of earning a livelihood.

Tourism generates enormous demand for scarce water resources, in direct com-

Harvesting melons (*above*) on Spain's Costa Blanca, on the northeast coast of Catalonia. Tourist and agricultural needs peak during the summer when water reserves are at their lowest. Land for farming has been lost to high-rise apartment blocks and hotels.

The impact on fishing (*left*) Spain and Portugal's picturesque fishing ports are themselves a tourist attraction. While some fishermen have benefited from tourist demand for fresh fish, others have abandoned their traditional way of life for higher wages in the tourist sector.

petition with agriculture. This is particularly important along the southern coasts where irrigation is essential for crop and fruit production. Here, tourist demands peak in July and August, when water supplies are at their lowest.

Tourists create extra demand for food, and larger or more enterprising landowners in the hinterlands of the tourist areas have exploited the situation by changing to fruit, vegetable or livestock production to satisfy the tourists' needs. Tourism has brought a better standard of living to working villages that are also small tourist resorts. The extra income generated from tourism can be used to fund improved community services such as schools and health care.

Even farmers in rural areas can benefit from the tourist trade: government grants are available to enhance their buildings in order to provide accommodation for tourists seeking more peaceful vacations. Tourism has also influenced the demand for Iberian farm products in the tourists' home countries. While some visitors prefer to stick to the food and drink they are familiar with, those who try the local produce – particularly wine – often want to obtain it at home.

The living past

Animal-driven water wheels – or *noria* – are still found dotted around the Iberian countryside. They are a very early form of irrigation, first developed by the Persians in what is today Iran. Their use was later spread throughout the Middle East and into northern Africa, Spain and Portugal during the expansion of the Arab civilization that took place in the early medieval period. Buckets attached to a wheel, which is turned by a blindfolded horse or mule, are used to raise water from a well supplied by groundwater; these underground reserves are recharged by rain falling during the short wet winter season.

This ancient technology survives because of the small scale of much traditional farming. The costs of installing mechanical pumps cannot be justified by the amount of income that the small plots of irrigated land generate, and so the old methods continue unchanged. There are sound environmental reasons for their remaining in use: although mechanical pumps would enable larger areas of crops to be irrigated, they run the risk of exhausting groundwater supplies, thus lowering the water table so that neighboring wells dry up.

Turning the wheel A mule drives a *norià* to irrigate a farmer's fields in La Mancha, central Spain.

MEDITERRANEAN FARMING

A HARSH ENVIRONMENT · THE FARMING LANDSCAPE · THE PURSUIT OF PROFITS

Agriculture is still a significant source of employment in Italy and Greece. However its importance is not matched by favorable natural conditions, and consequently farmers have devised ingenious methods of overcoming the difficulties of relief, climate and poor soils to improve overall productivity of the land: carving out elaborate hillside terraces, creating irrigated meadows, and reclaiming lowlying marshes. Largescale cultivation is only possible in a few favored areas of good land. Most farms are small and fragmented. Farmers usually cultivate six or seven different crops; because pasture is generally poor, livestock farming is limited. There has been a strong drive in recent years to improve the profitability of agriculture. Specialist commercial farming, particularly of fruit, is growing in importance.

COUNTRIES IN THE REGION

Cyprus, Greece, Italy, Malta, San Marino, Vatican City

Land (million hectares)

Total	Agricultural	Arable	Forest/woodland
43 (100%)	26 (61%)	12 (28%)	9 (22%)

Farmers

2.9 million employed in agriculture (11% of work force)
4 hectares of arable land per person employed in agriculture

Major crops
Numbers in brackets are percentages of world average yield and total world production

	Area mill ha	Yield 100kg/ha	Production mill tonnes	Change since 1963
Wheat	4.0	29.2 (125)	11.6 (2)	+9%
Grapes	1.3	101.2 (135)	13.1 (20)	+15%
Barley	0.7	32.7 (146)	2.4 (1)	+292%
Sugar beet	0.3	522.6 (151)	17.4 (6)	+112%
Vegetables	—	—	18.0 (4)	+58%
Peaches	—	—	2.1 (27)	+62%
Other fruit	—	—	8.2 (3)	+17%

Major livestock

	Number mill	Production mill tonnes	Change since 1963
Sheep/goats	27.9 (2)	—	+21%
Pigs	10.8 (1)	—	+101%
Cattle	9.7 (1)	—	−6%
Milk	—	11.6 (3)	+19%
Fish catch	—	0.7 (1)	—

Food security (cereal exports minus imports)

mill tonnes	% domestic production	% world trade
−4.8	21	2

A HARSH ENVIRONMENT

The climate is predominantly Mediterranean in type, with hot, dry summers and mild, wet winters; the north is generally cooler. In most of the region the period of summer drought is exacerbated by the hot, drying sirocco wind, and the benefits of occasional heavy showers in fall and spring are lost through surface evaporation before the water has penetrated the soil. Only a few plants, such as the olive, have adapted naturally to these conditions. Otherwise crop cultivation is reliant on methods of preserving soil moisture or on irrigation.

Physical relief also poses a challenge to agriculture. About 80 percent of the region is hilly or mountainous, leaving only a little cultivable lowland. The rocky skeleton breaks through the thin skin of topsoil at every opportunity. Fast-flowing mountain streams cause rapid erosion of the steep limestone mountain slopes of the central Apennines in Italy and the Pindus Mountains in Greece.

Despite generally poor soils and the threat of landslides, most hillsides are heavily cultivated. The soils, reflecting the underlying rock, are chiefly limestone. Where climate permits olives and vines are grown on the coarser soils. The finer marls are better for cereals, and pockets of terra rossa soils (red soils developed on the limestone bedrock) in the south of Italy and the basins and valleys of Greece are ideally suited to the interculture of cereals and tree crops (*coltura promiscua*). The alluvial deposits of the Po valley in northern Italy and the clays and loams of the lowlands of Thessaly, Macedonia and western Thrace in northern Greece are highly productive too, supporting largescale cultivation.

An early center of agriculture

The traditional agriculture of the region has roots that stretch back nearly 3,000 years, and was shaped by a succession of trading and colonizing peoples – the Phoenicians, Greeks, and Romans – who introduced wheat, barley, vines, olives and citrus fruits. The Greek farmers of classical times grew wheat and other cereals on the plains, where winter rains

A landscape of infinite variety A hillside in Tuscany, in central Italy, shows the characteristic mix of arable land, with fruit trees, vines and olives, that has predominated in the region since classical times. Tall cypress trees, typical of the Mediterranean, act as effective windbreaks.

Map of agricultural zones Despite the hilly nature of the countryside, land in Italy and Greece is heavily cultivated. Dairying is limited to irrigated pasture in lowland areas, and there is some rough grazing for sheep and goats.

Agricultural zones

- arable with some pasture
- fruit, vines and vegetables
- pasture
- rough grazing
- woods and forest
- nonagricultural land
- ▲ mountain peak (meters)

allowed seasonal use of land that was otherwise parched, olives on thin soils and rocky slopes, and vines on alluvial slopes with abundant subsurface water – a threefold arrangement that still continues today. The Romans left their mark on the Italian farming landscape with the practice of *coltura promiscua*. The Arabs, who colonized Sicily and various parts of southern Italy between the 9th and 11th centuries, introduced some important agricultural practices, such as techniques of irrigation, as well as a range of new crops (rice, cotton, apricots and mulberries), and extended the use of terracing.

Land reclamation projects

In more modern times, the most important agricultural developments have been the extension of irrigation and reclamation of marshland. About 30 percent of Italy's agricultural land is irrigated, mainly in the valley of the river Po, which provides 70 percent of the country's irrigation water. In Greece, 23 percent of the cultivated area is irrigated, these lands supporting 85 percent of all flower and vegetable cultivation, and 21 percent of the tree crops.

Reclamation of marshland has only been carried out to a significant extent in Greece in the period since World War II. In Italy, it has been in progress since the 10th century, and its single most striking success took place in the 1930s when the lowlying Agro Pontino in the central west coast of the country was transformed from an undeveloped area of marsh and dunes into a thriving agricultural landscape of more than 3,000 farms.

THE FARMING LANDSCAPE

Mediterranean agriculture is based on three types of crop – those, such as wheat, that are grown in the period of winter rain; those, such as the vine and the olive, that can survive the summer drought; and those, introduced from outside, that have to be irrigated. These include temperate fruits such as apples or pears and subtropical fruits such as peaches or citrus fruits, as well as vegetables. This intensive cultivation has earned the Mediterranean peasant the description of gardener rather than farmer.

Cereal cultivation takes up roughly half the arable land. Of the remainder, 40 percent is in fallow (uncultivated), and 10 percent under fruit, vegetables or other crops. Cereal crops are generally sown with the arrival of the fall rains and are harvested in early summer. The use of fallow is a distinctive feature of cereal cultivation in both countries of the region: once the crop has been harvested, the land is left fallow throughout the next two summers in order to conserve soil moisture. Because there are few cattle, very little manure is available, but artificial fertilizers have been increasingly used over the past few decades.

Cattle farming is limited by the prevalence of summer drought, which causes a lack of good pasture. Transhumance (moving herds from winter grazing to summer pastures in the mountains) was once widely practiced in the Alps and other upland districts, but is now far less common. Dairying is restricted to irrigated areas such as the north Italian plain. Sheep and goats are kept on rough grazing, particularly in the south.

Family farms and large estates

Almost all agricultural land in Greece and Italy is farmed by owner-occupiers; in Italy family farms account for 90 percent of all holdings and two-thirds of cultivated land. These traditional farms are widespread, but are most common in the north. They usually practice mixed farming, but where conditions are favorable, such as the Conca d'Oro around Palermo in Sicily, commercial specialized fruit

Rocks and red earth Stone walls describe an intricate pattern over a hillside on the Mediterranean island of Malta. Cultivable land is patchy in this rocky limestone terrain, but tiny terraced plots like these are a hindrance to farm modernization.

Commercial sponge fishing is a traditional means of support for fishermen in the Greek islands. The sponges – the skeletons of certain aquatic animals – are harvested with hooks or harpoons from the bottom of the sea. The industry is declining.

production has developed. In both countries, many family farms on marginal land have been abandoned as younger people drift to the towns in search of more lucrative employment.

Other systems of landholding remain locally important. A form of sharecropping known as *mezzadria*, once dominant throughout central Italy, is now confined to the Marches area in the northeast of the peninsula. Under this system, an estate is divided into a number of compact holdings, each with its own farmhouse. The owner of the estate, or his agent, supervises the operation from the home farm – the *fattoria* – and receives rent in kind from the tenants. Originally, produce was divided equally between the owner and the tenants, but legislation after World War II shifted the balance progressively in favor of the tenants. A further law (1978) banned the agreement of new *mezzadria* contracts.

Commercial farms are concentrated in northern Italy. They are capital-intensive enterprises run by a small team of full-time workers, supplemented with hired labor as necessary. These farms are extremely efficient and make use of high levels of mechanization. They are usually associated with well-developed processing and marketing systems, often organized on a cooperative basis.

In marked contrast to these efficient farms are the large estates, or *latifundia*, in the south. These were more important in the past than they are today, though many survive despite attempts to reform them. Frequently owned by absentee landlords, the *latifundia* made use of cheap seasonal labor supplied by landless peasants, and productivity remained low. They were traditionally devoted to the cultivation of durum wheat (used for making pasta) and sheep rearing.

Fragmented farms

Most of the region's farmers are disadvantaged by the small size of their holdings, or by their fragmented nature. This often results from the broken nature of the landscape, which leads to fields being divided by rocky outcrops, but it is exaggerated by inheritance laws that require equal division of land among heirs. Government policy has also played a part: in Greece a feudal system, based on large estates, has been transformed through legislation over the past 100 years into one dominated by owner-occupiers so that today small to medium-sized farms account for 90 percent of all landholdings. A typical farm has less than 5 ha (12 acres), split into 6 or more small plots, and large farms of over 50 ha (123 acres) are rare. (In the rest of the European Community as a whole farms of this size occupy more than 40 percent of agricultural land.)

In the last 30 years attempts have been made to consolidate these fragmented plots, with little success. So long as the inheritance laws remain unchanged, such plots are liable to be broken up again. Land consolidation schemes have benefits in increasing the economic viability of individual farms, thereby maintaining traditional agricultural communities. But unless they are accompanied by efforts to increase farm size, most farms will remain unable to provide a reasonable income. Consolidation of land and farm enlargement can only be effective within the much wider context of regional rural development to provide non-farm employment for farmers who are compelled to leave the land by these policies.

LAND REDISTRIBUTION

Great disparities in wealth and power in the rural societies of Italy and Greece concentrated ownership of land in the hands of a very small minority. Land reform programs – the radical attempt by government to reorganize the rural economy to redress the balance have been undertaken in both countries. However, such sweeping attempts to redistribute land to landless peasants, tenants or small farmers have not always brought longterm advantage. In Greece a land distribution scheme after World War II assigned between 4 and 18 tiny plots of land to each recipient in order to produce an equitable spread of land types, and the maximum size for individual holdings was set at 30 ha (74 acres). This program was thus a major contributory factor to the problems of farm fragmentation.

Italy has seen one of the biggest land reform programs ever undertaken by a Western capitalist government in the postwar period. Begun in 1950, the intention was to break up the *latifundia* by expropriating and improving the land, and redistributing it to landless peasants. Eight land reform agencies acquired a total of 767,000 ha (1.89 million acres). Eventually, 681,000 ha (1.68 million acres) was assigned to 113,000 families.

The overall impact of the reform was mixed. Little was achieved in Sicily, for example, where the activities of the Mafia criminal organization interfered with the program. But in other areas, particularly the coastal plain of Apulia in the southeast of the country, the reforms released the productive potential of the land. As a result, the quality of life of numerous small farmers was greatly enhanced.

THE PURSUIT OF PROFITS

Farming provides employment for 16 percent of Italy's labor force and 28 percent of Greece's. Revenue from agricultural products accounts for 7 percent of Italy's gross national product and 16 percent of Greece's. Agriculture thus makes a significant contribution to the economies of both countries, and recent government agricultural policy in each has been directed almost entirely toward improving the profitability of the farming sector.

The pressure for change

Italy has been a member of the European Community (EC) since its foundation in 1957; Greece joined later, in 1981. Working within the framework of its Common Agricultural Policy (CAP), both governments have fostered greater farming productivity by subsidizing improvements to rural services such as transport and electrification, and by providing price support and agricultural credit at low interest rates. Both countries are now agriculturally self-sufficient, except in animal products.

The pressure to modernize has seen an increase in mechanization wherever there is sufficient natural potential to justify the capital investment required. The number of tractors in Italy rose from 57,000 in 1950 to over 1 million in 1980; in Greece the number leapt from only just over 1,000 in

Riviera reflections *(above)* Greenhouses are packed close together along this coastal hillside in the Italian Riviera. Intensive flower and vegetable production, for the early season market in northern Europe, makes maximum use of available land and sunlight.

The marmalade mountain *(below)* A truckload of oranges, surplus to the requirements of the European Community, is added to a dump in Sicily. Membership of the EC gives farmers the support of subsidies and guaranteed prices, but encourages overproduction.

FIVE O'CLOCK FARMERS

Part-time farming has now become an increasingly important alternative to those who can no longer find full-time employment on the land. The majority of agricultural households in Italy and Greece now derive additional income from off-farm work. This typically involves commuting to work to an urban factory, or activities related to tourism. Part-time farming has become more and more common over the past few decades. It is partly a response to the persistent rise in costs (without equivalent increases in prices) faced by farmers, but also reflects the aspirations of most agricultural workers for higher incomes and a better quality of life.

Part-time farming has many advantages, both for the individual and in a wider social context. The extra income it generates may be used to buy consumer goods to improve the family's standard of living or it may be invested in the farm. This type of farming also helps to keep up sufficient levels of population in rural areas to support the local services, and contributes labor for local industries without adding to urban congestion. It also provides rural families with a safety net in times of industrial recession.

These advantages must be set against several disadvantages. Land may be left unused because "five o'clock farmers" do not have time to work it. This is not only a waste of resources; such land may become a haven for weeds and pests that spread to land under cultivation. Parcels of unused land belonging to part-time farmers may also obstruct efforts to consolidate and amalgamate smaller plots and farms. Nevertheless, part-time farming is likely to continue to increase; for many small farmers it offers the most practical method of survival.

1938 to more than 200,000 in 1980. The application of chemical fertilizers, pesticides and insecticides has also shown a dramatic increase.

The gradual commercialization of agriculture has been accompanied by increasing specialization. In most of Italy and Greece this has taken the form of the intensive production of one or two crops within existing farm systems. Specialist production of fruit and vegetables, particularly apricots, peaches and tomatoes, has been the area of greatest growth in Greek agriculture. Specialist agricultural areas in Italy include rice production in the upper Po valley, flower growing on the northwest Mediterranean coast (the Italian Riviera), tomato growing and processing around Naples in the southwest, and bergamot cultivation (for essential oils used in perfumery and food flavoring) in Calabria in the southernmost tip of the peninsula.

Balancing the two systems

Capital-intensive commercial farms are geared to realizing maximum productivity and profit. Small owner-operated farms are more attuned to the goal of long-term stability and survival. The coexistence of both these systems of farming within the region – and the contrasting attitudes they represent – creates difficulties and stresses for agriculture.

The push for greater productivity, with its reliance on crop specialization and the use of mechanization and agricultural technology, demands radical changes to traditional farm practice. It also involves the development of expensive processing methods and more efficient marketing and distribution networks. But problems arise when these techniques, aimed at the mass production of standardized products, are applied to the prevailing system of growing many different crops (that is, polyculture) practiced by the region's small farmers, particularly those in difficult physical environments.

It is impossible to achieve economies of scale on small, fragmented holdings and their existence continues to place severe constraints on the development of a more commercial agriculture. Plot fragmentation imposes significant extra costs. The value of time lost in moving labor or livestock between the scattered plots, the need for fencing, the difficulty of mechanization, all feature on the balance sheet. Disputes over access to plots and contested ownership, leading in some cases to litigation, are an additional social disadvantage. In Greece, 47 percent of plots and 33 percent of farmland are accessible only by trespass on a neighbor's property. The inadequacy of existing marketing and distribution services, and the presence of many small merchants and middle men, are further obstacles to greater efficiency.

Commercial methods, however, do not always offer the best strategy for farming in the region. In many areas of Italy and Greece, peasant farmers have adapted ecologically sound methods to cope with variations in the quality of land. The multiplicity of plots creates an intermixed pattern of farming. This can halt the spread of crop or animal disease. It also diffuses the risk of crop failure arising from drought or some other environmental hazard. Polyculture additionally offers peasants some protection against fluctuations in demand for a particular product.

The commercialization of agriculture causes unfortunate economic and social side effects. There has been a sharp reduction in the number of subsistence farmers, growing food for their own consumption. The overall number of people employed in agriculture has also plummeted. Many of the younger members of the rural population have emigrated to the towns, leaving an aged farm population behind. In Greece, 50 percent of all farmers are over 55 years old.

An early morning market at Epirus in northwest Greece. Local farmers display surplus vegetables and eggs produced on their small farm plots for sale. The commercialization of farming is threatening the livelihood of thousands of farmers such as these.

The versatile olive

The ancient Greeks believed that people could live off the products of the olive tree (*Olea europaea*) – a key element in Mediterranean agriculture for over 2,000 years. For centuries, olives – high in food value – together with goat's milk and bread, provided the basic sustenance for peasant farmers, and this hardy plant can be put to many other uses.

Crushed, the olive fruit gives oil that, when refined, can be used as a cooking medium or medicinally as a laxative or skin emollient. The residue of the refined oil can be boiled down to make soap, and the remaining pulp then returned to the land as a fertilizer or used as domestic fuel. The value of the olive tree does not end with the fruit and oil it produces. Annual pruning of the smaller branches provides fodder and bedding for sheep and goats; the hard, close-grained wood from larger branches provides timber for dwellings, furniture, tools and utensils; any waste wood is used as fuel for heating and cooking.

The olive tree ranges from 3–12 m (10–40 ft) in height. It has leathery leaves, whitish flowers and fruits that are a shiny purple-black when ripe. The wood is resistant to decay; if the top dies back a new trunk will often arise from the roots. Some trees can consequently achieve a very great age. The olive tree prefers an average temperature of 14°C (57°F), though it can also tolerate a minimum of −10°C (14°F). Though the olive prefers hot climates, it is otherwise remarkably adaptable. It can flourish on many different types of terrain and soil, and its gray-green foliage is often seen growing wild among the shrubby Mediterranean vegetation (maquis) or in natural groves.

Olives are frequently cultivated on what would otherwise be unproductive or uneconomic land: it prefers poorer, light soils, and become more susceptible to disease on richer ones. The mature tree develops an extensive root system that enables it to withstand summer drought. The roots also benefit farmers by binding fragile hillside soils together.

Although the olive tree does not bear fruit for seven years after planting, it subsequently yields fruit for long periods with comparatively little attention. For table use, the fruit is harvested either when it is green, and less oily, or when it is ripened to the purple-black color of the fully mature fruit. The bitter taste of the fruit is removed by soaking them in a hot, weak alkali solution that also softens the skin. The fruit are then pickled in salt water or preserved in their own oil. Sometimes the pits of green olives are removed and the centers of the fruit stuffed with anchovy, nuts or pimento.

If the fruits are to be harvested for processing into oil, they are left until they are fully ripe, when as much as 65 percent of their content may be oil. The oil is obtained by pressing or crushing the fruit, and is classified into five grades. The finest – "virgin" oil – is obtained from the first pressings and has to meet defined standards of quality to be so named. The next grade, of "pure" or edible oil, is a mixture of virgin and refined oils and is also sold for cooking; subsequent grades have commercial and industrial uses.

The olive economy

Italy and Greece contribute one third of the world's olive area. Olive trees cover 2.3 million ha (5.7 million acres), worked by 1.2 million families, in Italy and 0.52 million ha (1.28 million acres), worked by 400,000 families, in Greece. While the area under olive cultivation remains fairly constant, annual production can fluctuate markedly as the trees do not fruit every year. Output is also affected by the high proportion of old trees, imperfect pruning or insufficient pest and disease control. Average production is 450,000 tonnes per year in Italy and 250,000 tonnes in Greece. A great deal of the produce is consumed domestically, but Greece is a net exporter. Italy, though a major olive producer and exporter, is also an importer of significant amounts to satisfy domestic demand.

The number of olive trees in cultivation is likely to decline in future. This is mainly due to competition from cheaper alternative vegetable oils, such as soybean and sunflower. The EC has provided some protection to the olive economy. Producers are offered subsidies based on a flat rate per tree. Refiners are also subsidized in order to keep down the cost to the consumers. The result is an expensive support system that, so far, has remained politically acceptable because of the small share of olive oil in the EC's total agricultural production.

Harvesting the olives Nets are spread beneath the trees, and the branches are then shaken to loosen the fruit. Trees are often in inaccessible places, and the work is laborious. Olives are harvested either in their green, immature state or fully ripened, when their oil content is at its highest.

A VARIED FARMING LANDSCAPE

POSTWAR CHANGES IN FARMING · A REGION OF SMALL FARMS · THE QUEST FOR SELF-SUFFICIENCY

The pattern of farming found in Central Europe, which extends from the Baltic and North Sea coasts to the southern Alps, reflects wide differences in soils, climate and relief. The glacial soils of the northern and central lowlands provide fertile land for arable farming; as the country rises to the south, the landscape forms a colorful patchwork of mixed farmland and forest. Livestock and dairying predominate on the mountain slopes in the Alps, stretching across southern Germany, Switzerland and Austria, with arable farming in the major valleys and in Austria's northeastern lowlands. Although productivity has risen, agriculture contributes less than 4 percent of the national product in each of the countries (in the east of Germany it is 11 percent), and nowhere does it account for more than 10 percent of the workforce.

COUNTRIES IN THE REGION

Austria, East Germany, Liechtenstein, Switzerland, West Germany

Land (million hectares)

Total	Agricultural	Arable	Forest/woodland
47 (100%)	24 (50%)	14 (29%)	15 (31%)

Farmers

2.4 million employed in agriculture (5% of work force)
6 hectares of arable land per person employed in agriculture

Major crops
Numbers in brackets are percentages of world average yield and total world production

	Area mill ha	Yield 100kg/ha	Production mill tonnes	Change since 1963
Barley	3.1	46.0 (198)	14.2 ((8)	+162%
Wheat	2.8	56.1 (240)	15.9 (3)	+126%
Rye	1.2	36.6 (174)	42.1 (12)	−19%
Potatoes	0.7	291.5 (186)	21.1 (7)	−45%
Oats	0.7	42.8 (232)	2.9 (7)	−14%
Sugar beet	0.6	457.4 (132)	29.7 (10)	+59%
Rapeseed	0.6	28.3 (198)	1.7 (8)	+508%
Grapes	0.2	108.9 (145)	1.8 (3)	+74%
Other fruit	—	—	3.9 (1)	−18%
Vegetables	—	—	4.5 (1)	−64%

Major livestock

	Number mill	Production mill tonnes	Change since 1963
Cattle	25.6 (2)	—	+17%
Pigs	43.1 (5)	—	+44%
Milk	—	41.2 (9)	+27%
Fish catch	—	0.4 (—)	

Food security (cereal exports minus imports)

mill tonnes	% domestic production	% world trade
−3.5	8	2

POSTWAR CHANGES IN FARMING

Historical, as well as environmental, factors have contributed to regional differences in Germany's agriculture. Farming in northern Germany benefited from the agricultural innovations of the 18th century, which spread eastward across northern Europe from Britain and the Low Countries. The ending of the open-field system of land tenure brought farmland into unitary, more efficient holdings. Fodder crops were introduced, making it possible to keep animals indoors in the winter. The northern lowlands became an important area for sugar beet, which was grown first as a fodder crop and later for its sugar content. Elsewhere in the region peasant holdings remained small; there was no shortage of labor, and most farmers grew crops primarily for their own use rather than for surplus.

The aftermath of World War II affected Germany's agriculture drastically. Many farms went out of production, and there was a marked loss of labor to the towns. Much valuable farmland lay in the Soviet zone of occupation in the eastern half of the country. When it became the German Democratic Republic (GDR) in 1949, its agriculture was reorganized on socialist principles of state control.

In the immediate postwar years much effort went into modernizing and reorganizing West Germany's farming more efficiently. As a result, agricultural productivity began to rise steadily, and there was a marked shift toward livestock farming, especially beef and pig production and also dairying. Wheat, rye, oats and potatoes are grown, particularly on arable land in the north. The wine and beer industries are expanding parts of the agricultural sector, both for domestic use and for export – the west of Germany grows the largest quantity of hops (used to add flavor in brewing) of any country in the world. In the east the most significant farming area is in the south, where the majority of the country's cereals are grown on loess soils.

West Germany became a founder member of the European Community (EC) in 1957. The EC's Common Agricultural Policy (CAP) – designed to protect farmers' livelihood, increase production, and ensure reasonable food prices – had far-reaching effects. Unwanted surpluses were created in many commodities, especially milk, cereals, beef and table wines. As a result fixed commodity quotas and other stratagems to reduce production were introduced, though they were not always welcomed by the farming community.

Alpine farming communities

In Austria less than half the land, and in Switzerland only a quarter, is available for farming. Dairying and beef production are the mainstay of the agricultural economy of the mountain areas. There has been considerable modernization of farming in the Danube plain in the northeast of Austria, which has experienced a "grain boom" as a result, producing 80 percent of the country's sugar beet and 70 percent of its cereals. Vine growing is important in some parts.

Haymaking in full swing in an undulating valley floor in Austria. The comparatively level ground in lowland areas allows some degree of mechanization, but traditional methods of drying and storing the hay are still commonly used.

Map of agricultural zones Relief clearly determines the pattern of agriculture in Central Europe. Fertile lowland plains in the north provide excellent land for arable farming; as the land rises toward the Alps livestock and forestry predominate.

Arable farming in Switzerland is concentrated in the rolling hills and meadows of the narrow plateau known as the Swiss Midlands, between the Jura Mountains and the Alps; two-thirds of all farms combine grass and grain cultivation. Livestock production provides the main source of revenue on most farms: animal products account for over 75 percent of Swiss agricultural income, with milk alone supplying a third of the total figure. There are important areas of fruit and vine production in the south.

Forestry is important throughout the region. A quarter of the land area in Switzerland is forested, between a quarter and a third in Germany, and 40 percent in Austria. For centuries exploitation of the woods in Switzerland and Austria was secondary to farming – felling was unregulated, with large tracts often being removed at a time. With the mountainsides deprived of protective stands of trees, avalanches would fall right down into the valleys. Since the beginning of this century the management of alpine woodlands has been more strictly controlled, particularly with regard to planting, felling and access.

One of the most serious problems to face the forestry industry is the damage being caused to trees by sulfur dioxide emissions from the east. With the reunification of the two Germanies (and with West Germany's greater legislative concern for environmental issues) it is hoped that the problem will be partly brought under control.

A REGION OF SMALL FARMS

The number of people employed in farming in the west of Germany has fallen by a third since 1950, though agricultural output is twice as great. However, there is generally an inefficient use of land, capital and labor. One reason for this is the large number of small, highly fragmented farms – the legacy of the transfer of land by divided inheritance that formerly prevailed in much of the country, particularly the south and west.

In some districts a holding of only 20 ha (50 acres) may be divided into 200 or more parcels of land, some of which may lie several kilometers away from the farmer's home. The consolidation of farmland has been a priority for regional governments; since 1950 over 5 million ha (12 million acres) of land have been reorganized in this way. The majority of farms nevertheless remain small: almost a third are under 5 ha (12 acres), while 60 percent are less than 20 ha (50 acres).

The larger farms are generally found in the north, particularly in Schleswig-Holstein and Lower Saxony. In these two areas the average farm size is over 35 ha (86 acres), while in Bavaria in the extreme south it is less than 10 ha (24 acres). The larger the farms the more labor efficient they tend to be. Nearly 10 percent of Bavaria's workforce is engaged in agriculture; in Lower Saxony the figure is 7 percent. The farming population tends to be distributed in large numbers of widely scattered small settlements. This can hinder consolidation and make it difficult for neighboring farms to cooperate in the use of machinery and marketing facilities.

Changing patterns

The marked rise in livestock numbers that has taken place since 1945 has come about partly through the greater affluence of the population, which has increased the demand for meat. West Germany is the leading meat producer in Europe: pigs outnumber all other farm animals in the country, and the consumption of pork is among the highest in the world. The imposition of the CAP has also been responsible for shifts in the relative importance of crops, encouraging the production of grain by guaranteeing relatively high prices on the market. This in turn has led to specialized production on many holdings, especially in northern Germany.

As elsewhere in Western Europe, agricultural production has increased substantially in Austria since World War II

Harvesting sugar beet *(above)* in northern Germany. Sugar beet is a major contributor to Germany's agricultural exports. Yields have been increased by developing varieties of seed that achieve a high rate of germination and produce seedlings that do not need thinning.

The ubiquitous cow *(below)* Even the smallest farms usually keep some dairy cattle. The higher milk yields of recent years have led to large production surpluses. Both the Swiss and the German governments have introduced milk quotas in an attempt to curb overproduction.

Rhine valley wines Winemaking in Germany is very efficient: from just 1 percent of the world's vineyards, it produces 13 percent of the world's wines. German wines are graded according to the natural sugar content of the grape, the most superior being the sweetest.

MAKING THE MOST OF THE LIE OF THE LAND

In vine growing, perhaps more than in any other agricultural activity, the combination of warmth, soil, slope and aspect – determining the intensity and duration of sunlight on the vines – plays a crucial role. The steep-sided slopes of the river Rhine and its tributaries, particularly the Mosel (Moselle), in West Germany are well suited to vine cultivation. When terraced, the slatey soils of the steep valleys provide a well-drained foundation for the vines, while the constant movement of air in the valleys prevents rises in the levels of humidity that cause mildew to form on the grapes. Vineyards cling to every possible face of the sunnier slopes along the Rhine. Special machinery has been developed to tend and harvest the vines on the slopes; where they are too steep even for this, the grapes are picked by hand.

About 90 percent of German wines are white; the hardy Riesling, Sylvaner, Müller-Thurgau and Gewürztraminer grapes create soft, fragrant wines. Sugar may be added to some categories of wine to compensate for the short, cool growing season, but wines of the highest category (*Qualitätswein mit Prädikat* – QmP: "quality wine with special attributes") must be fermented from natural sugars only. Long hours of sunshine in late summer and early fall prolong the ripening process, raising the sugar content of the fruit, while the heat is not so strong that the skin of the grapes is damaged.

despite the large number of workers who have left the farm sector. A rationalization of production, brought about in part by difficult farming conditions, particularly in the higher mountainous areas, has led to farmland being abandoned and replaced by large tracts of forest.

As much as 80 percent of all farms are less than 20 ha (50 acres) and almost half are smaller than 5 ha (12 acres). Such fragmentation makes farming for many a part-time activity; less than 40 percent of Austrian farms are run on a full-time basis. In the mountainous areas both forestry and tourism (particularly skiing) provide alternative sources of income in winter when farmwork is low.

Considerable public money has been made available for farm improvement schemes. More than 100,000 farms have been enlarged by consolidating holdings, mostly in the Danube plain, which has facilitated the expansion of crop production in that area.

A shortage of land

Although lowland farming in Switzerland is extremely productive, it is not without handicaps. Farmland in the Swiss Midlands is at a minimum altitude of 300 m (1,000 ft) and rises rapidly toward the Alps. The average farm size on the plain is 16 ha (40 acres), compared with 9 ha (22 acres) in Switzerland as a whole. Although farms are increasing in size, albeit slowly, the low availability of agricultural land is a major constraint upon the development of the farm sector.

In Switzerland some 1 million ha (2.5 million acres) are used for agriculture. A third of this land lies in the mountains; another million hectares of alpine pastures come into use during the summer. However, farmland is under considerable pressure; since 1945 100,000 ha (247,000 acres) – 80 percent of which were in the lowlands – have been taken out of agricultural production and used for urban development. Switzerland has only 0.05 ha (0.12 acres) of arable land per head; comparable figures in the west of Germany and Austria are 0.13 ha (0.32 acres) and 0.21 ha (0.52 acres) respectively.

In addition to the extreme fragmentation of holdings and the small size of

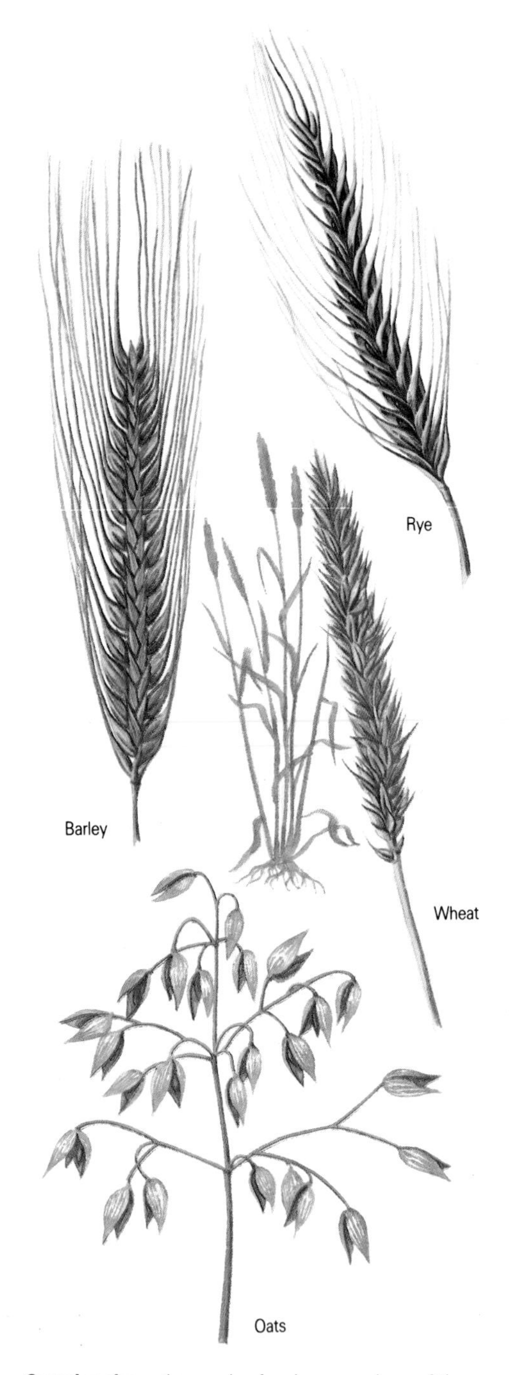

Cereal grains – the seeds of various members of the grass family – are the world's major staple foods, usually milled into flour. Four types – wheat, barley, oats and rye – are known to have been cultivated in northern Europe since prehistoric times.

farms, farming in Switzerland suffers from an extreme shortage of labor. Farm workers have been drawn away from the land by the incentive of higher wages in manufacturing and the service industries. This is a problem that affects the farm sector in the west of Germany and Austria as well, and to a lesser extent in eastern Germany. In an attempt to achieve economic viability some farmers have purchased additional land, but this has had the detrimental effect of burdening many of them with debts.

THE QUEST FOR SELF-SUFFICIENCY

In the past all the countries of the region had to rely on extensive imports of cereals and other commodities to feed their rising industrial populations. Largely as a result of the region's involvement in two devastating world wars, the principal concern of governments throughout this century has been to achieve agricultural self-sufficiency – an aim that is now approaching realization, largely through government intervention.

In contrast to other sectors of the West German economy, agriculture was well supported by both federal and regional governments. Since the 1950s policies aimed to increase the productivity of the farm sector, often at great cost to the taxpayer. In 1990 West Germany was able to produce most of its own foodstuffs: it was more than self-sufficient in sugar (131 percent), milk (123 percent), beef and veal (119 percent), and nearly self-sufficient in cereals (95 percent). Agriculture accounted for 6 percent of total exports.

In the wider European context, West Germany produced 18 percent of the EC's total production of barley, 21 percent of its beef, 24 percent of its sugar, 28 percent of its pigmeat and 70 percent of its rye.

Cereal growth Fields of oats, wheat and barley contribute to the stripy, patchwork effect of this farming landscape. The creation of larger fields and more open farmland helped to increase grain production in the west of German after 1945.

MERGING TWO SYSTEMS

At the end of World War II agriculture throughout Germany was still largely based on peasant farming, and was in disarray and decline. After the post-1945 division of the country, widely contrasting systems of agriculture emerged: in West Germany one of the main problems facing farming was how to dispose of surplus agricultural production; in the East there was a constant effort to increase production to avoid food shortages.

The reunification of the two Germanies posed a challenge to agriculture. On joining the economic structure of the West, farmers in East Germany became subject to EC regulations for agricultural production. For many of the more efficient farms, the CAP was likely to increase their incomes through the setting of price levels higher than those currently existing in the East. Farmers were also set to benefit from EC aid, especially if their farms were situated in disadvantaged areas such as uplands, were on poor soils, or were serviced by inadequate road and rail links.

The other side of the picture was less rosy. After reunification farmers in the East had to pay more for their fertilizers and machinery. Consumers in the East showed a preference for higher-quality food from the West. With the introduction of high CAP prices the gap between producer and consumer prices widened, raising the cost of maintaining an artificially cheap food supply in the East, where consumer prices had been protected. For many, German reunification seemed likely to offer less than positive benefits.

After the introduction of the CAP, West Germany proved, along with France, to be one of its most enthusiastic supporters. However, it proved to be a very costly policy, and its unwanted surpluses have had to be countered by lowering the prices for market support, by only guaranteeing subsidies for a fixed amount of production in each commodity and, in the case of cereals, by paying farmers to take their land out of production (a policy known as "set aside").

Upheaval in the East

Agriculture was the economic sector that experienced the most sweeping changes following the introduction of communist rule in East Germany. Responses to the socialist upheaval varied in different parts of the country, but in general state control was not easily imposed on the deep-rooted traditions of peasant independence. More than 40,000 farmers, many of the most able and successful, fled to West Germany in the years before the frontier was closed in 1961.

The massive reorganization of farming that took place in the 1950s absorbed substantial state investment, but this was not matched by major increases in agricultural productivity. It brought visible changes to the rural landscape. Holdings of over 180 ha (250 acres) were confiscated without compensation and pooled into a land fund for redistribution to peasant farmers. Specialization and cooperation were encouraged, leading to the development of farms based on only one or two lines of production.

Lands flowing with milk

Although agriculture is no longer as important to the economy of Austria as it was in the past, it is nevertheless able to produce most of its food requirements; in some commodities, such as milk, regular surpluses are produced. More than half of the country's gross income from farming is derived from livestock rearing. Wine is also an important export – vast surpluses were produced in 1982 and 1983, when climatic conditions were favorable, but an early frost in the fall of 1985 destroyed as much as 20 percent of the expected grape harvest, causing longterm damage to the Austrian wine industry.

Switzerland's determined pursuit of food self-sufficiency is largely the consequence of its role as a neutral country. In order to minimize its level of reliance on others, government has played a key role in managing the farm sector through a range of policies including fixed prices on most crops, protected markets, import fees and quotas, restrictions on farm size, and subsidies for mountain farmers.

As a result, Switzerland has achieved striking agricultural success, producing 60 percent of its domestic food needs on only a quarter of its total land area. It is able to supply all its milk and 83 percent of its meat, and is nearly self-sufficient in pork and veal. Given the limited land for arable farming, it has to depend much more heavily on crop imports: 50 percent of its grain, 60 percent of its vegetables, 80 percent of its sugar and 90 percent of its vegetable oils are imported.

Since 1970 increased milk production through higher yields, coupled with a marked fall in domestic demand, has regularly created large surpluses – the origin of the Swiss chocolate and cheese-processing industries. In common with the EC, when faced with escalating costs in its support of the milk sector, the Swiss government in 1977 was forced to introduce a quota system to limit production. Although this was originally brought in as a short-term measure, it was never withdrawn and has now become an established instrument of Swiss agricultural policy.

Cheese making has long been practiced in Switzerland as a way of consuming surplus milk production. Gruyère cheeses, here being washed, are world famous. Their characteristic holes result from the rapid fermentation of the curd.

Mountain farming

For centuries pastoral farmers in the Alps have practiced transhumance – the regular movement of herds of cattle, sheep and goats from their winter grazing land on the lower mountain slopes to higher alpine pastures between the months of May and September. In Switzerland it is estimated that 80 percent of all goats, two thirds of all sheep and 20 percent of all cattle are still moved to summer mountain pastures each year.

The grazing period on these upper pastures, which come into use as the winter snow melts, is often quite short, between 75 and 120 days, its duration depending on altitude and weather. Barns for storing hay, animal shelters and even farmsteads have been built at higher levels for temporary summer use, and some members of the farming family may spend days, even weeks, high up the mountain. Before the days of easier road transportation, milk would be turned into cheeses and then taken down the mountain to local markets in the valleys.

The grazing on these high alpine pastures in Austria can provide enough fodder to feed 100,000 cattle. However, difficulty of access, a dwindling labor supply and competition from tourism means that the use of alpine pastures has been on the decline since the 1970s. In Austria alone more than 300,000 ha (740,000 acres) – that is, one in every five mountain pastures – have now been abandoned.

The first pastures to be given up are, not unnaturally, those high up on steep slopes where the terrain is too inaccessible for machinery and the amount of labor needed to use the land is disproportionate to the returns. The growing unprofitability of mountain farming has forced farmers to look for alternative sources of income. Forestry provides

Summer in the mountains Cows are moved to high pastures after the snows have melted. In the past, when roads were scarce, cowherds spent weeks alone with their livestock. Bells were hung around the animals' necks to stop them from wandering.

between 12 and 30 percent of the total income of mountain farmers in Austria. In Switzerland the figure is much lower, perhaps as little as 5 percent.

The dramatic rise of tourism and related leisure activities in recent years allows many farmers to supplement their incomes in other ways. Farm buildings on the upper pastures, no longer used for transhumance, have been converted for letting to holidaymakers, for example. Although income from tourism may act as a disincentive to farming in marginal mountain pastureland, leading to a reduction in the number of livestock kept, the converse is also true: the fresh injection of supplementary income may spell survival for such farmers, even enabling some to modernize their holdings. However, the uncertainty of the tourist trade, vulnerable as it is to variations in snowfall from year to year, means that at best it can only offer temporary respite from the difficulties of alpine farming.

Preserving the alpine environments

The governments of both Austria and Switzerland have recognized the important role of mountain farmers in preserving the alpine environment. One of the essential goals of the Swiss government's agricultural policy is to ensure that farmers are rewarded for protecting the rural landscape, especially in areas that are popular with and much visited by tourists. This aim justifies the expenditure of large amounts of public money in mountain localities. For example, farmers in these areas receive income supplements based on the number of livestock maintained, as well as subsidies for growing feed grain, guaranteed bread-grain prices, a milling bonus, and grants for potato growing. They may additionally receive subsidies for structural improvements that are more than one and a half times as great as those given to farmers in the lowlands. As a result of these measures the number of animals kept on alpine pastures in Switzerland has increased in the last decade, reversing the trend of previous years, when the number of cattle fell from 141,000 in 1955 to less than 100,000 by the late 1960s.

Uneconomical farms The inaccessible, sloping land on steep alpine farms makes the use of machinery virtually impossible. Although this limits their efficiency, it ensures that farmers preserve the natural beauty of the mountains – one of the great attractions for tourists, who are a source of extra income.

FARMING AT THE CROSSROADS

TWO AGRICULTURAL ZONES · STATE FARMS AND PRIVATE ENTERPRISE · SUCCESSES AND FAILURES

Agriculture has had a long history in Eastern Europe; the relatively warm and dry Balkan lands in the south served as a corridor for the spread of crops from the Middle East, where farming began before 5000 BC, into the rest of Europe. The region contains extensive fertile lowlands in the Danube basin in Hungary, Yugoslavia and Romania, and in the North European Plain in Poland. In the southeast of the region climatic conditions become very dry, and irrigation is a critical factor in supporting agriculture. Just over a fifth of the land is more than 500 m (1,500 ft) above sea level, limiting agriculture to the hardiest cereals and to rough grazing. Despite the constraints imposed by the rigid communist system that dominated all the countries in Eastern Europe until recently, there is great variation in styles of farming.

COUNTRIES IN THE REGION

Albania, Bulgaria, Czechoslovakia, Hungary, Poland, Romania, Yugoslavia

Land (million hectares)

Total	Agricultural	Arable	Forest/woodland
115 (100%)	69 (60%)	46 (40%)	36 (31%)

Farmers

12.2 million employed in agriculture (20% of work force)
4 hectares of arable land per person employed in agriculture

Major crops
Numbers in brackets are percentages of world average yield and total world production

	Area mill ha	Yield 100kg/ha	Production mill tonnes	Change since 1963
Wheat	9.8	40.3 (172)	39.6 (8)	+132%
Maize	7.1	53.5 (147)	38.0 (8)	+122%
Barley	3.4	39.7 (171)	13.5 (7)	+143%
Rye	3.0	25.6 (122)	7.7 (22)	−14%
Potatoes	2.8	177.9 (113)	50.6 (18)	−11%
Sunflower seed	1.4	20.1 (141)	2.9 (14)	+147%
Sugar beet	1.2	326.1 (94)	39.4 (13)	+39%

Major livestock

	Number mill	Production mill tonnes	Change since 1963
Pigs	61.5 (7)	—	+60%
Sheep	45.7 (4)	—	+12%
Cattle	31.7 (2)	—	+13%
Milk	—	36.7 (8)	+50%
Fish catch	—	1.2 (1)	—

Food security (cereal exports minus imports)

mill tonnes	% domestic production	% world trade
−1.3	1	1

Traditional methods In much of Eastern Europe farming has been slow to give way to change. Flax, being harvested here, is grown in parts of the region where conditions are suitably cool and moist. The stems produce fiber that is woven for linen cloth, and the seeds yield linseed oils.

TWO AGRICULTURAL ZONES

Eastern Europe began to be opened up to agriculture in prehistoric times by people spreading northward and westward from Mediterranean lands. The Danube and other river corridors gave them access through the Balkans to more fertile land farther north, and woodland was quickly cleared for farming, especially on land with a light loess soil. Early settlement in the south tended to be in the mountains, which offered protection from marauding armies and from diseases that were endemic in lowlying areas, and some remarkably large agricultural communities grew up here.

Although routes along the northward-flowing rivers of Eastern Europe were established very early by traders dealing in amber, a valued commodity found along the Baltic coast, the dense coniferous and deciduous forests of northern Europe posed a severe challenge to people who wished to settle. It was only during the early medieval period (500–1200 AD) that German and Slav colonists began to open up the lands in the north to agricultural and economic development.

Expanding markets

Gradually these northern farmers came to challenge the superiority of the Balkans, and then to eclipse it. Access to the Baltic along the rivers Elbe, Oder and Vistula allowed food surpluses to be traded with Western Europe from the cereal lands of the north. Those of the south, on the other hand, were more inaccessible; a long sea journey was required to reach the Danube delta, and the towering Dinaric Alps along the coast of Yugoslavia cut off access from the Adriatic Sea. From the 14th century onward the Balkan states, including much of Hungary, came within the orbit of the Ottoman empire centered on Istanbul, severing trading relations with the West for several centuries. The ending of the Ottoman trading monopoly in the mid-19th century and the coming of the railroads and steamships subsequently helped to make Romania one of the world's leading cereal exporters.

As industrial growth spread throughout the region, more land was required to feed the population and to create surpluses for agricultural exports. In the Danube basin agriculturally unproductive land, such as the grasslands of the central Hungarian plain (the puszta) mostly supports poor grazing; in recent years, however, the supply of irrigation water has allowed some land to be more intensively farmed. Irrigation schemes have also enabled agriculture to take place in the very dry areas in the southeast of the region, which receive an average annual rainfall of below 500 mm (20 in) and have high summer temperatures.

Nevertheless, the demand for food has outstripped agricultural production. In the mid-1930s net exports equaled 6.4 percent of total production in the region; by the early 1960s there was a deficit of 5.9 percent. Pressure to provide better living standards make any return to Eastern Europe's traditional exporting role difficult to foresee, despite the strenuous efforts that are being made to reduce the level of cereal imports to the region from North America.

Some of the deficit in food production is met by fish, which are caught by trawlers operating from Polish and other ports in distant waters. The catch of sea fish is supplemented by the output from fishponds, of which there are a particularly hugh number in Czechoslovakia.

In most areas cereal production is complemented by both sugar beet and fodder crops. In the mountain districts there is greater specialization in cattle, sheep and pigs, with some cereal production (mainly rye and maize). Stock rearing and fattening have expanded in the main arable zones in the lowlands. Fodder cereals are grown, and animals are raised and slaughtered, within the same district, an arrangement that considerably reduces transportation and marketing costs.

A few specialized crops, such as flax, hemp, sugar beet and sunflowers, are grown in some areas. Fruit, flowers, vegetables and vineyards are also important where soils and temperature permit. Forestry in the wooded mountains is commercially important too, producing timber for board, plywood and furniture.

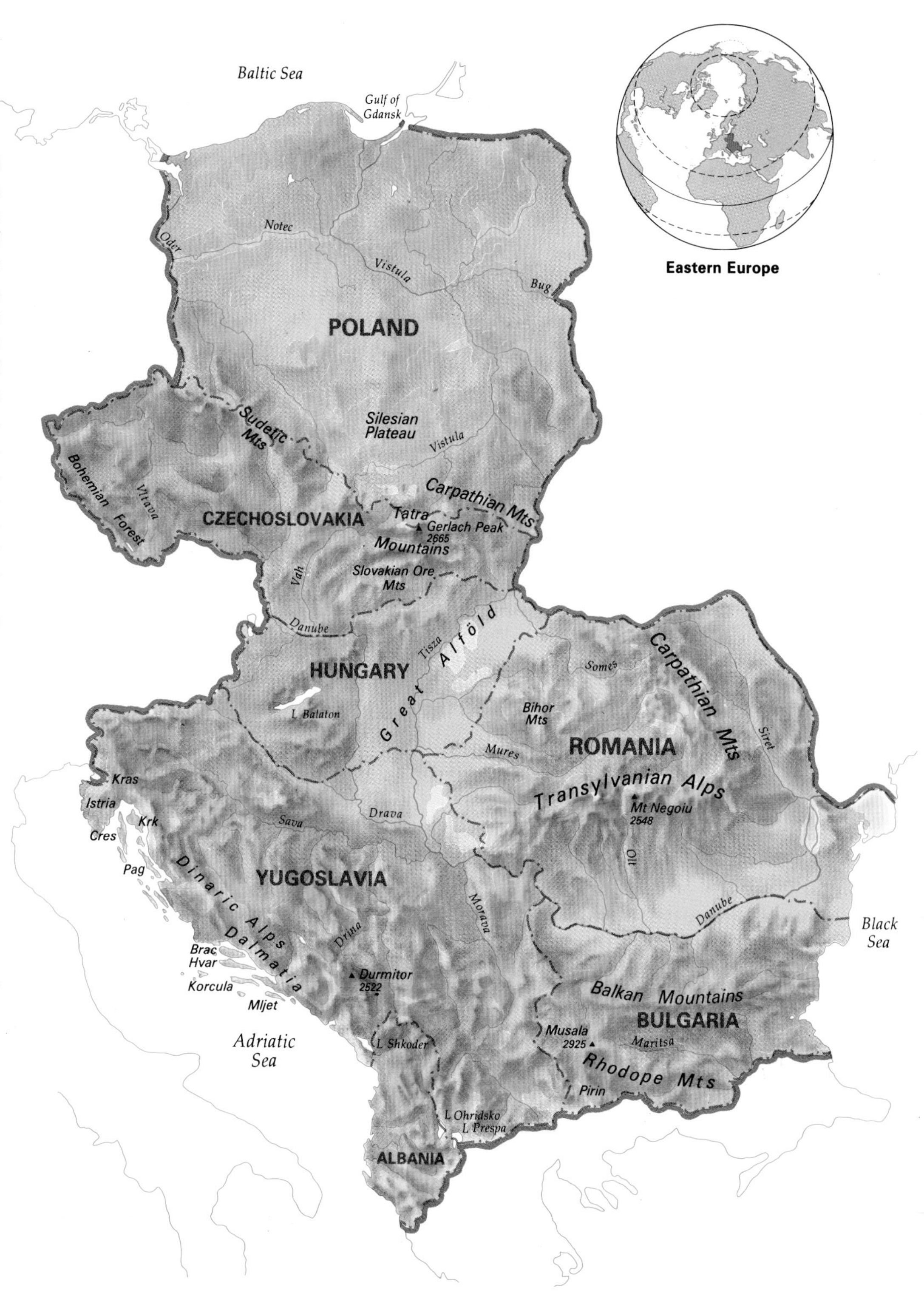

Agricultural zones

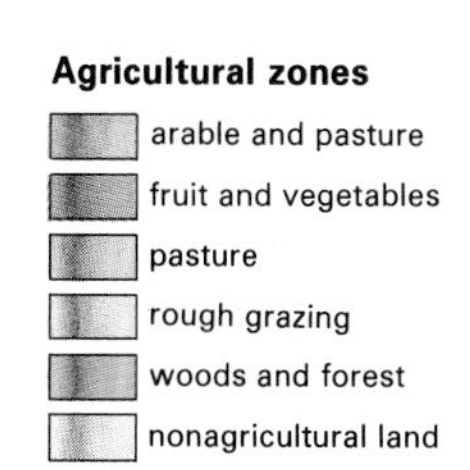

▲ mountain peak (meters)

Map of agricultural zones Fertile river plains, divided by heavily wooded mountain ranges, provide land for arable farming. Livestock predominate in upland areas and where conditions are arid; irrigation has extended the amount of cultivable land in the southeast.

STATE FARMS AND PRIVATE ENTERPRISE

Before events in the aftermath of World War II brought about wholesale political change to Eastern Europe, farming almost everywhere was predominantly peasant in character. Although feudal obligations binding peasants to landowners had been swept away in the last century, there was no shortage of labor; lack of employment opportunities elsewhere meant that large landowners, paying low wages, were able to impose burdensome labor contracts on farm workers. This had prompted some prewar governments to introduce land reform programs to give peasants greater security and independence, at the same time encouraging the acceleration of industrialization in the region.

The imposition of communism in one country after another in the years after 1945 as they came under the influence of the Soviet Union brought about the complete reorganization of agriculture in the region. Efforts were made to bring the peasantry into a Soviet-modeled state system, while ensuring adequate farm output at prices determined by the state. Farms were of two kinds: state farms, under direct central control; and collectives, created through the pooling of peasant land, livestock and equipment, by coercion if necessary. The Communist Party hierarchy controlled the setting of production targets and ensured delivery of produce to the state marketing organizations. Workers on the collectives were allowed to retain small private plots to support their families and to create small surpluses for sale in local markets.

Tinkering with the system

This picture of a uniform, state-controlled agriculture is inevitably generalized. Nowhere in the region was collectivization ever absolutely complete; some farms remained where existing field systems and settlement patterns made cooperation difficult. In Poland and Yugoslavia the political will, even at the height of communist control, was not forthcoming to collectivize all the peasants in all areas.

As the socialist experiment was seen to have failed in Eastern Europe during the 1970s and the 1980s, many modifications were made to the system. The young male population increasingly opted for work in the towns, so the agricultural labor force fell, becoming predominantly elderly and female. A greater degree of investment in fertilizers, machines and irrigation schemes was necessary to increase output. Regular wages for members of collectives, along with improved welfare benefits, became commonplace so as to encourage people to stay on the land.

In order to raise productivity still further, attempts were made to increase individual commitment and motivation by allocating specific tasks to individuals and groups in return for payments by results. Numerous *ad hoc* procedures were worked out to allow cooperatives, state farms and private plotholders to collaborate to make fullest use of land and labor, and farms were enlarged by amalgamation to allow greater economies of scale. Nevertheless, decision making remained very much concentrated at the center, so any local initiative was subordinated to the state agricultural program. For example, a cooperative with sustained success in stock rearing might be called upon to lose a large part of its herd in response to a sudden demand for a cutback in grain consumption.

The trend in a reforming Eastern Europe must be to harness the energies of the peasant workforce in a way that is likely to stimulate production and increase exports. Farming within the state sector guaranteed work, but left little scope for initiative and payment according to results; productivity consequently remained depressed. Independent farms, however, although they allowed farmers greater choice, were limited in output: as a result, farming had become a part-time activity for most people.

Developments in Hungary, as it moved away from Communist Party control, may have pointed the way ahead. Although all land remained in the hands of cooperative and state farms, individual enterprise was encouraged and integrated with state agriculture. The role of the state in fixing prices was reduced in favor of free market forces, and there was an improved supply of machinery to small farms.

Many examples can be found of successful and imaginative schemes to increase productivity. A cooperative at Ocsa near Budapest, for instance, allocates

Scything an upland meadow for hay Centralized agricultural policies, deployed through state farms and collectives, contrived to expand cattle production in lowland arable areas, but very few efforts were made to increase farming output in the mountain areas.

These little pigs went to market A busy scene in a Polish market, where young pigs are being sold for fattening on private plots. Agriculture in Poland was never centralized to the extent that it was in other Eastern European countries.

private plots, the size of the plot depending on the number of cows each member agrees to look after. In some of the villages in northeastern Hungary cooperatives have planted orchards and arranged for periodic spraying while contracting out other routine tasks (including harvesting) to individuals, who can work in the knowledge that there will be a guaranteed outlet for their produce. A cooperative near Sopron on the border with Austria deals with several hundred individual gardeners and smallholders to produce grapes, vegetables and poultry. These are then supplied to markets on the other side of the border.

A GROWING MARKET FOR FLOWERS

Flowers have traditionally played an important part in the lives of the people of Eastern Europe; the specialized cultivation of flowers to meet the local and national market is an exception in a region that has, for the main part, discouraged specialized production in order to achieve self-sufficiency in its staple foods. Quite often, the initial stimulus from a local market has provided the basis for successful entry into a wider commercial market. Horticulturalists at Turany, near Brno in central Czechoslovakia, for example, have derived great benefit from a strong local market for flowers to extend their trade into Western Europe.

One particular specialization is found in Bulgaria (also an important area for tobacco production). Here roses have been grown for their oil-bearing properties since the days of the Ottoman empire. They yield attar of roses, which is used in the manufacture of perfumes.

The demand for flowers is particularly strong in the northern cities of Eastern Europe; a flourishing center of production has grown up in Poland, for example, in the immediate area around Warsaw, especially in Jablonna. Here carnations are very successfully grown under glass, benefiting from a supply of pure underground water and carefully prepared manures. There is good local organization to support an industry that is able to employ one person for every 2.5 ha (6 acres) throughout the year, providing an example of the sort of enterprise that will have to become more widespread if the agriculture in Eastern Europe is to change to meet the challenges of the future.

SUCCESSES AND FAILURES

Even though agricultural production has registered some spectacular increases in Eastern Europe, planned targets are frequently missed, sometimes by considerable margins. While natural disasters may be blamed in part for this – the severity of a summer drought in nonirrigated areas of the Balkans, for example, can have a devastating effect on cereal yields – the main problem in every country has been the mistaken expectation that the mere setting up of a socialist farm structure would enhance output, compounded by the assumption that political education would overcome adherence to traditional peasant proprietorship.

All the same, investment has increased and progress been made in several areas. Greater attention to the use of fertilizers, to pest control, and to animal and plant breeding has brought important benefits. There has been interregional cooperation in agricultural research; Bulgarian sheep, for example, have been crossed with the Soviet Romanov breed, and the Soviet wheat variety called Bezostayal 1 has been widely used in Bulgaria as a basis for new, shortstalked hybrids.

A considerable amount of effort has also been expended in improving farm machinery and storage facilities in order to prevent heavy losses through inefficient harvesting and transportation. Not every innovation, however, has brought with it unmixed blessings: the excessive application of nitrate fertilizers to soils brings diminishing returns, and the land can become compacted through the use of heavy machinery.

Modifying the environment

Irrigation to increase the amount of land available for agriculture has been very successful in the Balkans. Bulgaria's irrigated area increased from 0.13 million ha (0.32 million acres) in 1950 to 0.72 million ha (1.78 million acres) in 1960 and almost 1 million ha (2.5 million acres) in 1968. However, this figure was well short of the target of 2 million ha (5 million acres) set in 1959 for achievement in 1965, and the present figure of 1.2 million ha (3 million acres) is far below the target of just under 3 million ha (7.4 million acres) set for 1980 by the Party Congress in 1962. Irrigation is very expensive when the costs of both artificial water storage and a distribution system are taken into account. Two harvests a year are necessary to justify the capital investment and compensate for the land that is lost, as a result of the

Aerial crop spraying in Hungary *(above)* Hungarian farms are among the most modern and efficient in the region. Farmers have responded to the relaxation of state control with innovations such as contracting out crop spraying or harvesting, and importing new techniques from Western Europe.

State-run agriculture *(left)* Not all collective farms benefited from attempts to modernize farming. This outdated threshing machine is at work in the fertile farmlands of northwest Poland. Inefficient harvesting methods and inadequate storage and transportation facilities meant that yields often fell short of targets.

installation of the system, but output has rarely met this target.

Other forms of land improvement offer further scope for increasing production. Flooding or erosion have rendered considerable areas of land in the region unusable for agriculture. Although longterm solutions to these problems are not yet forthcoming, steps have been taken to protect river floodplains by constructing dikes so that former permanent grasslands can be drained and then converted to arable farming. Much of the lower Danube has been reclaimed in this way. Risk of damage from flooding is reduced if water storage facilities along the river's course (linked perhaps to hydroelectric generation schemes or to navigational canals) are adequately controlled to prevent any sudden rise in river levels after heavy rain or snowmelt.

Erosion, usually in the form of gullying but sometimes involving landslides and mudflows, is widespread in the hill and plateau country of Eastern Europe. The damage was caused in the past by woodland on unstable slopes being cleared to provide more land for subsistence farming. The only longterm remedy may be to return the slopes to forestry, though this will reduce the amount of available farmland. But the terracing of eroded hillslopes has met with some success in preventing further soil loss. Orchards and vineyards may be established on these terraces, but without adequate marketing and food processing facilities, opportunities for such specialist fruit production cannot be fully exploited.

Unrealized potential

In many respects, and by comparison with other parts of the world, agriculture in Eastern Europe is highly successful. What places it at a disadvantage is the lack of an effective trading system. It is arguable that the region's potential (the physical resources assessed in the context of available technology) has not been fully realized, especially in the Balkans, and this is reflected in the trading pattern of surpluses and deficits. Eastern Europe was obliged to import 8 percent of its total cereal requirement of 62.8 million tonnes in 1965, but by 1980, with total consumption at nearly 100 million tonnes, imports had risen to just over 12 percent.

Greater involvement by Western enterprises benefit the region's agriculture. Already there have been important outside investments in Hungarian farms and food processing concerns, operating to standards set by the foreign partners. State farms in Hungary have imported dairy cattle from North America and Western Europe. As a result of the radical changes of 1989–90, virtually all the ideologs, who insisted on the primacy of socialist farm structures despite their proven lack of success, were eclipsed by the pragmatists, who seek the best standard of farming attainable in the region through greater efficient investment and increased incentive.

NEW SCOPE FOR AGRICULTURE

For centuries agriculture in the lowland areas of the Balkans was severely constrained by the low summer rainfall, though the high temperatures offered great potential for cereal and other kinds of farming. For example, farming in the steppelands of the Baragan and Dobrogea districts in the southeast corner of Romania was limited to providing winter grazing; the shepherds brought their animals down from the Carpathian Mountains when fodder was no longer available on their high summer feeding grounds. Despite the proximity of the river Danube, these grasslands were subject to severe summer drought, making cereal farming very hazardous.

Although the lifting of Ottoman restrictions on trade with Western Europe provided a stimulus to create cereal surpluses for sale to places elsewhere in Romania, these districts – lacking irrigated water – were unable to benefit, and it is only since the 1960s that the full potential of this area has begun to be exploited. A small irrigation system was built in the early 1950s and later augmented by the completion of a larger scheme in 1984. This increased the area of land under irrigation to 200,000 ha (500,000 acres).

A byproduct of this scheme has been the utilization of spoil from the canal excavations to reclaim waste land in the area. This has provided an additional 2,800 ha (6,900 acres) of agricultural land, which supports fruit trees, vines and vegetables under irrigation.

The legacy of rural resettlement in Romania

Reform and reorganization in Eastern Europe will be a painful process; the communist past will haunt the new democracies for a long time to come. Perhaps the most distasteful legacy will be the rural resettlement program, or *sistematizare*, which was launched by Romania's president Nicolae Ceaucescu (1918–1989) in the late 1980s.

While most governments in Eastern Europe were allowing farmers greater initiative at this time, Ceaucescu remained firmly opposed to liberalization and continued to seek to strengthen the power of the state. Determined use of the state security apparatus allowed him to persevere with social projects that had their origin in the Stalinist policies of the 1950s, and have been paralleled in recent programs in some communist countries in Southeast Asia. As economic growth slowed down in the 1980s, full employment was maintained in the construction industry by ordering the rebuilding of selected villages to create *agrogorods* with much of the housing being provided, inconveniently, in apartment blocks.

Her way of life destroyed *(above)* A bulldozer moves in to flatten what was once this woman's home to make way for new apartment blocks. The anguish on her face reveals only too clearly the human cost of the rural resettlement program.

A tranquil way of life *(right)* The apparent prosperity of Romania's private farmers incurred the jealousy of the Romanian Communist Party, and led to increasing restrictions on them. Under Ceaucescu, whole villages were destoyed to make room for state agriculture.

Destroying a tranquil way of life

After the early traumas of adapting to state-controlled farming, with delivery quotas and compulsory collectivization, Romanian agriculture had settled down by the late 1960s. There was a small private sector, but enlargement of these farms was prevented by prohibiting the use of nonfamily labor and by imposing high taxes on land. All the same, produce raised on private farms was able to attract high prices in the market because of its superior quality, and families living close to small towns were able to find additional employment in industries such as mining or wood processing.

Life in such households was therefore fairly comfortable: they were largely self-sufficient in food, and were able to enjoy the proceeds of regular nonagricultural work and the occasional sale of livestock, providing adequate means to spend on consumer goods. This included the ability to build larger and more modern houses for themselves. Young families were offered a choice: to concentrate on farming and remain in their traditional homes conveniently situated for access to the fields, or to look for ancillary employment and perhaps build a new house closer to transportation routes for the daily journey to and from work.

This tranquillity was upset in the 1980s when coercive measures were introduced to maintain (if not increase) agricultural production at low prices. The apparent prosperity of private farmers aroused fears within the Communist Party of the rise of a rich class of peasant farmers (kulaks) and led to the fixing of maximum prices for produce sold on the free markets; the surpluses available for sale were reduced by obliging farmers to sell most of their produce to the state. To ensure that livestock other than poultry was not disposed of illegally, all animals had to be registered and the births of young animals notified immediately – the period allowed was shorter than that required to register human births.

By the end of the decade these policies had hardened to include complete resettlement, to the even greater consternation of Romania's rural populations. Justified on the grounds that land would thereby be freed for state agriculture, the plan was to consolidate scattered villages and hamlets into huge units of faceless apartment blocks. Whole communities might be ordered to transfer from their

Beyond the reach of the state *(below)* Upland farmers were fortunate in being able to survive the sweeping resettlement programs of the 1980s unscathed. Their remoteness and the difficulty of the terrain made them unpromising candidates for collectivization.

outlying settlements to a key village, or individuals simply moved within a village. Compensation paid for properties compulsorily purchased was minimal; the private plots fell into the state agricultural system as the population was moved, and the newly settled people were unable to restart their private agriculture. Production from intensively cultivated private plots was consequently lost.

During 1988–89 the authorities started rehousing villages in Bucharest's agricultural sector (called Ilfov), and it was expected that this would be repeated throughout the rest of the country in the 1990s. This was social engineering on an almost unprecedented scale. It was widely condemned for its social and economic costs, as well as for its devastating impact on a rich cultural heritage. One of the first changes put into effect by the interim government after the collapse of the Ceaucescu regime in December 1989 was the abandonment of resettlement in favor of a policy giving more land to individual farmers. This was good news for the majority, but for those already trapped in substandard accommodation in one of Ceaucescu's *agrogorods*, the future was less certain.

AGRICULTURE UNDER STATE CONTROL

THE CHALLENGES TO AGRICULTURE · GROWING FOOD FOR THE PEOPLE · IMPROVING THE SYSTEM TO MEET DEMAND

The Soviet Union is the largest country in the world, but size alone does not bring agricultural success. Only about 10 percent of the land is suitable for arable farming; the remainder is too dry, too hot or too cold. Just under half the total land area is covered by coniferous forest, and in the northern tundra zone (about 20 percent of the land area) the subsoil is permanently frozen, making it too cold for crop cultivation. Useful agricultural land is squeezed into a central band (the black earth belt), which is wider to the west, narrower and more broken to the east. These physical limitations are compounded by inefficient state-controlled agriculture, exemplified by the fact that there are 24 million farmers, more than in the whole of the West and Japan together, but they produce only about a quarter as much.

COUNTRIES IN THE REGION

Mongolia, Union of Soviet Socialist Republics

Land (million hectares)

Total	Agricultural	Arable	Forest/woodland
2,384 (100%)	729 (31%)	230 (10%)	959 (40%)

Farmers

20.7 million employed in agriculture (14% of work force)
11 hectares of arable land per person employed in agriculture

Major crops
Numbers in brackets are percentages of world average yield and total world production

	Area mill ha	Yield 100kg/ha	Production mill tonnes	Change since 1963
Wheat	47.2	17.8 (76)	83.9 (16)	+30%
Barley	30.8	19.0 (82)	58.5 (32)	+188%
Oats	11.8	15.7 (85)	18.5 (43)	+205%
Rye	9.7	18.6 (89)	18.1 (53)	+20%
Potatoes	6.3	121.7 (77)	76.1 (27)	−7%
Dry peas	5.2	16.1 (99)	8.4 (54)	+56%
Maize	4.6	32.4 (89)	14.8 (3)	+13%
Sunflower seed	4.2	14.6 (103)	6.1 (30)	+20%

Major livestock

	Number mill	Production mill tonnes	Change since 1963
Sheep	155.4 (14)	—	+6%
Cattle	124.6 (10)	—	+46%
Pigs	79.6 (9)	—	+38%
Milk	—	103.1 (22)	+61%
Fish catch	—	11.2 (12)	—

Food security (cereal exports minus imports)

mill tonnes	% domestic production	% world trade
+32.1	16	15

THE CHALLENGES TO AGRICULTURE

Serfdom was well established in Russia by the mid-16th century. This system made the peasants who worked the land the property of the landowning classes, who could buy and sell them as they pleased. It was formally ended in 1861, when the land was divided between the original landowning class and the freed peasantry, organized into communes.

In the communes land was divided into small strips often less than a meter wide and several hundred meters long. A peasant family might use its strips for a year or two only, before they were reallocated to others. This system, which provided no incentive to improve the land, was a major contributing factor to the backwardness of Soviet agriculture at the beginning of the 20th century.

Reform and collectivization

In the years before World War I reforms were introduced with the aim of creating a class of landowning farmer peasants (kulaks). By 1915 about half of the rural households owned strips, and over a million had been able to consolidate their landholdings. Migration eastward was encouraged, helped by the opening of the Trans-Siberian Railway at the end of the 19th century: 3 million peasant farmers settled in Siberia between 1907 and 1909. After the communist revolution in 1917 these peasant farmers were absorbed into large state-owned farming enterprises, and agricultural settlement continued to extend eastward until the 1950s.

In the early years after the revolution force was often used to extract sufficient food for the urban and industrial populations from the peasant communes and farmers. It is hardly surprising that the majority produced just enough for themselves; there was little point in producing a surplus for confiscation. The food shortages in the cities became acute, and in October 1929 Stalin (1879–1953) started a violent campaign to compel the kulaks to amalgamate their holdings into collective farms, which meant that production was brought under direct state control.

Checking the work tally Workers on a collective cotton farm in the southern Soviet Union queue to have their sacks of picked cotton weighed. Under the Soviet system, all aspects of farming are regulated and controlled by the state.

Today vast tracts of the Soviet Union remain virtually empty of agriculture. In the northern tundra the only agriculture that takes place is based on hunting, fishing and reindeer herding. In the broad belt of coniferous forest (known as taiga) agriculture is confined to clearings where potatoes and small amounts of rye, oats and barley are grown.

The mixed deciduous and coniferous forests to the south of the taiga were the first to be cleared for permanent cultivation: their fertile soils (the black earths or chernozems) make them very suitable for agriculture. By the beginning of the 19th century wheat was being grown here for export to Western Europe. Large areas of this productive land in the Ukraine were destroyed by the nuclear accident at Chernobyl in 1986.

Still farther to the south are the natural grasslands of the great Russian steppe. The traditional pastoral economy that has been practiced here for hundreds of years has given way this century to cultivation. At their dry southern margins, however, where the grasslands merge into semidesert and desert, migratory agriculture based on herds of sheep still survives. Most of the herdsmen, though, are now permanently housed in settlements, and only the herds move.

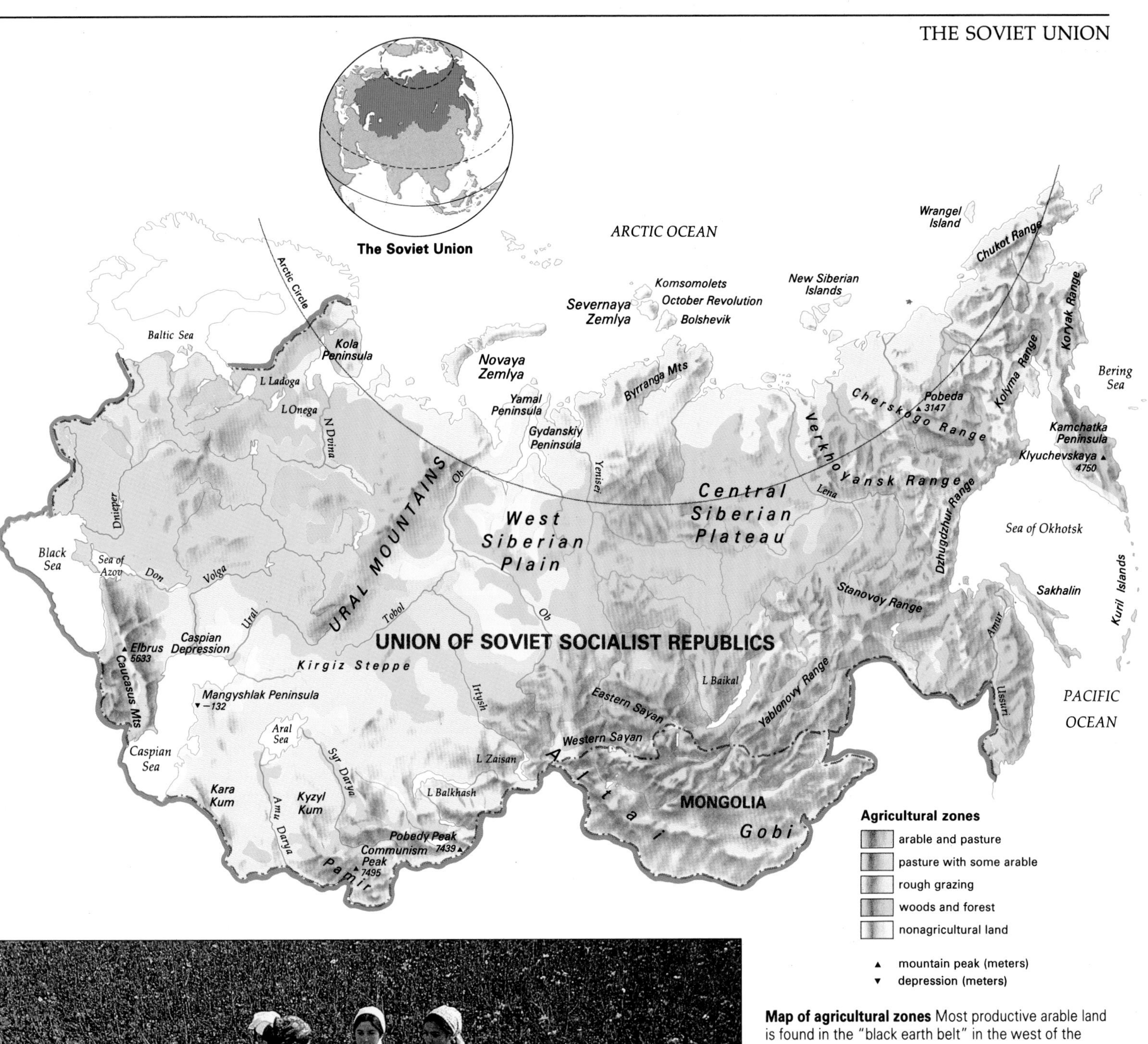

Map of agricultural zones Most productive arable land is found in the "black earth belt" in the west of the Soviet Union. North and south of the vast taiga forest conditions for cultivation become too cold or too arid.

Cotton and rice are grown in the valleys of the rivers that flow into the Aral Sea. So much water is extracted for irrigation that the Aral Sea has shrunk to nearly half its former size. In the 1960s the lake's area was 66,820 sq km (25,800 sq mi). It is now about 40,000 sq km (15,500 sq mi), and in places the old sea bed forms a salt desert. The lake once supported many fishing communities, now abandoned.

A large area of northwestern European Russia is characterized by heavy, acidic, poorly drained soils called podsols. Here cultivation is difficult, and the application of lime and extensive soil drainage is generally necessary. Potatoes remain the staple food crop, though recent heavy investment in agricultural improvement has allowed a sharp increase in cereal yields, especially near the large cities.

GROWING FOOD FOR THE PEOPLE

The state manages all aspects of agriculture in the Soviet Union – what is grown, how it is grown, when it is harvested, and how most of it is distributed through state marketing organizations. It does this by setting the production plans and by controlling the farming units at local level. These are of two kinds. The collectives, or *kolkhozy*, were the instruments of Stalin's coercive policies against the kulaks and remain a cornerstone of Soviet agricultural organization. The state appoints the farm managers who implement the production plans, which are set centrally within the context of a state agricultural program. The state farms, or *sovhkozy*, in which workers are employed directly by the state, were set up after 1918 to parallel state-owned industrial ventures.

In recent years the numbers of state farms have grown rapidly as the collectives have amalgamated or have been transferred into state farms. The expansion of state agriculture into areas east of the river Volga, where there were few peasants to collectivize, has contributed to the growth of the state farms.

The organization of farming

The collectives and state farms are both very large units. The former have, on average, 3,485 ha (8,610 acres) of cultivated land and 44 tractors; the latter are even larger, with 4,767 ha (11,780 acres) and 57 tractors. Farm workers live in small villages and towns, and commute to work on the farms. Some of the state farm enterprises are intensely specialized, producing, for example, only irrigated cotton, sugar beet or tobacco for export, or tomatoes and cucumbers under glass for the Moscow market. A "farm" of this sort may have 250 ha (620 acres) under glass and a small power station to provide heat early in the season.

Inflexible pricing policies, centralized planning and the power of local party officials have left little incentive for raising productivity in state farms and collectives. The prices charged to consumers have been set unrealistically low without serious consideration of production costs;as a consequence the volume of production of different crops is controlled by imposing quotas. Since the mid 1970s state farms have been given some control over their investments, but without any marked increase in production levels.

The principal crop of both the collectives and the state farms is wheat. This occupies about half of the total area under cereals. Winter wheat, planted in the fall, is grown in the western Soviet Union, and spring wheat over the whole of the cereal belt. The amount of land planted to both wheat and potatoes has been falling since 1960, while production of fodder for livestock has been increasing, especially barley, oats and maize. After a period of rapid expansion in the early 1960s sugar beet production has stagnated; it is very variable from year to year.

Change in diet

As the standard of living of Soviet people has risen since about 1960 there has been a trend away from staple foods such as bread and potatoes toward more meat. All the same, the consumption of meat per person remains very low by Western standards. In the past, following poor harvest years (1963, 1966) livestock, particularly cattle, were slaughtered to match herd sizes to the available fodder. Meat, milk and eggs were often unobtainable. After 1970, however, growth in livestock numbers became more or less continuous. Even the dreadful harvest of 1975 brought only a small dip in the upward trend of cattle, pig and sheep production, which was made possible not only by the increased national production of fodder grains, but also by large imports of grain from outside the Soviet Union.

Milk production is more volatile and is very sensitive to the availability of feed for dairy cattle, so any shortage shows up in reduced milk production well before animals have to be slaughtered. In addition little milk is produced by the state sector. Most of it comes from small private plots, and these have few resources to see them through periods of feed shortage.

The Soviet Union's vast size and its relatively undeveloped system of transportation mean that regional food shortages are common. Some high-value crops are moved in small volumes by air, but most marketing and distribution has to rely on a limited rail network and truck movement over very poor roads.

The Soviet grain harvest Despite mechanization, production levels have not risen as high as targeted. Even when harvests are good, the lack of bulk road and rail containers means that grain may go moldy before it can be moved from the farms.

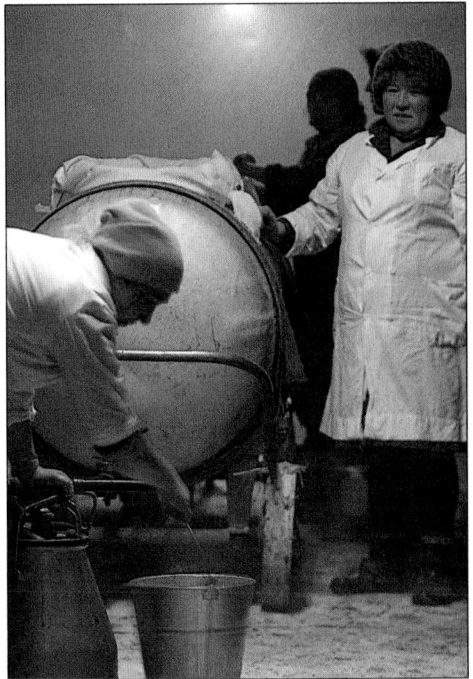

A milking shed on a collective farm in Ulyanovsk, Lenin's birthplace, on the river Volga. Cattle are kept indoors during the long winters; however, milk supplies can be still very variable, as they are dependent on the availability of fodder.

THE PATTERN OF GRAIN HARVESTS

The expansion of agriculture east of the Ural Mountains that has taken place this century means that grain is now grown in areas that have high temperatures in summer, and soils that do not retain moisture well. In the traditional grain-producing areas to the west conditions are generally much wetter. It is only in exceptional years that adverse weather occurs throughout the entire grain belt. This happened in 1975, when hot dry conditions prevailed throughout the whole country, resulting in a disastrous harvest. More often, however, unfavorable weather in one area, giving rise to a poor harvest, is likely to be balanced by better weather, and healthier harvests, elsewhere.

Climatic patterns are consequently important in helping to prevent great fluctuations in grain availability from one year to the next. However, yields per hectare are increasing much faster in the western half of the grain belt than in the climatically more difficult east. If this trend continues, the west will come to dominate the total Soviet production once again. Overall harvest sizes will be dependent on whether there is a good or bad year in the west, giving rise to much greater variability in supply, and the level of Soviet grain imports will become more unpredictable. Although stocks are now being built up to provide some security for poor years, two or more disastrous harvests in succession could rapidly increase the demand for imported grain, and destabilize grain markets throughout the world.

"Agricultural technology of the future" (*left*) So reads the legend on this sign outside a state farm. Soviet investment in farm machinery such as combine harvesters and tractors has been considerable. But failures in planning – which the reforms initiated under the *perestroika* movement have not effectively been able to overcome – means that there is a very low level of return. Machines often break down, and cannot be easily repaired.

Branching out (*right*) This privately run flower nursery, on a collective farm, exemplifies the sort of changes in agriculture that *perestroika* seeks to bring about. The reforms have given farmers far greater control over their operations than they had before. Some of them have diversified from basic food production into commercial enterprises such as this. Greater financial independence may prove to be a spur to increased productivity.

IMPROVING THE SYSTEM TO MEET DEMAND

As personal incomes in the Soviet Union began to rise in the 1960s there came an increase in the demand for consumer goods, and in particular for higher quality foodstuffs. Their continued absence inhibits the country's economic development and labor productivity. Why should people work harder if there is nothing for them to spend their wages on?

The fertile conditions of the black earth grain-producing belt had allowed surplus wheat to be exported since the ending of serfdom. Except for a brief period in the mid 1960s the Soviet Union continued to export grain even during times of shortage at home. The decision of the Soviet government in 1972 to import grain in order to meet the domestic demand for higher quality foodstuffs therefore came as a great shock to the international grain market. It had profound effects on the price and availability of food for much of the rest of the world.

Great secrecy surrounded the Soviet Union's dealings on the international grain market. Most of its purchases were made from grain companies based in the United States. In 1971–72 the Soviet Union imported 7.8 million tonnes of grain; this shot up to 22.8 million tonnes in 1972–73, and of this 13.7 million tonnes came from the United States. The world price of wheat increased almost fourfold between July 1972 and February 1974. Although other factors, including a rapid rise in the price of oil and fertilizers, were involved, much of this price rise was attributable to the Soviet Union's unexpected decision to buy grain.

Soviet grain imports have remained high as domestic grain harvests continue to be disappointing and the demand for meat remains largely unmet. Annual grain imports have risen as high as 55 million tonnes, close to the maximum that can be handled by port facilities in the Soviet Union. Many exporting countries, including Argentina, Canada, the European Community and Thailand, as well as the United States, now rely on exports to the Soviet Union.

The situation remains precarious, both for the Soviet Union and for those parts of the world that rely on food imports and food aid. The Soviet Union is able to buy imports by selling both gold and natural gas, but continued Soviet demand, if it is combined with a world grain shortage following droughts and harvest failures elsewhere, will leave other food importing countries short, especially those without foreign exchange resources.

High investment, low returns

The problems of low productivity that continue to beset Soviet agriculture are not a result of a lack of investment in land improvement and farm technology. On the contrary, the scale of investment is even higher than in the United States. Good investment intentions, however, tend to be frustrated by the practical difficulties of centralized planning and supply over such a vast country.

The money is not well spent. Heavy seasonal demand for fertilizers can create bottlenecks in their manufacture, storage and transportation. Deliveries to farms often arrive too late in a crop's growing cycle, and are left to lie by the roadside. At one time tractors and other machinery were held in centralized "tractor stations" from where they were supposed to be delivered to farms as required, but rarely were. Machinery is now supplied direct to each collective and state farm. However, the machines themselves continue to be extremely unreliable, and it is

THE OBSTACLES TO REFORM

Perestroika has been likened to a stone dropping in a pond; by the time the ripples reach the great central plains of the Soviet Union they have died away and do not disturb the accepted way of doing things. Many state and collective farm managers have been quietly replaced with younger people, who are in sympathy with reform, but memories are long in country areas. Some 75 years of state control, including the devastating purges of the Stalin era, have left the mass of the rural population unwilling to trust edicts from Moscow, and ill-prepared to take responsibility for their affairs. If the reforms are to take hold and win widespread acceptance, results in the form of more and better consumer goods will rapidly have to follow the rhetoric. It gains a farm worker little to be paid more when there is nothing for the family to buy with the extra roubles.

Creating a further stumbling block to reform is the reluctance of the local and central party bureaucracy to surrender its authority over the detailed implementation of state planning. In 1987 President Gorbachev merged six ministries, all concerned with agricultural planning and organization, cutting 47 percent of the staff at the same time. Local reorganization, however, is not as easy to achieve, and has not been as successful. While agricultural reforms remain vital to the growth and security of the Soviet economy, many of those with power and influence remain devoted to the status quo.

difficult to acquire spare parts – yet another example of high investment producing a low rate of return. Agricultural investment increased by 190 percent between 1965 and 1980, yet output rose by only 17 percent. Labor productivity remained only a tenth of that of agriculture in the United States. Moreover, it was significantly higher on the private plots than on the state-controlled farms.

Prescribing remedies

In the Soviet Union change must come from the top. The package of economic reforms (known as *perestroika*, or "reconstruction") launched in 1986 by President Mikhail Gorbachev set out to raise levels of productivity throughout Soviet industry by loosening the stifling grip of central planning. In agriculture this has meant giving farm managers more control over what they grow and how they market it. This extends even to some aspects of international trade, so that decisions about exports of crops such as cotton can be taken locally. Farms are allowed to retain a greater amount of production for their own use, rather than delivering it compulsorily to the state, and they also have had to become more self-financing, paying for fertilizers and machinery from their income alone.

The stifling effects of working for the vast state farms has been countered by dividing the work force into small teams, each responsible for all stages of production on an area of farmland. The team is paid according to its production success. In some areas the results have been remarkable, increasing labor productivity to ten times the national average. Hardline conservatives who oppose reform see this development as spearheading an undesirable return to the class of rich peasants, or kulaks.

A complete transition to a market-based system of agriculture, however, requires that the prices paid by customers in the shops more closely reflect production costs and provide a reasonable profit for the farmers. Attempts at price reform have met stiff opposition in the Soviet Union, as in Poland and elsewhere in Eastern Europe. For market incentives to work and lead to increased productivity in Soviet agricultural production, prices will have to be more than doubled, no matter what political difficulties they will cause for the reforming elements of the Soviet government.

Farming the private plots

The state owns all the land in the Soviet Union and controls the use of the bulk of it. Many Soviet people are, however, permitted to control the use of a small private plot of land, to grow food for their own consumption.

Private plots were originally granted as compensation for those peasants whose lands were amalgamated into the collective farms. Of the 35 million private plots today 13 million are still worked by the collective farmers, or *kolkhozniks*, and a further 10 million are worked by the employees of the state farms, the *sovkhozniks*. The remaining 10 million are divided among the employees of various other state organizations. The plots that belong to the *kolkhozniks* are the largest in size (an average of 0.31 ha/0.76 acres).

Having a plot involves control of its use, but not ownership. Control can be passed on within a family, but if the family moves or does not use the land it reverts to the state. Great care is lavished on the private plots, especially by grandparents and children, the incentive being the high prices that are available in the private markets, and the pleasure of growing quality produce for the home.

By Soviet standards productivity is very high, but because the plots are so small the type of produce is limited. A cow, a pig or two and a few chickens are usually combined with a number of different vegetables, the choice depending on the season and climatic zone. Some 45 percent of the Soviet Union's vegetable and potato production, 31 percent of dairy cows, but less than 1 percent of the cereal production comes from private plots.

Supplying the private markets

Because so much state produce is of poor quality and is subject to irregular supply, fresh food sold through private markets, which are found in every town and city, can be a very valuable source of income. Domestic transportation is cheap, so producers may travel long distances to sell relatively small quantities of produce from their private plots. A producer living in one of the southern areas of the Soviet Union may undertake a round trip of 1,000 km (over 600 mi) or more to sell fresh vegetables and fruit in the private markets of Moscow. Sales have to be made directly by the producing family without employing middlemen or traders of any sort, so the private markets are noisy and colorful, with very large numbers of sellers from many parts of the country, each trying to dispose of small quantities of produce.

The success and importance of this private agriculture remains something of an indictment of the failures of the state sector, and the private plots remain more tolerated than liked. All the same, cooperation between the private and the state sector has been encouraged by allowing the release of fodder from state and collective farms for private livestock. Moves have also been made to put private plot production under contract, using seed, fodder and young livestock as well as the marketing channels of the state sector, harnessed to the productivity of the private sector. State and collective farms are allowed to sell up to 30 percent of their produce in private markets, as the incentives for private production slowly spread into the state sector.

Marketing the produce A woman in a mountain village in the Caucausus displays the tomatoes, apples, pears and grapes grown on her private plot for sale. Markets like this are found in every town and village in the Soviet Union.

The fruits of their own labor A couple at work on their private plot. The onion-domed church behind them is under repair – a sign of the changing political climate, which has also grown more tolerant of private farming.

The private sector's share The small size of most private plots means that they are limited in what they can produce, but in some items, notably potatoes, milk and vegetables, their percentage contribution to overall Soviet production is considerable.

Percentage per year / Product	1940	1960	1970	1975	1980	1985
Meat	72	41	35	31	31	28
Milk	77	47	36	31	30	29
Eggs	94	80	53	39	32	28
Wool	39	22	20	20	22	26
Grain	12	2	1	6	1	1
Cotton	0	0	0	0	0	0
Sugar beet	0	0	0	0	0	0
Sunflower seeds	11	4	2	3	2	2
Potatoes	65	63	65	59	64	60
Vegetables	48	44	38	34	33	29

TRADITION AND CHANGE

A CRADLE OF AGRICULTURE · FARMERS, HERDERS AND NOMADS · HIGH TECHNOLOGY FARMING

Farming in the Middle East has undergone major changes in recent years. High-technology farming now coexists alongside the surviving traditional systems. Climate and relief are the major constraints determining the types of farming undertaken, though land tenure and the farming structure that has developed also have a significant effect. Peasant arable farmers and pastoral nomads live a precarious existence – the outcome of a fragile balance between terrain, crops and livestock. The availability of water is a crucial factor, and extensive irrigation schemes – a feature of the region's agriculture since ancient times – have helped to reduce the uncertainty associated with rainfed cultivation. Nonetheless, less than 10 percent of the region supports arable farming – the remaining areas are either too mountainous or too arid.

COUNTRIES IN THE REGION

Afghanistan, Bahrain, Iran, Iraq, Israel, Jordan, Kuwait, Lebanon, Oman, Qatar, Saudi Arabia, Syria, Turkey, United Arab Emirates, Yemen

Land (million hectares)

Total	Agricultural	Arable	Forest/woodland
680 (100%)	265 (39%)	61 (9%)	47 (7%)

Farmers

23.7 million employed in agriculture (38% of work force)
3 hectares of arable land per person employed in agriculture

Major crops
Numbers in brackets are percentages of world average yield and total world production

	Area mill ha	Yield 100kg/ha	Production mill tonnes	Change since 1963
Wheat	21.2	16.6 (71)	35.2 (7)	+119%
Barley	8.5	13.3 (57)	11.3 (6)	+76%
Lentils	1.1	9.4 (119)	1.1 (41)	+373%
Cotton lint	1.0	8.8 (153)	0.9 (5)	+45%
Grapes	1.1	60.2 (80)	6.5 (10)	+45%
Other fruit	—	—	12.7 (5)	+138%
Vegetables	—	—	31.3 (7)	+138%

Major livestock

	Number mill	Production mill tonnes	Change since 1963
Sheep/goats	170.2 (10)	—	+10%
Cattle	28.6 (2)	—	+15%
Milk	—	7.7 (2)	+80%
Fish catch	—	1.2 (1)	—

Food security (cereal exports minus imports)

mill tonnes	% domestic production	% world trade
−23.0	43	11

An ancient farming tradition Peasant farmers in Iran winnowing grain by hand, so that the wind blows away the chaff. Farming methods have remained unchanged for thousands of years.

A CRADLE OF AGRICULTURE

The Middle East is dominated by conditions of extreme aridity. Turkey and Lebanon are the only countries that receive more than 240 mm (9.5 in) of rainfall – the minimum required for rainfed cultivation – over their entire area. In other parts of the region, rainfall is extremely variable and irrigated agriculture consequently becomes increasingly important. Water for irrigation is unevenly distributed. Iran has over 40 percent of the region's irrigated area, though its cultivated area is no more than 22 percent of the total land area. Less than 21 percent of all agricultural land in the Middle East is under irrigation.

Climate and terrain contribute to agricultural diversity. The Middle East is a transitional climatic zone between subtropical Africa and the Mediterranean. Savanna conditions, similar to those found in Africa south of the Sahara, give rise to coffee growing in northern Yemen. Livestock pasture in Oman, on the southeast tip of the Arabian Peninsula, benefits from monsoon rains during the summer. Mediterranean conditions prevail in Israel, Lebanon and Turkey. Rainfall in the highlands of Turkey, Iran and Afghanistan feeds the region's river systems: the Amu Darya in the far northeast, and the Tigris and Euphrates rivers in the west. These cross Syria and Iraq, uniting to form the Shatt al Arab as they enter The Gulf. The Tigris–Euphrates

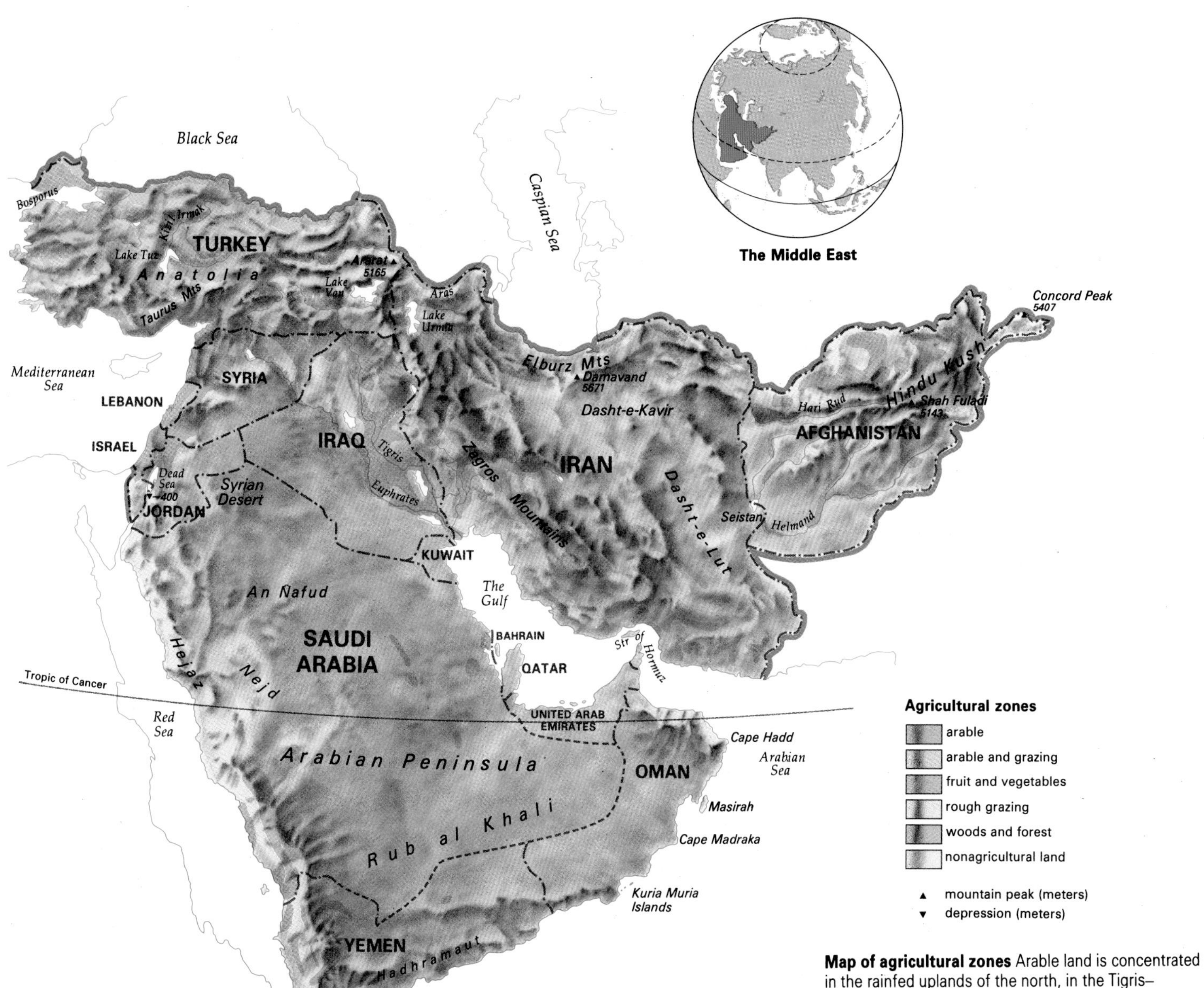

Map of agricultural zones Arable land is concentrated in the rainfed uplands of the north, in the Tigris–Euphrates river valley, and wherever seasonal rainfall can be collected and stored for irrigation. Livestock is grazed on less productive land.

basin – the largest lowland belt in the region – is irrigated and forms one of the great fertile areas of the Middle East. To the south, the arid plateaus of the Arabian hinterland, touching on Syria, Jordan, Iraq and the whole of Saudi Arabia, form the major deserts of the region.

Ancient farming techniques

Farmers in the region have displayed considerable ingenuity in coping with a range of harsh environments. The Middle East is one of the world's oldest agricultural areas, and has seen a great amount of agricultural experimentation and innovation. The dry farming of early domesticated cereals, such as einkorn and emmer, along with techniques of grain storage in pits and granaries and the domestication of animals, first evolved in the Middle East before 5000 BC. The region saw the development of a system of subsistence farming based on cereals, cattle, sheep and goats that was later spread into Europe and the Indian subcontinent. As the population increased, very possibly as a consequence of these developments, new farming systems – based on the use of extensive irrigation – evolved. Irrigated agriculture probably originated in simple form in the Jordan valley. It later provided the basis for the largescale irrigation that developed in the Euphrates and Tigris valleys in the 3rd millennium BC.

Accelerating change

For much of the region's long history, agriculture has remained largely unchanged. The majority of farmers have eked out a marginal existence as subsistence peasants, as settled livestock herders, or as nomadic pastoralists. In recent years, however, the introduction of machinery, fertilizers and high-yielding seeds, along with the exploitation of the region's oil reserves and consequent industrialization, have rapidly changed the face of Middle Eastern agriculture.

Although improvements have been made in farming methods, opening up the region's agricultural potential, great social and political changes have diminished the importance of agriculture, and many countries are now witnessing the disintegration of their traditional farming systems. In war-devastated Lebanon, for example, it is estimated that 23 percent of the farming population left the land between 1975 and 1985. In Jordan, farming increasingly relies on imported labor, now accounting for nearly 60 percent of the agricultural workforce.

FARMERS, HERDERS AND NOMADS

Farming in the Middle East falls into two broad categories: arable cultivation, which uses either dry farming or irrigation techniques, and the rearing of livestock. The latter may involve settled herding, where livestock are taken out from the village to graze land each day, or nomadic pastoralism, where both herders and animals are continually on the move.

The distinctions between the different types of farming systems are often blurred. Arable farmers may use both irrigation and dry farming methods, and pastoralists, herders and arable farmers are often interdependent, moving between one activity and the other. In the past, cultivators have often taken to pastoralism after crop failure. In more recent years, pastoralists have begun to settle as permanent farmers or herders.

Dry farming is of great importance to the countries in the northern part of the region – Jordan, Lebanon, Iran, northern Iraq, Syria and Turkey. The main crops are wheat and barley, though some rye and oats are grown in parts of Iran and Turkey. Olives, fruits and vegetables are important in the upland areas of Jordan, Lebanon, Syria, Turkey and the occupied West Bank.

Elsewhere in the region, irrigation is crucial to agriculture, and a range of important crops are successfully grown under irrigation. The monoculture of citrus fruits – oranges and grapefruits – as well as other fruit, vegetables and salad crops, takes place along the Mediterranean coast in Israel, Lebanon and parts of Turkey. Cotton is important in Syria and Turkey. Where conditions are hot and humid enough rice is grown in southern Turkey, Iraq and Iran, while southern Iraq, Oman and Saudi Arabia are important date-growing areas. Other cash crops grown in the region under irrigation include maize, sorghum, millet, sugar cane and beet, oil seeds, pulses (lentils), alfalfa and tobacco.

Herding and irrigation

Livestock and herding activities have traditionally taken place in marginal lands, either as a complement to dry farming or in areas too arid or mountainous for cultivation. Nomadic pastoralism in Afghanistan, Iraq, Iran, Saudi Arabia and Turkey has developed largely as a response to seasonal and altitudinal variations in the availability of pasture and water. Tribes such as the Bakhtiari, Basseri, Quashqai and Lur in Iran often cover 1,000 km (650 mi) during their annual migrations. In winter they graze their sheep and goats on the grasslands of the central plateau. When these become parched in the summer heat, they move their herds up to the rich pastures of the Zagros Mountains, in the southwest of the country, which have come into growth following the melting of the winter snows.

Similar seasonal migrations are undertaken by the Kurds and Yürüks of Turkey and the Baluchi, Durrani and Ghilzhai of Afghanistan. Tribes in the sandy deserts of Saudi Arabia, such as the camelherding Al-Murrah of the Rub al Khali (or Empty Quarter) in the south and the Mutair in the northeast, depend on wells and oases for summer water, and only venture out into the interior desert after the scanty winter rains.

Land ownership

All farming systems in the region involve some form of organization based on the use and ownership of land. Traditional grazing lands (*dira*) were often established informally among desert tribes. In Saudi Arabia, tribes had exclusive

Productive farmland (*above*) More than 42 percent of the region's cultivated land lies in the fertile plains of Anatolia in Turkey. Wheat and barley are the main arable crops grown throughout the north of the region.

A fruitful harvest (*below*) These grapefruit have been grown on one of Israel's collective farms. Irrigation and intensive production methods transformed agricultural output in the early years of the Israeli state, and valuable export markets were created.

Sheep at a desert well in Syria. Government policies have forced many nomads, such as these Bedouin herders, to abandon their traditional migratory way of life for settled agriculture.

COOPERATIVE FARMING IN ISRAEL

Unique in the Middle East are the forms of cooperative farming settlement developed in Israel. The two main types are the *moshav* and the *kibbutz*. Agricultural settlement in the early half of the 20th century was an important means for Jewish colonization of what was then Palestine and provided a source of inspirational value for the founding of the state of Israel in 1948. The early pioneers saw agriculture as a "way of life", a symbol of a society dedicated to making the "desert bloom".

The *moshav* is a cooperative smallholders' village. The village is made up of farm holdings of equal size, small enough to be cultivated by the farmer and his family. The cooperative structure of the village provides the farmer with loans for investment, with inputs (fertilizers and machinery) and with marketing facilities.

The *kibbutz* is a collective enterprise based on the common ownership of resources and on the pooling of labor, income and expenditure. No wages are paid because the *kibbutz* provides all the goods and services needed by its members. In recent years activity on most of the *kibbutzim* has diversified into areas such as manufacturing; only 28 percent of working time within the *kibbutz* sector is now spent in farming.

This reflects Israel's growing role as an industrial country. Only 5 percent of Israel's labor force is actually engaged in agriculture, and agricultural products account for less than 6 percent of the country's national product. The *moshav* and the *kibbutz* have correspondingly declined in importance. Estimates suggest that 83 percent of *moshav* are no longer economically viable without substantial government support. The *kibbutz* sector faces similar problems. However, there is still considerable attachment to these early pioneering forms of agricultural settlement.

rights to water sources, while in the Bakhtiari regions in western Iran, pastoralists were also large landowners, able to secure winter pasture through control of village areas. Localized grazing and water rights were often secured through agreements that were mutually beneficial to both settled farmers and pastoralists. The latter provided labor in the fields in exchange for cash and cereals; farmers relied on pastoralists for meat during winter and livestock during spring.

The cultivated lands of the region were traditionally owned by the state or sultan – a practice known as *miri* in Arab regions and in Turkey, and *khaliseh* in Iran. Although farmers did not own the land, they had rights to use it, for which they paid a tax or tithe.

This system of land ownership, which

Money into food Oil wealth allows the states of The Gulf to overcome the limitations of their arid environment by developing costly methods of high-technology production. Vegetables in this greenhouse in Abu Dhabi, one of the United Arab Emirates, are being grown by hydroponics, without soil.

has concentrated large amounts of land in the hands of relatively few landowners, is responsible for much rural poverty and low productivity. In Iraq, as in many other countries in the region, approximately 50 percent of the farming population own less than 8 percent of the land under cultivation. Sharecropping is the dominant tenure system: small farmers are allowed to farm a plot of land and, in return, give up between a half and two-thirds of the harvest as rent. This type of arrangement leaves little scope for farm improvement since the benefits of modernization go directly to the landlords.

A number of countries have attempted land reforms of varying degree; yet holdings in most of the Middle East remain fragmented and small – 1.5 ha (3.7 acres) is normal, but they may sometimes be as small as 0.2 ha (0.5 acres). It is difficult to modernize holdings of this size profitably: most are operated as peasant concerns, with the family providing the workforce and growing only sufficient for basic needs; the small surplus is sent to the local markets.

HIGH TECHNOLOGY FARMING

High-technology farming systems are now as much a feature of the region's agricultural landscape as its traditional smallholdings. Largely because of the landholding system, however, adoption of new technologies has been uneven. Many groups of farmers have not benefited from the changes, and in many cases the environment has also suffered.

Despite major efforts to achieve self-sufficiency, agricultural output continues to fail to meet demand. All Middle Eastern countries, with the exception of Turkey, are net importers of foodstuffs: Jordan imports 50 percent of its food needs; and Saudi Arabia, despite its remarkable success in achieving wheat surpluses through technological innovation, imports 70 percent of its overall food needs. This situation is likely to continue even though attempts are being made throughout the region to increase agricultural productivity.

In most countries in the region, farmers are making the transition from low-technology farming, with its reliance on simple implements, large inputs of labor and use of traditional methods such as crop rotation to restore soil fertility, to modern farming systems. In Israel and Lebanon, investment in mechanization as well as expansion of irrigated land meant this process was already well advanced by the early 1960s. Elsewhere developments took place later, but the rate of change was equally spectacular. For example, in the 20 years between 1961 and 1981, the number of tractors per unit area increased ninefold in Iran, fivefold in Syria and eightfold in Turkey.

Fertilizer consumption also increased dramatically and crop yields were further boosted by the introduction of high-yielding varieties. In Turkey, yields of wheat and barley have increased between

THE GREENING OF SAUDI ARABIA

Although conditions in Saudi Arabia are extremely unfavorable for agriculture, except in a few specialized areas, oil wealth has enabled the country's farming to flourish. As a result, a new type of farming enterprise has emerged in Saudi Arabia alongside the country's traditional agriculture. This was based on date cultivation in the oasis areas along The Gulf; cereals, coffee and bananas in the uplands of the southwestern coast along the Gulf of Aqaba; and pastoral nomadism in An Nafud and the Nejd in the north and center of the country. While all other Middle East countries are importers of wheat, Saudi Arabia is now able to export it. Wheat production grew tenfold between the mid 1970s and 1985 and this remarkable increase is the result of the use of new capital-intensive technologies, such as sprinkler, or center pivot, irrigation systems, within largescale commercial crop production.

This success has not been without its critics. The water that is used to irrigate the new wheat-growing areas is obtained by tapping supplies of "fossil" water trapped in underground reservoirs, or aquifers, of permeable rock. These water reserves are scarce and nonrenewable, and concern has been raised about the loss of this precious resource. Criticism has also been expressed about the size of the subsidy, which is five times that of the world market price, currently being paid to Saudi Arabia's wheat farmers.

Wheatfields in the desert Massive investment in irrigation and machinery means that Saudi Arabia is now the only Middle Eastern country to export wheat. Its boom in agriculture has raised concern over the rapid depletion of nonrenewable water reserves.

60 and 70 percent since the 1960s. New irrigation-related technologies have also increased crop yields, which are in many cases as much as five times higher than those in dry farming areas.

The drawbacks of modernization

Much of this development has taken place with little regard for its overall impact on agriculture. Mechanization in Turkey during the 1960s and 1970s displaced many smaller farmers, who were evicted to make way for larger farms. In dry farming areas, mechanization encouraged farmers to expand the amount of land under cultivation into more arid areas, receiving less than 240 mm (9.5 in) of rain a year. These had previously been used for grazing. In Jordan and Syria, a quarter of the land currently devoted to cereals is in such areas. The abandonment of the traditional rotation of cereal and fodder crops for continual cereal cultivation on former rangelands has led to declining yields, as well as to widespread soil erosion and degradation.

Another effect of these changes has been to push herders and pastoralists farther out into the desert fringe. The dwindling amount of rangeland left in Jordan means that livestock levels are four times higher than the land can sustain. Attempts are now being made to reduce the numbers of livestock and to introduce nitrogen-fixing forage crops such as alfalfa into the cropping cycles of the dry farming lands.

In The Gulf States and Saudi Arabia, oil wealth has led to the development of a highly specialized and mechanized farming sector. In many cases, the farms are owned by corporate and overseas business interests and operated by foreign workers. High-technology agribusiness – including American-style feedlot dairy farming, factory farming for poultry and eggs, fattening units for livestock, along with irrigated hothouse and hydroponic methods of vegetable cultivation – has been widely adopted in an effort to increase food security.

Despite the desire for regional food security, Gulf markets have become increasingly internationalized at the expense of the region's own agricultural producers. Saudi Arabia, for example, now obtains onions, potatoes, oranges and apples from over 28 different sources, though these crops were traditionally imported from Jordan, Lebanon, and Turkey. The production both of fruit and vegetables in The Gulf and Saudi Arabia is highly subsidized and protected, and this, too, has reduced the access of Jordan and other Middle East exporters to their largest export market in the region.

Coping with population growth

By the year 2000, it is estimated that the population of the Middle East will have grown by 55 percent, from 164 million in the early 1980s to 254 million. There is a growing awareness that agricultural resources in the region are insufficient to feed an extra 90 million people. Grazing land and dry farming areas are already suffering acute environmental stress and the high cost of irrigation, as well as the exhaustible nature of underground water supplies, place limits on how much more land can be irrigated. The challenges to agriculture in the harsh environments of the Middle East seem more formidable than ever over the coming decades.

Watering the dry land

From the distant past, water for agriculture in the Middle East has been obtained by collecting floodwater (runoff) from hills and slopes, and distributing it to areas suitable for cultivation. Systems of dams, canals and terraces were built to create the catchment areas for this water. In the highlands of Jordan, Lebanon and Yemen, runoff terraces are still being used, and some experts believe that this ancient technology (known to have been practiced 4,000 years ago) holds the key to the better management of the region's drylands. In the Badia region of the desert between Iraq, Jordan and Syria, for example, water is being collected and distributed in this way to create small areas of irrigated pasture for the native Awassi (fat-tail) sheep.

Another ancient water supply system – the qanats, developed in Iran about 2,500 years ago – are still used in Afghanistan, Iran and the mountains of Oman today. Underground water courses, filtering through alluvial gravels at the foot of mountain ranges, are tapped, and gravity carries the water downhill through a manmade network of tunnels and wells to irrigate cultivable land some considerable distance away – sometimes as much as several kilometers.

A patchwork of green (*right*) Intensely cultivated fields at the foot of the highlands of Oman, at the tip of the Arabian Peninsula, benefit from the monsoon rains that touch the mountains in summer and are trapped and stored for irrigation.

Traditional irrigation (*below*) A blindfolded donkey turns a wheel to raise water from a bucket well in Iraq. This ancient technology is used to tap sources of subterranean water in aquifers just below the surface.

Other traditional methods of surface irrigation are also widely used in the region. Water is either flooded into fields ("basin" irrigation) or channeled between crops ("furrow" irrigation). Another more modern system known as trickle or drip irrigation consists of applying the water directly to the soil through smallbore pipes, often in combination with black polythene coverings and mulches to prevent evaporation.

Increasing irrigation

Since the early 1960s, the use of irrigation in the Middle East has risen dramatically

– the area under irrigation has grown by as much as 30 percent, with the result that the vagaries of seasonal rains no longer threaten farmers with the total calamity of crop failure. Irrigation also offers the opportunity to improve crop yields and extend the growing season – double-cropping or even triple-cropping become a possibility. In many parts of the region there is an acute shortage of new land to bring under cultivation, and irrigation provides a means for intensifying production and for utilizing existing crop areas more efficiently.

Much of the expansion in irrigation has been as a result of largescale government projects, though smaller schemes are also significant. Farmers in Syria have used mechanized water-pumps to turn marginal rainfed lands into the country's principal cotton-growing area. Overhead sprinklers, or center-pivot systems, are also widely used in a number of countries. The water is led into a horizontal pipe pivoted at one end. As it moves in a circular motion, the water is drip-fed through holes along the pipe's length to soak a circular area of crops.

The increase in the use of irrigation has undoubtedly raised levels of production, but irrigation is an expensive business. Small family farms consequently choose to grow high-value crops such as vegetables and fruit, often for the export market, in preference to less profitable cereal staples, such as wheat and barley. The region's reliance on grain imports therefore remains undiminished.

Farmers also face major difficulties with waterlogging of the land and increased levels of salinity as a result of irrigation; in Iraq, large areas of farmland have been abandoned because they have become too saline for use. Declining reserves of groundwater have become a problem particularly in Israel, Jordan, Saudi Arabia and The Gulf states, while ambitious irrigation schemes along the Euphrates require complex water-sharing arrangements between Iraq, Syria and Turkey. The scarcity of water is increasingly important in determining the role to be played by irrigation in solving the Middle East food equation.

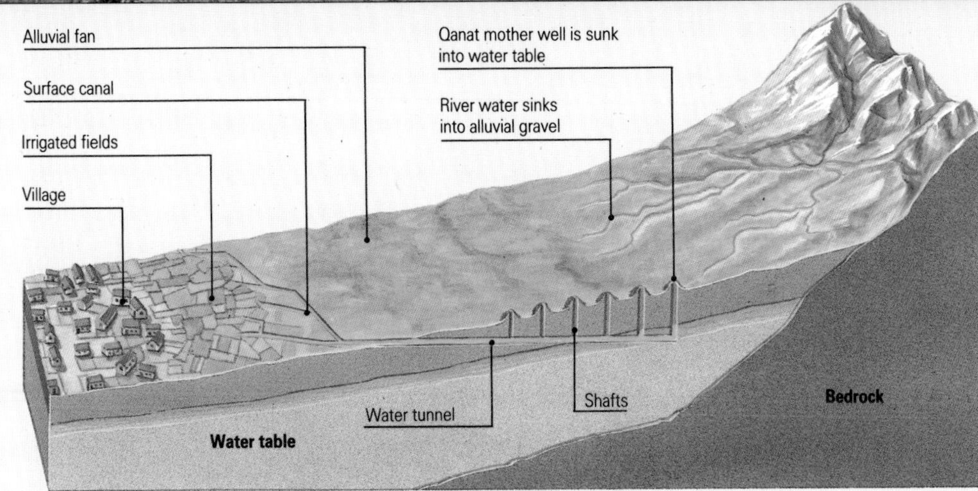

The qanat irrigation system A river carrying rainfall collected in the mountains disappears into alluvial gravels, and the water gradually filters down to the watertable. Wells are sunk down to tap it, and it is led downhill through a system of tunnels to irrigate fields lying some distance below.

The fruit of the desert

The date palm has long been cultivated as a source of food in the Middle East. Its long roots are able to reach sources of water deep below the surface, and it can be grown where other forms of agriculture are difficult to carry out. It is thus a traditional crop of desert oases, but in Israel it has been transformed from a traditional staple crop into a plantation crop.

The trees grow up to 23 m (74 ft) tall; they begin to bear fruit after 4 or 5 years, reaching full maturity after 10 or 15 years. The yields slowly decline thereafter, but the trees may continue to fruit for as long as 150 years. One tree may produce more than 1,000 dates in a single bunch. The ripe fruits, which contain more than 50 percent sugar by dry weight, are packed and exported throughout the world, or they are sold in a compressed form after the seeds have been removed.

The date palm traditionally has many other uses. The trunks are an important source of timber, while the leaf ribs are used for furniture, and the small leaves for baskets. The fruits themselves are a source of syrup, alcohol and vinegar; the seeds may be ground and used for animal feed. Finally, when a tree is felled, the growing tip provides a luxury element in salads.

Tending the crop A date plantation in Nazareth, Israel.

FARMING IN AN ARID LAND

HISTORICAL INFLUENCES · MAKING THE MOST OF WATER · DEBT, WAR AND FAMINE

The traditional subsistence farming of northern Africa – settled agriculture or migratory pastoralism – has largely been overtaken by the need for modern farming systems to feed the rapidly expanding population. Intensive farming requires water, and the extraction and supply of this resource dominates the region's agriculture. Although the colonial powers that ruled northern Africa for so long imposed a measure of stability on the region, they encouraged the cultivation of crops for the European market. In the post-colonial era, demand for many of these crops has declined, creating a heavy debt burden. Civil wars have severely disrupted the agriculture of many countries, and areas that were already experiencing difficulties through drought and desertification have been precipitated into famine.

HISTORICAL INFLUENCES

Agriculture has a long history in northern Africa. Wheat, barley and flax were being cultivated in the Faiyum depression of northern Egypt between 4000 and 5000 BC; and the great civilization of ancient Egypt depended, from at least 2000 BC, on irrigated agriculture made possible by the seasonal flooding of the river Nile. Cereals, including wheat, barley, millet and sorghum, were being cultivated in Ethiopia about the same time. When the Romans occupied northern Africa, they turned it into the great wheat-growing area of their empire.

In the centuries that followed, agriculture in the northern part of the region, stretching from Morocco, Algeria and Tunisia (known collectively as the Maghreb) to Libya and Egypt, developed away from the coastal plains, where malaria was endemic. The great Arab trading cities, such as Cairo, Kairouan, Tunis and Fès, depended upon local agriculturalists to provide for their increasing populations. In areas where the soil retained sufficient moisture from the sparse winter rains to support the dry farming of cereals, farmers combined crop cultivation with seminomadic pastoralism; they moved their herds of sheep and goats between areas of seasonal pasture.

The colonial period

European colonial occupation during the 19th and 20th centuries had a profound effect on farming in the Maghreb and neighboring areas. Much of the best land

COUNTRIES IN THE REGION

Algeria, Chad, Djibouti, Egypt, Ethiopia, Libya, Mali, Mauritania, Morocco, Niger, Somalia, Sudan, Tunisia

Land (million hectares)

Total	Agricultural	Arable	Forest/woodland
1,464 (100%)	383 (26%)	58 (4%)	133 (9%)

Farmers

39.3 million employed in agriculture (57% of work force)
1.5 hectares of arable land per person employed in agriculture

Major crops
Numbers in brackets are percentages of world average yield and total world production

	Area mill ha	Yield 100kg/ha	Production mill tonnes	Change since 1963
Millet/sorghum	13.0	5.1 (44)	6.6 (7)	+4%
Wheat	6.4	13.9 (60)	8.8 (2)	+67%
Barley	5.0	8.1 (35)	4.1 (2)	+17%
Maize	2.4	23.3 (64)	5.7 (1)	+76%
Cotton lint	1.2	5.8 (105)	0.7 (4)	+6%
Vetetables	—	—	17.6 (4)	+149%
Fruit	—	—	8.7 (3)	+49%

Major livestock

	Number mill	Production mill tonnes	Change since 1963
Sheep/goats	175.7 (11)	—	+36%
Cattle	77.9 (6)	—	+40%
Milk	—	6.3 (1)	+147%
Fish catch	—	1.2 (1)	—

Food security (cereal exports minus imports)

mill tonnes	% domestic production	% world trade
−20.3	63	9

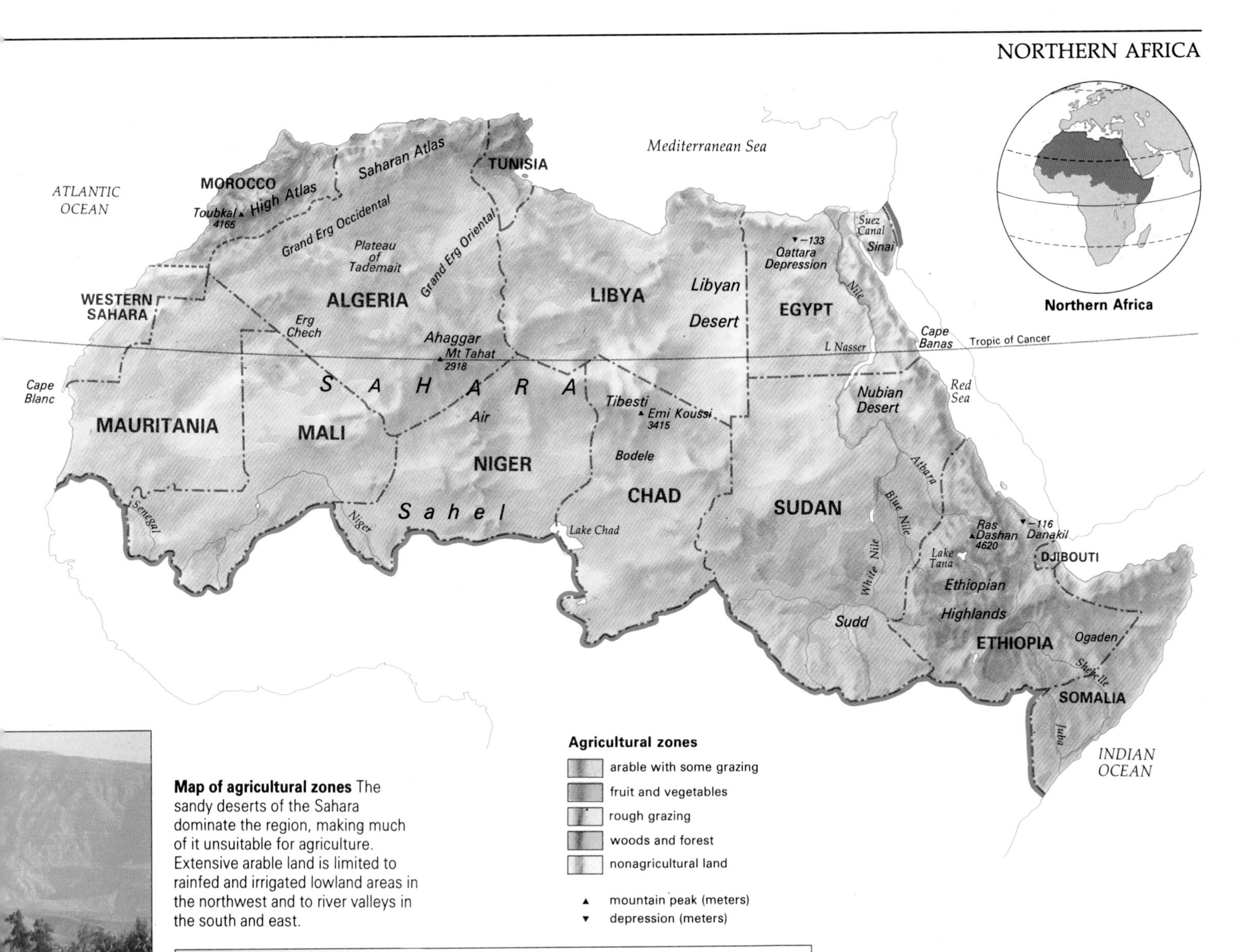

Map of agricultural zones The sandy deserts of the Sahara dominate the region, making much of it unsuitable for agriculture. Extensive arable land is limited to rainfed and irrigated lowland areas in the northwest and to river valleys in the south and east.

Arduous agriculture (*above*) Farm workers in Chad use hand tools to prepare the fields for planting millet, which is the staple cereal crop of most of sub-Saharan Africa.

The mighty Nile at Luxor, Egypt (*left*) People have farmed in the Nile valley since earliest times; the reliable flooding of the river renewed the fertility of the land; the waters were used to irrigate the crops.

was appropriated by settlers who introduced largescale methods of farming with the aim of producing agricultural goods for export to Europe. Morocco, for example, supplied France with both citrus fruits and wheat, while Tunisia specialized in olive production. Algeria – the most heavily colonized country in the region – became a major exporter of wine and cereals. Similarly, the growth of the textile industry of northwest Britain was made possible by supplies of cotton from Egypt and Sudan.

Colonial exploitation extended to the region's fishing and forestry resources. The vast fishing banks off the coast of Mauritania, which had traditionally been fished by a coastal people, the Imraghen, began to be fished on a far greater scale. The trend continued after independence, and traditional fishing today accounts for only 1 percent of Mauritania's total catch. Tunisia began to export coral in large quantities, and both Tunisia and Libya also became involved in the commercial fishing of sponges in the eastern Mediterranean. The enormous cork oak forests of Morocco, Algeria and Tunisia were harvested for their cork.

Traditional pastoralists

The raising of livestock never became a major colonial concern. The herding of camels, goats and sheep by nomadic pastoralists was, and still is, the primary agricultural activity in the increasingly arid savanna and semidesert areas north and south of the Sahara. Agriculture in the desert itself is limited to cultivation in the oases, with date palms, fruit trees,

The results of overgrazing (*above*) In sub-Saharan Africa overuse of grazing land has disastrous effects. The rate at which land is being lost through erosion due to the removal of vegetation cover by grazing and for burning as domestic fuel is rapidly rising.

An irrigation channel (*left*) conducts water to a date palm plantation in Morocco. It is one of many largescale irrigation schemes supporting commercial cash crop production in northern Africa.

vegetables and some cereals being grown wherever water lies sufficiently close to the surface to be used for irrigation.

As rainfall becomes less sparse in southern parts of Mali, Niger, Chad and Sudan (the Sahel), extensive settled agriculture, in combination with cattle herding, returns. Historically, crop cultivation in sub-Saharan Africa was rainfed, and depended on dry farming techniques, but irrigated farming was possible where large rivers could be utilized. There is evidence to show that Mali was producing rice in quantity in the 14th century, using the waters of the river Niger. Even today it is the second largest rice producer in northern Africa, after Egypt.

MAKING THE MOST OF WATER

The major challenge to farming in the north of Africa is the low and variable rainfall. Sizable areas receive less than 200 mm (8 in) in a year. The crucial problem for all farmers is to ensure as reliable a system of production as possible, faced with this climatic constraint. Where irrigation can be undertaken, the unreliability of rainfall may be wholly or largely mitigated; where it is not, other strategies for survival must be adopted.

In the past farmers in much of the region were able to make use of the large expanses of land available to move to a new area when drought occurred and crops failed. But the development of national boundaries and economies, and the need to support growing populations – the region has some of the highest birthrates in the world – has taken subsistence farming into ever more marginal areas as the better land has been taken over for commercial production.

Dryland farming techniques, practiced mainly in the northern part of the region and in the lowland areas of Ethiopia and the Horn of Africa, include the growing of a judicious mix of crops, with sowing carefully timed and spaced to reduce the risk of crop failure, as well as fallowing to preserve soil moisture. In upland areas,

Baling alfalfa In Tunisia a horse-turned wheel is the traditional means of providing power to drive an antique baling machine. Alfalfa has deep roots and is widely grown as a fodder crop in dry regions of the world. It fixes nitrogen in its roots, and consequently helps to return fertility to the soil.

hillsides are terraced and water runoff from seasonal rainfall is trapped and stored for irrigation. Farmers are mainly subsistence producers, growing barley and wheat as staples, and a range of vegetables, pulses and tree crops, including olives, nuts and various fruits, to provide dietary variety. Any crop surpluses are sold in local markets.

Rainfed agriculture is also practiced in the southern parts of the Sahel, with cattle farming providing a major proportion of household income. The main crops are cereals, especially sorghum and millet, which local farmers regard particularly highly as it is a drought-resistant crop. Although there has been some increase in mechanization, handtools – particularly the hoe – are still widely used. Irrigation takes place wherever water is locally available, beside rivers or wadis. Farmers in Niger and Chad are able to make use of moisture left in the soil by the contracting waters of Lake Chad.

People living around the oases of the Sahara have developed a complex pattern of intensive irrigated agriculture with limited pastoralism. The cultivation of the date palm is essential to oasis life. Dates are a highly nutritious source of food; the fronds provide shade and can be used for making ropes and baskets; the trunk provides wood for building.

The water on which the oases depend comes from subterranean sources, either from wells or delivered through *khattara* – underground tunnels that collect water from close-lying aquifers (underground reserves held in permeable strata of rock). Usually such water is found quite close to the surface, but sometimes it may be necessary to bore holes as deep as 90 m (300 ft) in order to tap it.

WATER FROM BENEATH THE GROUND

Northern Africa has massive subterranean water reserves. Some of these lie close to the surface, particularly in desert regions where they support the oasis farming societies of southern Morocco, Algeria, Tunisia and western Egypt. Other reserves of water are held in aquifers, which are replenished by the runoff from infrequent rainfall. Aquifers can be found on the Jifara plain and Jabal al-Akhdar areas of Libya. These Libyan reserves are often very fragile. If overused, they become contaminated with salt as seawater can permeate the porous rocks of the aquifers for some considerable distance. Around Tripoli, for example, seawater has penetrated 20 km (12.5 mi) inland.

The most striking examples of subterranean water reserves are found deep under the Sahara desert. Here rainwater from the Ethiopian Highlands takes up to 30,000 years to seep through vast water-bearing aquifers, which can be 2–3 km (1.25–1.8 mi) thick, extending as far as the Mediterranean, almost 3,200 km (2,000 mi) to the north. These aquifers have been tapped for modern intensive irrigated farming deep in the Libyan Desert. They also provide water for one of northern Africa's grandest development projects, the Great Manmade River, which is piping water for urban, industrial and agricultural use to the salt-contaminated plains along the coast of Libya.

Water rights

In an arid environment it is not surprising that rights over the use of water are meticulously defined. Frequently, the actual cultivation of the oases is undertaken by sharecroppers on behalf of those enjoying ownership rights, who are often wealthy pastoralists. The amount of time that a cultivator may draw upon the local water resource is carefully regulated.

In the Sahara and the Sahel, pastoralists frequently leave one area of pasture in order to move to a new one as the arrival of sparse rains brings it into growth. Consequently rights over access to water resources, particularly wells, need to be clearly established. Different family groups have evolved sophisticated ways of regulating access to water and pasture, which generally work to the advantage of all groups in the area.

Nomadic pastoralism depends upon the extensive use of land; as subsistence farmers move into marginal lands used for grazing, remaining pastures have to carry more livestock. Overgrazing is having serious environmental consequences in many parts of sub-Saharan Africa, leading to the loss of vegetation cover, and accelerating desertification.

Commercial farming

Largescale irrigation schemes – in the fertile plains of the Maghreb, for example – are characteristic of colonial and postcolonial agriculture. Many river systems, such as the Senegal river in Mauritania and the Niger river in Mali, have now been dammed to provide water for hydroelectric power and for intensive irrigation in vast modern agricultural schemes where cash crops for the urban market and for export, particularly to Europe, are grown. These vary in different parts of the region, but include potatoes and early vegetables in Morocco and Egypt, citrus fruits and grapes (for wine) in Algeria, rice in Egypt and Mali, cotton in Egypt and Sudan, and groundnuts in Sudan, Mali and Niger.

Storing grain (*above*) In Somalia, in the Horn of Africa, sacks of harvested millet are being emptied into a pit for storage. Millet stores well, but in periods of prolonged drought such small reserves will be powerless to prevent famine.

An ancient way of life (*right*) Nomads with their flocks gather at a trading post in the Sahara. Their traditional way of life is threatened by restrictions on grazing lands and other pressures.

DEBT, WAR AND FAMINE

Development of the farming sector in the 30 years since independence has been hampered by the stranglehold that the region's past colonial dependence imposes on its economies. In the Maghreb, modern irrigated commercial farms constitute 20 percent of the total arable land, and generate about 80 percent of agricultural exports. Government policies encourage development of this commercial sector so that output of high value crops such as vegetables has risen, at the expense of staple cereals. Responsibility for producing these cereals falls upon the rural labor force of small farmers.

In the face of competition from other sectors of the economy there has been inadequate investment in this system of farming. Many peasant farmers are constantly in debt as they borrow money to cover poor harvests when the rains fail. Unable to generate enough cash to pay off the loans, they are threatened with loss of land to meet the rising interest.

Massive food importers

In Algeria these developments were compounded by the imposition in 1971 of an unpopular and ineffective land reform program. As a result cereal production failed to increase, and food imports have risen. The country now imports 65 percent of all the food it consumes. Increasing social inequality has caused resentment of the "vegetable millionaires", and these feelings contributed in no small measure to the violent riots that swept the country in October 1988.

In much of the region a lack of available land for farming means that governments must in any case rely on imports to feed their rising populations. In Libya, for example, only 1 percent of the total land area is arable, and only a further 4 percent is usable for livestock grazing. Libya's population has tripled since independence in 1951, with the result that, despite massive investment, it still imports up to 60 percent of its food. Furthermore, much of what it does produce relies heavily on costly imports of fertilizers, which are necessary for the cultivation of higher-yielding crop varieties.

In Egypt the damming of the Nile by the Aswan High Dam in the early 1960s for hydroelectricity resulted in the loss of the fertile silt that the annual river flood had laid down on agricultural land for millennia. Fertilizer use has had to increase massively in order to preserve soil fertility. Although attempts have been made to expand the area under cultivation through projects such as the "New Valleys" irrigation scheme, Egypt's agricultural output cannot keep pace with population growth.

Fertilizers are available only to those who can afford them; for most subsistence farmers the costs are too high. Morocco is the world's largest exporter of fertilizer phosphates. However, the country's huge foreign debt forces it to export them rather than use them to improve domestic agricultural output.

Until recently subsidies were used in many countries to stimulate agriculture by maintaining artificially high prices for produce. Further subsidies prevented these prices being passed on to the consumer. The appalling problems of heavy debt experienced by many countries in

NOMADISM: A THREATENED WAY OF LIFE

Saharan Africa, with its huge expanses of arid and desert land, constitutes one of the last great domains of pastoral nomadism in the world. The way of life adopted by livestock farmers in these marginal areas of sparse vegetation, where conditions are difficult for settled agriculture, represents a very efficient means of producing food, based on extensive use of land. Camels, sheep and goats are the animals chiefly kept.

The pastoralism of these Saharan herders is of two kinds. True nomads, such as the Tuareg, are perpetually on the move from one pasture to another, following established customary routes and often traveling great distances. Seminomads migrate from a permanent settlement at certain times of the year to seasonal pastures some distance from their base, and often combine animal husbandry with the smallscale cultivation of crops.

Some nomad groups will occasionally cultivate cereals by making use of *maidars* – depressions in the surface of the desert that retain sufficient water for a meager harvest. They supplement their own meat and milk produce with the crops grown by the oasis populations who were traditionally their serfs, or leased land and water rights from them. In return, the nomads protected these people from attack.

Nomadism is now undergoing a serious decline. The creation of national borders and administration controls interfered with traditional patterns of migration. Severe droughts have decimated herds, and war has destroyed nomad social structure, particularly in Western Sahara. Modernization and other pressures have persuaded many to switch to settled farming; others have gone to the cities.

the 1980s, particularly Algeria, Egypt, Morocco, Sudan and Tunisia, led the International Monetary Fund to insist on the removal of all subsidies in return for economic aid. The result was to depress output and increase food prices.

The expansion of the European Community has also caused problems by closing off what was traditionally a major market for much of the region's agricultural goods. Alternative markets have been hard to find. In Morocco alone, it has been calculated that the livelihood of up to as many as 1.5 million people has been affected as a result.

Drought and its consequences

These problems are dwarfed by the largescale devastation and the suffering caused by war and famine elsewhere in the region. All the sub-Saharan countries experienced a series of terrible droughts in the 1970s and 1980s, particularly in 1972–73 and again in 1984–85. Human suffering was incalculable as crops failed and livestock died or were sold off at disastrously low prices.

In some countries, notably Ethiopia and Sudan, the drought led to widespread famine. Civil war and the incapacity or unwillingness of governments, and the international community, to deal effectively with the crisis ensured that it became a disaster, for which there were obvious underlying causes. In the Horn of Africa, a progressive shift away from subsistence farming toward cash crop production contributed to the precarious nature of the food economy, and elsewhere in the Sahel the expansion of cultivation into marginal areas meant the traditional way of coping with drought – by moving to areas of better pasture and greater resources – had been destroyed, robbing indigenous farming systems of their resilience.

Cotton – a major cash crop

Tending the cotton crop in Sudan (*above*) The creamy white flowers soon turn deep pink and fall off, leaving small green seedpods. These burst open upon ripening, to reveal the soft masses of cotton fibers that are attached to and surround the seeds.

Baling the cotton (*right*) After picking, the springy cotton bolls are pressed and packed down into bales to be transported to the ginning plant, where the seeds are mechanically removed.

Cotton is a white fibrous substance produced from hairs surrounding the seeds of various tropical plants. The cotton fibers develop within a closed seed pod, or boll, which splits open after a period of growth (between 50 and 75 days). In the past, the bolls were picked by hand; today they are harvested by machines that strip them from the plants after the leaves have been chemically removed. The cotton fiber (lint) is then detached by a machine known as a gin. It takes between 1.5 and 2 tonnes of cotton bolls to produce 1 tonne of raw cotton fiber.

The plant has additional uses. The very short cotton hairs that remain on the seeds after ginning are used in the manufacture of paper, cellulose-based chemicals, explosives and other products. The seed itself has a number of industrial uses: the hull is used in oilcake for cattle food, in fertilizers, or in fuel; the kernel yields oil. Cottonseed oil is the third-ranking edible oil in the world; it is used in cooking, and in the manufacture of margarine as well as soap.

Cotton, first used for papermaking in China in the 1st century AD, has long been known in the region. It became a major industrial crop in Egypt in the early 19th century. Later in the century large-scale farming schemes, particularly on the fertile plain of El Gezira between the Blue and White Nile, were introduced into Sudan by the British. Throughout the colonial period both Egypt and Sudan served as a major source of cheap cotton for the British textile mills. Today, the two countries generate between 2 and 3 percent of the world's total output of cotton, and about 35 percent of Africa's production. Cotton was introduced by the French into Chad in the 1920s to satisfy their national market and is a major cash crop there as well as in Mali, Morocco, Niger and Somalia.

Declining markets

In Sudan, 53 percent of export revenue comes from cotton, and in Egypt the crop forms the largest agricultural export. Up to 400,000 ha (1 million acres) produce an annual output of about 300,000 tonnes of raw cotton in Egypt, while in Sudan 300,000 ha (700,000 acres) yield 150,000 tonnes. However, a depression in world prices throughout the 1980s contributed

The cotton plant is a shrublike annual that is native to subtropical regions around the world. It is a valuable source of vegetable oil, which is extracted from the crushed seeds, as well as fiber.

to a crisis in the industry, which was also adversely affected by disease and by unfavorable climatic conditions. Attempts are now being made to reform the cotton industry in both these countries, and to introduce new marketing policies, in the hope that prices will recover as the developed world turns away from synthetic materials back to the use of more popular natural fibers.

Production of cotton in the former French colonies of the Sahel has increased considerably since their independence. In some cases output has risen from 200 kg per hectare (178 lb per acre) in 1961 to 2,100 kg per hectare (1,068 lb per acre) in 1987, largely as a result of improved production methods and the increased use of irrigation. In Mali, for example, annual output grew from 16,000 tonnes in the mid 1960s to about 183,000 tonnes 20 years later, and investment in two major development schemes is likely to increase production still further.

In Chad, cotton produced by some 60,000 small farmers is bought, ginned and marketed by a state monopoly – Cotonchad. During the late 1980s, as part of the conditions attaching to a loan package provided by the World Bank, Cotonchad underwent radical restructuring in order to increase productivity in areas where intense production is possible. One likely effect of this will be that large numbers of less economic small farmers in remote areas will be forced out of business.

TRADITIONAL FARMERS

THE COLONIAL LEGACY · FARMERS AND NOMADS · CASH CROPS AND FOOD SHORTAGES

Farming still dominates life in Central Africa, mostly carried out by smallscale producers who grow subsistence crops, though commercial farming is increasing in importance. The range of physical environments means that there is considerable diversity in farming : extensive shifting cultivation is practiced alongside labor-intensive permanent cultivation and plantations. With some exceptions, farm implements are generally unsophisticated and traditional methods are used. In drier areas nomadic herders move livestock between seasonal grazing lands. The farming calendar revolves around the seasonal pattern of rainfall; periods of exceptional drought and also pest attack (both commonly experienced) can wreak havoc. Overuse of land and overgrazing in many places is causing widespread soil erosion.

COUNTRIES IN THE REGION

Benin, Burkina, Burundi, Cameroon, Cape Verde, Central African Republic, Congo, Equatorial Guinea, Gabon, Gambia, Ghana, Guinea, Guinea-Bissau, Ivory Coast, Kenya, Liberia, Nigeria, Rwanda, São Tomé and Principe, Senegal, Seychelles, Sierra Leone, Tanzania, Togo, Uganda, Zaire

Land (million hectares)

Total	Agricultural	Arable	Forest/woodland
823 (100%)	218 (27%)	75 (9%)	393 (48%)

Farmers

84 million employed in agriculture (70% of work force)
0.9 hectares of arable land per person employed in agriculture

Major crops
Numbers in brackets are percentages of world average yield and total world production

	Area mill ha	Yield 100kg/ha	Production mill tonnes	Change since 1963
Millet/sorghum	15.2	9.9 (82)	15.0 (17)	+49%
Maize	8.0	11.7 (32)	9.4 (2)	+119%
Cassava	6.3	85.7 (92)	53.9 (40)	+100%
Groundnuts	3.7	9.0 (78)	3.4 (16)	−9%
Cocoa beans	3.7	3.2 (87)	1.2 (59)	+27%
Palm kernels	—	—	0.6 (22)	−21%
Palm oil	—	—	1.5 (18)	+59%
Bananas	—	—	21.2 (32)	+90%

Major livestock

	Number mill	Production mill tonnes	Change since 1963
Sheep/goats	110.5 (7)	—	+67%
Cattle	58.3 (5)	—	+40%
Milk	—	2.6 (1)	+36%
Fish catch	—	2.2 (2)	—

Food security (cereal exports minus imports)

mill tonnes	% domestic production	% world trade
−4.6	15	2

Sheep and cattle at a waterhole In the north of the region dry savanna and desert provide poor grazing for livestock. These Turkana herders in northern Kenya move their flocks between seasonal pastures. In the rainy season they return to a permanent site where they raise millet and vegetables.

THE COLONIAL LEGACY

The revolutionary transition in human society from food gathering to food production is thought to have begun between 4,000 and 5,000 years ago on the savanna grasslands of Central Africa. These extend across the north of the region south of the Sahara, and cover the vast plateaus of eastern Africa. This early farming focused on the cultivation of indigenous millets for grain; these were suited to the dry conditions and light soils. Where the vegetation was less dense, pests such as the tsetse fly, carrier of sleeping sickness (trypanosomiasis), were fewer; this encouraged the spread of domesticated livestock.

It was probably not for another 2,000 years that comparable development took place in the densely forested parts of western and central Africa. The transition to food production here depended on clearing dense bush on heavy soils rather than hoeing light soils, and on planting root crops rather than sowing and cultivating grains. These activities were combined with fishing to provide additional protein in the diet.

The forests yielded many kinds of food crops, including the oil palm, the shea butter tree, the kafir potato, a variety of other root crops and various beans and peas. However, the majority of Africa's modern forest foodplants have been introduced from other parts of the world. Bananas, yams and cocoyams (taro or dasheens) came from Southeast Asia, probably about 1,000 years ago. Maize and cassava (manioc) were brought by Portuguese colonizers from their empire

in South America in the 16th and 17th centuries, as was cocoa, which – grown as a plantation crop – has become the mainstay of several economies in the countries of western Africa.

The growth of commercial farming

When the slave trade to America came to an end in the 19th century, European traders and colonizers searched for other tradeable items from Africa. One way of recouping European investment in the administration and infrastructure imposed on the new African colonies was through taxation, but this was only possible if Africans had goods to sell. Crops that were of use to the European market were introduced and promoted, and this policy did much to change the agricultural geography of the region. Subsistence producers were drawn into commercial production, and the range and mix of crops altered.

Some areas were affected more than others. The southern part of western Africa provided valuable crops of palm oil, coffee, cocoa and rubber (both introduced from South America), while the savannas were a source of groundnuts and cotton. Coffee and cotton were exported from eastern Africa, and timber came from the densely forested areas.

Agriculture was most advanced in the coastal states of the region, which were well served by ports from which goods were transported to the European market. Farming in the interior developed more slowly, since much greater investment was required to open up communications. European involvement in farming itself was limited, an important exception being the Kenyan highlands where land was appropriated by Europeans and used successfully for largescale mixed farming. Much of this land has now been returned to the Africans.

When most of the states in the region achieved independence in the 1960s, agriculture was relatively prosperous and most of the new nations were virtually self-sufficient in foodstuffs. But there was a tendency in the early years of independence not to reinvest the profits from farming in the agricultural sector. Ghana, for example, was heavily reliant on revenues from cocoa exports, but yields fell as profits were transferred to industrial development; here, and in neighboring Nigeria, imports of staple foods rose as agriculture struggled through lack of investment. Attempts to increase agricultural productivity by replacing the workforce (increasingly in short supply as more people moved to the towns) with capital-intensive technology have met with only limited success. It is now appreciated that the longterm future of even mineral-rich states such as Gabon and Nigeria depends on the rural sector being developed, with more emphasis being put on encouraging traditional farming along sound ecological lines.

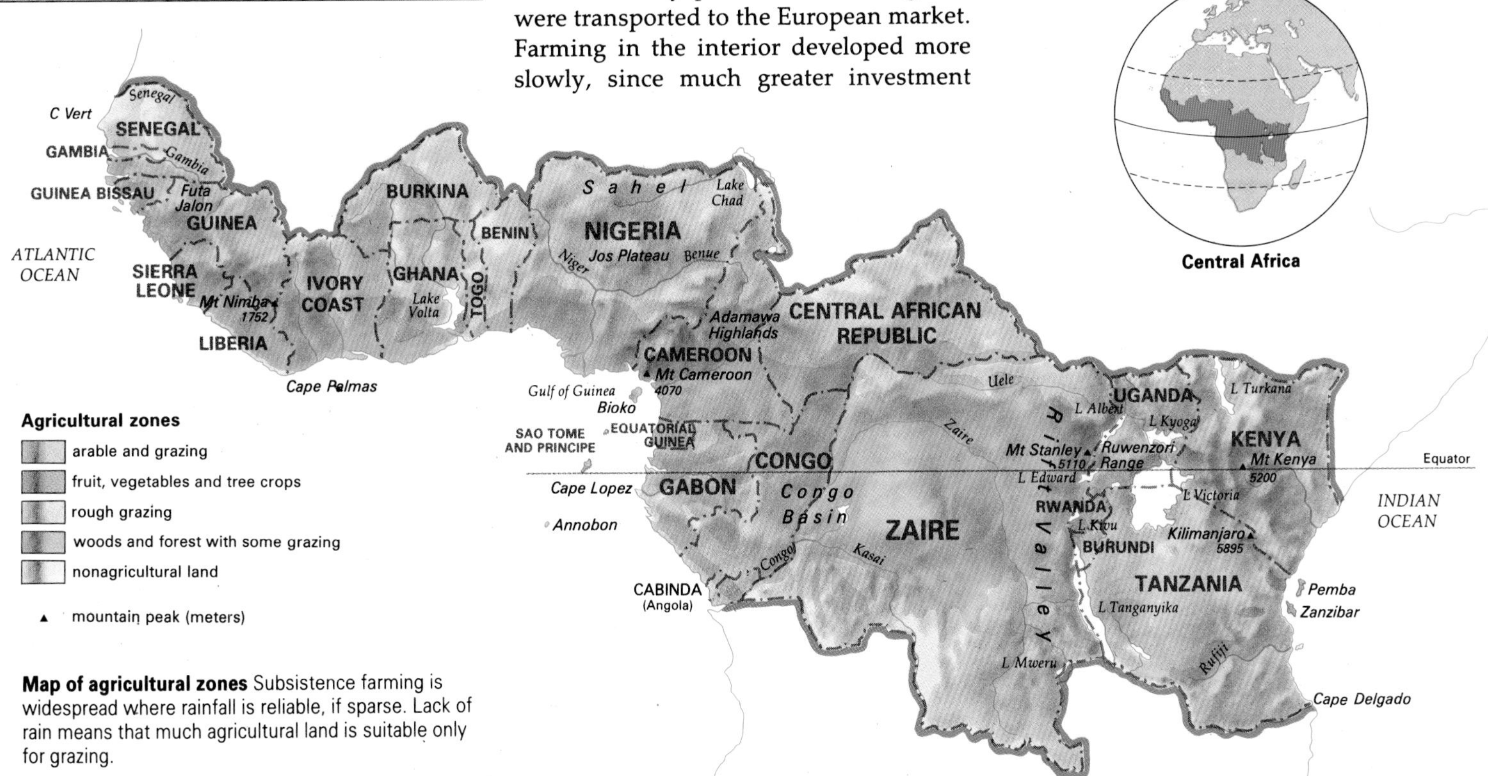

Map of agricultural zones Subsistence farming is widespread where rainfall is reliable, if sparse. Lack of rain means that much agricultural land is suitable only for grazing.

FARMERS AND NOMADS

The farming systems of western, central and eastern Africa range from extensive shifting cultivation to zones of permanent cultivation, large plantations and modern capital-intensive schemes. The savanna lands, with their marked seasonal rainfall, are generally the main cereal-producing areas. Millet, sorghum, maize and rice are the principal subsistence crops. Vegetables such as potatoes, tomatoes and eggplants form an important addition to the diet, and groundnuts and cotton are grown as cash crops. In the forests, cereals give way to cassava, yams, cocoyams, sweet potatoes and other root crops; citrus fruits, pineapples and bananas are also grown here.

The organization of traditional farming is broadly similar throughout the region and is carried out within the framework of the extended family. The men are largely responsible for cultivating the staple crops, while the women contribute significantly to cash crop production, grow vegetables, and gather resources such as wood for fuel, medicinal herbs, fruits, nuts and leaves from the bush. Children play an important role in looking after animals and scaring birds away from the crops in the fields. Cattle are frequently kept for milk, though the meat is rarely eaten; they are highly valued as a source of wealth and status. Goats and chickens are the most common form of livestock, found nearly everywhere. Bush animals are hunted for meat, and fish are caught in local rivers as an additional source of protein.

Bush fallowing

After crops have been grown on a field for four or five years, the fertility of the soil is greatly reduced; weed growth may also threaten to stifle the sown crops. Fertilizer and manure are seldom available in sufficient quantities to restore nutrients to the soil, and traditionally the land is left to lie fallow for several years, allowing it time to recover. This system of shifting cultivation, rotating crops with fallow on fields around a fixed settlement, is known as bush fallowing. It is practiced throughout the region.

Crop rotation is important, as planting the same crop on the same plot year after year reduces soil fertility. In Senegal, for example, millet, the main staple crop, is

Thatched huts *(above)* in the savannas of northern Cameroon. Bulrush millet – a staple cereal, valued for its drought-resistant qualities – is being dried and stored on a platform. It is usually intercropped with groundnuts or maize.

Masai cattlemen *(right)* Herds of cattle represent wealth and prestige to the nomadic Masai people of Kenya and northern Tanzania. They live in areas occupied by settled farmers, trading cattle with them in exchange for crops.

rotated with groundnuts, the main cash crop; sorghum, maize or beans may also be planted. When the land lies fallow – rarely for more than six years – weeds and woody shrubs regenerate. When it is cleared again for cultivation, any trees or saplings of economic value are allowed to stand. Tree stumps are often left in place: they are difficult to remove, and their roots help prevent soil erosion. The cut vegetation is burned, but overuse of fire in the savannas is hindering the longterm regeneration of vegetation. In some areas the quality of the land has been reduced so much that grasses are the only species that now grow readily.

In more densely forested areas the even distribution of rainfall means that cultivation is possible throughout the year. As forest clearance is difficult, the cultivated plots tend to be quite small. Toward the end of the cultivation period larger spaces are left between the rows of plants so that growth of weeds and regeneration of the bushes can commence even before the fallow period begins.

Bush fallowing remains a viable and ecologically sound system of preserving soil fertility so long as the length of the fallow period is maintained. In many parts of western and eastern Africa fallows are declining for a variety of reasons; in the absence of other methods of improving the soil, such as the use of organic manures, improved crop rotations and the mixing of tree and field crops, the land is inevitably becoming degraded. Without the widespread adoption of soil improvement schemes this deterioration will continue. The closely settled zones around Kano and some of the other towns in northern Nigeria show how farmers using alternative methods of soil improvement, such as composts and intercropping, can modify their traditional farming systems to cultivate land successfully on a permanent basis.

DROUGHT AND SALINITY

The floodlands of many of the rivers in western Africa are highly prized areas for cultivation. They are particularly used by women for growing rice and other crops during the rainy season, which falls between May and October. The persistent drought that has affected this part of the region since 1968 has had a serious impact on these areas of riverbank cultivation.

The subsequent fall in the volume of water washing down rivers such as the Senegal, the Gambia, the Casamance and their many tributaries has caused a considerable amount of saline water from the coastal estuaries to penetrate up the rivers: the Gambia river and its feeder streams are now saline for about half their length in Gambia, and the network of rivers in Guinea Bissau has been equally badly affected. As a result the cultivation of swamp rice has suffered badly in Gambia. Rice plays an important part in the local diet – if domestic production falls, demand has to be met with costly imports.

In an attempt to keep the saline water at bay, the Senegal river has been dammed at its mouth. At a local level mud dikes have been constructed where the streams and river meet. It is not known just how effective these measures are, or whether they will, in fact, create further environmental problems. It is more certain that it will take a very long time for saline soils to regain their fertility. The future of women's rice cultivation in the lower Gambia and elsewhere therefore looks extremely bleak at present.

Nomadic herding

Cattle, goats, sheep and even camels are kept in the drier parts of the region, which are free of tsetse fly. There is not enough rain to sustain permanent pastures, and seminomadic herders migrate in search of forage and watering places for their animals. These pastoralists trade their goods with the settled farmers: the Fulani of western Africa and the Masai of eastern Africa trade with the Hausa and Kikuyu respectively, exchanging meat and dairy products for cereal and pulses. The settled farmers may also lease cattle to the nomads.

Herd sizes have increased in spite of recent years of drought, placing greater pressure on limited grazing lands. Both overgrazing and burning the ranges to promote the growth of fresh grass has led to deterioration in the quality of the soil, environmental degradation and a further decrease in the amount of land available for grazing. South of the Sahel the reduction of vegetation in formerly wooded savanna areas of western Africa has led to permanent nomadic herding. However, the migration routes are not as marked as they are farther north, as rainfall is more even, reducing reliance on seasonal pastures. Overgrazing is likely to be a problem here, too.

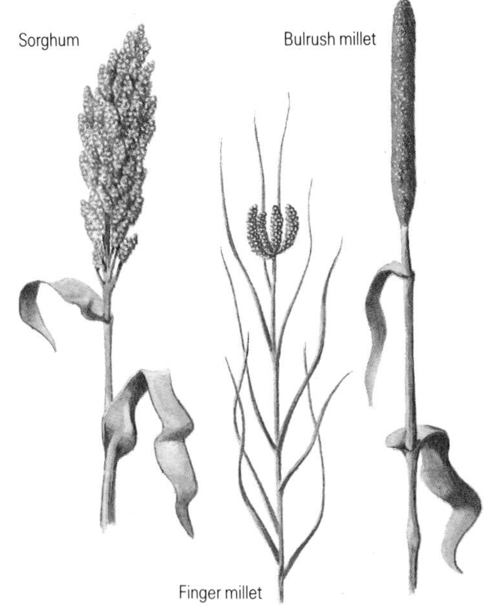

Tropical grain crops Millet and sorghum are grown widely throughout Central Africa. Finger millet has the highest nutritional value, being rich in minerals, and can be stored for as much as five years. Millets are usually eaten in the form of porridge; they are also used in the production of beer.

CASH CROPS AND FOOD SHORTAGES

Farmers throughout the region face a variety of pressures. Droughts during the 1970s and 1980s significantly reduced crop production and contributed to food shortages. Areas where crops have failed are vulnerable to erosion – drought thus contributes to degradation of the soil. Pressure on the land also comes from other sources. Population increase in the region places a great strain on food production systems. In much of the region there is an annual population increase of over 3 percent (the figure is higher still in Kenya), and domestic food supplies fall drastically short of demand.

Of greater significance than the natural increase in population is the migration of rural people to the towns and cities. The traditional farming systems are labor-intensive, particularly at peak periods of activity when the land is being cleared, sown and harvested. Migration depletes the village labor supply, and in parts of western Africa farm production has stagnated or even fallen as a result. Seed is sown late, and the crop may be past its best before labor is available to harvest it. Those who are left to work the land – mainly the very old and the very young – are unable to clear new land and so tend to overcultivate accessible land lying close to the village.

Sacrificing food for foreign exchange

Overemphasis on export crops such as cocoa in Ghana and Nigeria and groundnuts in Gambia has resulted in a shortage of food crops. As a result, valuable foreign exchange earned from agricultural exports often has to be spent on food imports. Self-sufficiency seems to be an increasingly distant goal for many countries – even for those, such as Gabon and the Congo, where population levels are not posing such an acute problem.

Tanzania is an exception, managing to produce most of its own food. However, farming here is adversely affected by poor road and rail links, and this holds up the supply of rural produce to the urban areas. Inadequate transportation of goods is endemic in all the large countries of the region, particularly affecting those that do not have access to coastal ports.

Changing traditional farming

Agricultural production throughout the region could be far higher than it is. African farming systems are well adapted to the environment, but economic pressure caused by inflation and the rising cost of labor and resources has resulted in farmers abandoning ecologically sound methods for short-term gains in productivity. The gains cannot be maintained unless traditional practices geared to extensive farming are modified to suit changing circumstances.

Farmers are accustomed to burning off vegetation prior to sowing their land. After two or three seasons of burning, the natural vegetation finds it difficult to recover. Furthermore, the ash, which is generally regarded as a form of fertilizer, is very often useless as it blows away, leaving a black, baked surface to hoe. Rather than burning the dry vegetation, it would be better to work it into the soil.

The use of crops such as pulses (particularly pigeon peas) that fix atmospheric nitrogen in the soil, improving its nutrient status, should also be encouraged. Improved soil has the additional benefit of making more efficient use of artificial fertilizers, as it retains both moisture and nutrients. Such changes would be relatively simple to introduce.

Major efforts have been made to increase agricultural output through farm settlement and resettlement schemes in Kenya and Uganda, and through the

River transportation *(above)* Mangoes are being transported by dugout canoe along the Tana river in southeast Kenya. The lack of roads in the larger countries of the region restricts the movement of food supplies.

Threshing millet by hand *(right)* The harvest is a time of peak activity in the countryside. Traditional farming methods make heavy demands on labor, and the migration of many men to the towns means that productivity is falling in some places.

Crop failure Drought has caused this field of maize to fail. Periods of extended drought during the 1970s and 1980s brought disastrous food shortages to some areas. Drought also exacerbates the worsening problem of soil erosion.

THE IMPORTANCE OF FENCING

The absence of good fencing is a major problem for farmers in much of Africa. Inadequate fencing results in widespread destruction of crops. The main culprits are domesticated animals – cattle, sheep and goats – that are left to wander during the dry season in search of forage. Wild animals also cause much damage. Without fencing, considerable labor is needed to keep birds and other animals from eating the crops.

Fence posts are made from trees such as ironwood, the stem of the borassus palm and sometimes the oil palm. Ironwood is resistant to insect attack by termites, but posts made from other tree species have to be replaced every two or three years. Dried palm fronds, Andropogon grass, local types of bamboo and sometimes corrugated iron are also used for fencing. None of these materials can withstand damage from animals – even corrugated iron, which is often very thin.

Live fences of climbing shrubs, such as bougainvillea, or cacti have been used, but have not always grown successfully. Farmers are keen to replace traditional fences with barbed wire or some other permanent alternative, but the costs of such materials are prohibitive. Without adequate fencing, however, demands on labor for unproductive tasks remain high, and the felling of rapidly declining numbers of trees continues.

establishment of rural cooperatives, or *ujamaa*, in Tanzania. Elsewhere largescale capital-intensive developments and integrated rural development projects have been attempted, but none of these has had the success predicted.

The reasons for this are varied; they include managerial inefficiency, a shortage of capital to buy machinery and costly inputs of fuel, fertilizers and pesticides, and in some cases the use of methods that have increased environmental damage. There is a growing appreciation that agricultural development must respect the traditional knowledge and skills of the peasant farmer. Projects to improve productivity and develop the farming sector must be much more closely tailored to local environmental conditions.

Cassava: Africa's staple root crop

Cassava is widely grown throughout the region for its starchy roots, which form an important part of the staple diet. Cassava is high in carbohydrates but low in protein. A diet that contains too much cassava can result in malnutrition. This is particularly common among young children who are often weaned straight onto it without being given other sources of protein in their diet.

The plant is not native to Africa, but was probably introduced from South America by the Portuguese 300 years ago. There are many advantages to growing it: it tolerates both poor soils and drought, and can be left in the ground for up to four years before being harvested. Plants are easily propagated from cuttings, and little skill is needed to raise them.

Although cassava is an easy crop to grow, preparation of the roots for eating is laborious. They contain enzymes that can convert certain compounds into hydrogen cyanide, which can kill people who eat it. In western Africa, to avoid this, cassava is safely prepared by staking and grating the roots. The pulp (known as *gari*) is then left under weights for a few days, where it ferments. Heating the fermented cassava drives off the cyanide. This heavy work of preparation is traditionally done by women, who are usually also responsible for cultivating and harvesting the crop.

Cassava deteriorates very rapidly after harvesting, and it is difficult to keep stocks for even a few days. It is grown principally by small farmers, and a few roots are harvested at a time for the farmer's own use or for sale locally or to a processing plant. It is thus an important subsistence and cash crop, but has little value as an export crop, though efforts have been made to increase exports of cassava starch – most familiar to consumers outside the tropics as tapioca.

Attitudes to cassava differ across the region. In the forested areas of western Africa it is seen as a valuable protection against famine, but is less popular in times of plenty as it is not particularly palatable or convenient to prepare. It is not a favored crop in eastern Africa, as people there were pressurized into growing it as a reserve against famine, and it lacks status because of the ease with which it grows – one does not need to be a "real" farmer to cultivate it. In the savannas of western Africa, however, cassava is becoming popular among people who have migrated to towns, but still have access to village land. A crop can be planted over a weekend and left to grow without needing further attention.

Improving cassava

Although development projects in the past tended to ignore cassava at the expense of cereals, much recent research has been done to improve the production and quality of cassava, as well as to find ways of prolonging the time it can be kept. Unfortunately, varieties that have been bred for greater yields tend to have inferior storing qualities – the tubers rapidly become fibrous, making them even less palatable. Attempts to reduce the toxicity of cassava have rendered the crop vulnerable to attack by rodents, which eat the roots, destroying the crop before it has been dug up.

Some types of cassava have been affected by disease. Varieties resistant to mosaic disease have been bred, but these are proving less successful in areas where the soil is poor, thereby negating a major reason for growing cassava in the first place. Attempts are being made to halt the recent spread of two pests that attack cassava – the cassava mealybug and the green spidermite – accidentally introduced to Africa from South America in the 1970s. Natural predators are used to keep the insects in check.

Drying the tubers (*left*) Cassava does not deteriorate when left in the ground, so it can be harvested at any time of year. This makes the crop good security against famine. However, cassava has poor storage properties and must be used quickly once harvested.

Laborious preparation (*right*) Cassava is exceptionally easy to propagate and grow, but it requires a good deal of work before it is edible. This work – as well as the cultivation and harvesting – is usually performed by the women and children.

Traditional fishing

In many parts of the world, where people have access to coastal waters, or to inland lakes and rivers, the hunting of fish is still carried on by individuals or family groups using minimal equipment. People living beside the sea may fish with nets or lines from small craft, or from the beach itself, using hand nets to take crustaceans such as shrimps at low tide, or wading out to sea to catch shoals of small fish carried on the incoming tide. Larger crustaceans, such as crabs and lobsters, may be caught in pots or baskets. Fish caught in these ways provide an important source of protein in grain-based diets, as well as adding an element of variety and interest to otherwise monotonous foods.

In many places such supplies of fish are being reduced by largescale fishing operations. Marine pollution is also responsible for dwindling fish stocks. Yet traditional smallscale fishing still accounts for the greater part of the total world fish catch.

In countries where electricity is limited for refrigeration, the freshly caught fish either have to be eaten immediately or preserved by drying or smoking. It is common in Africa to see fishermen selling their catch by the roadside. The fish are kept alive in nets or strung together through their gills until a purchaser is found.

Making the catch Hundreds of people in Nigeria wade out to sea with nets and pots.

LOW TECHNOLOGY AGRICULTURE

THE OLD AND THE NEW · FARMING TO LIVE AND FOR PROFIT · THE STRUGGLE TO EXPORT

Farming is the main economic activity of most southern African countries, with the exception of Zambia and South Africa. The majority of farmers are smallscale peasant producers whose main concern is subsistence production of maize – the staple food crop. Given favorable conditions, these farmers may be able to produce surplus maize for sale, and grow cash crops such as groundnuts, tobacco or cotton. In many parts of the region, however, rural households are unable even to feed themselves, making them dependent either on migrant relatives who have found paid work, or on food aid. The causes of this problem are complex, but include lack of access to land, military activity and also drought. Largescale commercial farming predominates in South Africa and is also important in Zimbabwe and Namibia.

COUNTRIES IN THE REGION

Angola, Botswana, Comoros, Lesotho, Madagascar, Malawi, Mauritius, Mozambique, Namibia, South Africa, Swaziland, Zambia, Zimbabwe

Land (million hectares)

Total	Agricultural	Arable	Forest/woodland
650 (100%)	366 (56%)	34 (5%)	160 (25%)

Farmers

23.3 million employed in agriculture (56% of work force)
1.4 hectares of arable land per person employed in agriculture

Major crops

Numbers in brackets are percentages of world average yield and total world production

	Area mill ha	Yield 100kg/ha	Production mill tonnes	Change since 1963
Maize	8.9	13.0 (36)	11.6 (3)	+36%
Wheat	2.0	17.0 (73)	3.4 (1)	+270%
Roots/tubers	2.0	55.5 (44)	11.0 (2)	+79%
Groundnuts	0.9	6.1 (52)	0.5 (3)	−17%
Sugar cane	0.6	724.3 (121)	42.9 (4)	+101%
Tobacco	0.2	11.8 (83)	0.2 (4)	+45%
Fruit	—	—	5.9 (2)	+105%

Major livestock

	Number mill	Production mill tonnes	Change since 1963
Sheep/goats	55.7 (3)	—	0%
Cattle	42.1 (3)	—	+24%
Milk	—	3.5 (1)	+13%
Fish catch	—	1.8 (2)	—

Food security (cereal exports minus imports)

mill tonnes	% domestic production	% world trade
−0.2	1	0.1

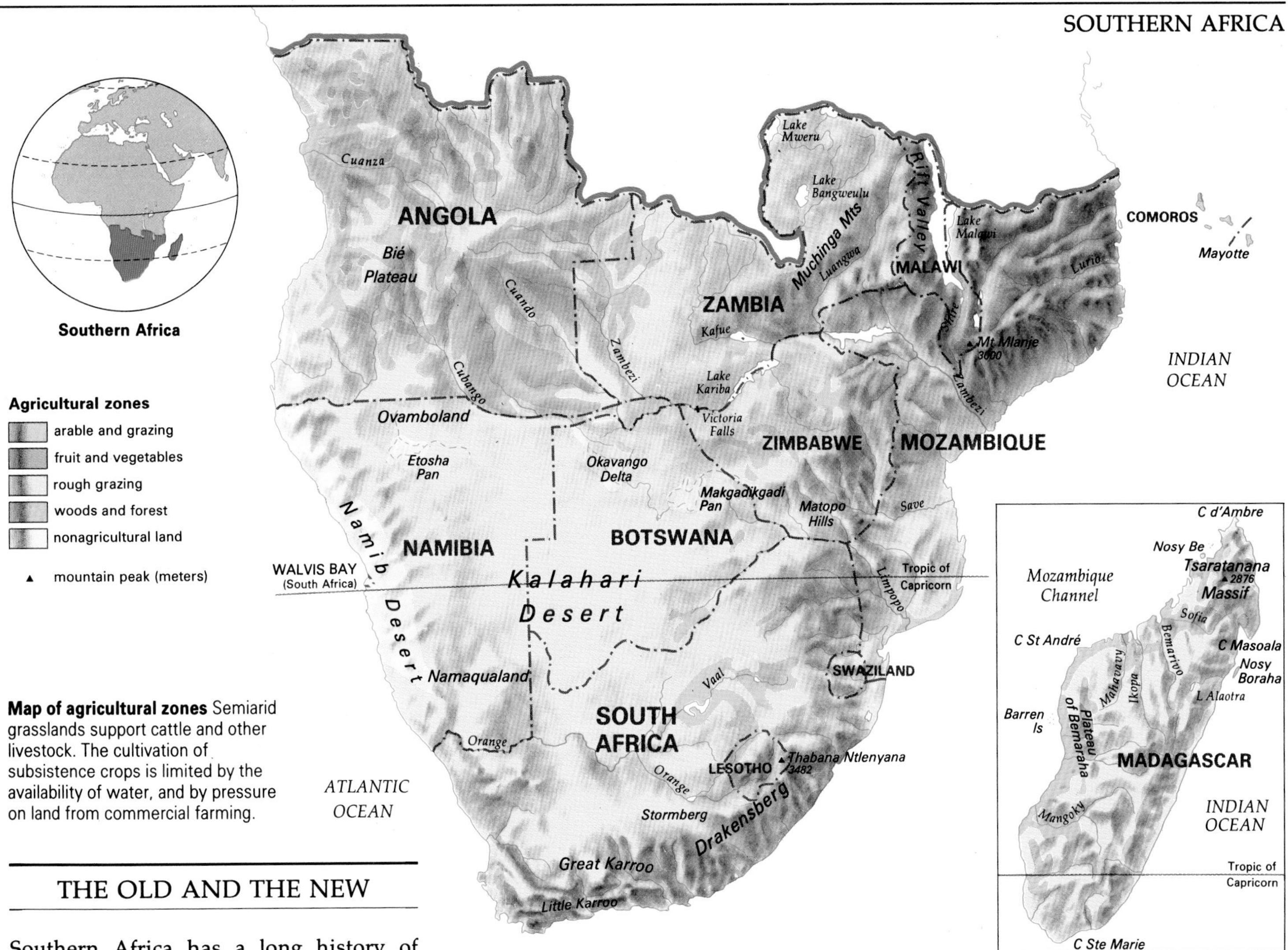

Map of agricultural zones Semiarid grasslands support cattle and other livestock. The cultivation of subsistence crops is limited by the availability of water, and by pressure on land from commercial farming.

THE OLD AND THE NEW

Southern Africa has a long history of human settlement, but a relatively short history of farming activity. Evidence of human occupation of the region dates back at least 2 million years: the earliest known people were hunter–gatherers, some of whom are still found in Namibia and Botswana. The transition to the use of domesticated animals by the indigenous people began about 2,000 years ago with the introduction of sheep from grasslands farther north. At about the same time cultivated crops were brought into the region by Bantu-speaking peoples from central Africa.

The major crops at this time were sorghum and millet. Animal husbandry was also important, and cattle played a significant economic and social role. By the time the first European settlement was founded in Cape Town in the mid-17th century, the original herders and hunter–gatherers had largely been replaced by Bantu farmers and herders.

Soil degradation and drought

Farming was originally practiced by a small population using extensive areas of land. Farmers and herders were able to live in relative harmony with their environment, choosing the soils most suited to their crops. Many of the soils in the region need careful nurturing to prevent leaching and erosion. The soil was traditionally cultivated with hoes. Although labor intensive, this causes less moisture to be lost from the soil than deep tilling.

In drier areas seminomadic herding over large expanses enabled both cattle and smaller stock to be moved to different grazing sites, limiting the environmental impact of the herds. Pastoralism was, and still is, limited to areas that are free of tsetse fly, the carrier of sleeping sickness (trypanosomiasis), which affects cattle as well as humans.

The most significant environmental constraint on farming in southern Africa is availability of water. The rains usually fall between November and April, and without irrigation very little can be grown for the rest of the year. Where the annual rainfall is less than 750 mm (30 in), rainfed agriculture becomes extremely difficult. Much of southwest Zimbabwe, southern Angola, and the northwest of the Transvaal in South Africa receive less; Botswana and Namibia and the southern interior and west of South Africa receive less than 500 mm (20 in) on average.

Furthermore, the rainfall tends to be unreliable: a very severe and prolonged drought in the early 1980s caused immense problems for both the commercial and peasant farming sectors. As the rain often comes in intense downpours farming techniques that reduce its erosive power and store the moisture are vital. The southern tip of South Africa experiences a different climate, similar to that of the Mediterranean, which supports orchards and vineyards.

Until the advent of European settlement, agricultural production for commercial reasons was limited. Colonialism introduced taxation and new consumer products. Where markets developed for livestock, or for crops such as cotton and tobacco, many peasants began to meet their cash needs from farming.

However, market production by indigenous farmers was frequently discouraged so that young men would have to migrate to the mines and European farms – where there was a severe labor shortage – to earn the necessary cash. Much of the marketed produce therefore came from European farms using African labor. In colonial Angola and Mozambique the use of forced labor on agricultural and other projects was widespread.

Livestock farmers with their herds in Botswana, where cattle herding is the chief agricultural activity. Vegetation is sparse, aggravated by overgrazing, and many cattle die in times of drought.

FARMING TO LIVE AND FOR PROFIT

The traditional arable farming system in southern Africa was based on shifting cultivation, which allows soil fertility and vegetation cover to regenerate. When land was cleared – principally a male task – the vegetation was burnt, the ash providing some nutrients for the new crops. In most societies in the region all further labor was done mainly, or exclusively, by the women, whose work is of crucial importance to farming. Planting was done at the beginning of the rains – timing the planting was critical as sometimes the rains come late. The major grain crops were then harvested between March and May.

Except where the prevalence of tsetse flies discouraged herding, cattle husbandry dominated the rural economy of most groups. It was usually an exclusively male preserve and tended to have much higher status than cultivation. Livestock were herded during the wet season to keep them out of the fields, but were left to wander in the dry season when little cultivation took place.

This traditional pattern of farming has been altered by a variety of factors. In South Africa, Zimbabwe, Namibia and Swaziland the most significant change was the appropriation of large amounts of land by European farmers: 87 percent of the land in South Africa and 50 percent in Zimbabwe was reserved for white ownership. This unequal division left African farmers with far less land than their farming systems had been devised for, to the detriment of their livelihoods.

Population pressure on the land has altered established agricultural practices: the land is left to lie fallow for shorter periods and is sometimes cultivated continuously, resulting in soil degradation and erosion. The combination of smaller plots and poorer soils frequently reduces productivity to below subsistence levels. The loss of many able-bodied men to the migrant labor system means that farming has become subsidiary in many rural households as their remittances often now form the major part of the family's income. Productivity has become increasingly dependent on the money they send back to purchase inputs of fertilizers and pesticides, and the men's absence means that the women have to carry an even heavier burden of herding and farming than under the traditional system.

Hoeing a maize field (*above*) Women were traditionally responsible for most crop cultivation, but with more and more men migrating to work elsewhere, their role as subsistence farmers has grown in importance.

Vineyards in Cape Province, South Africa (*right*) South African agriculture is dominated by white commercial farmers and largescale farms. The Mediterranean climate supports fruit and vegetable production, which is geared to supplying European markets during the winter, when their own crops are out of season.

The main grain crop grown in southern Africa is maize, which was introduced in the 17th and 18th centuries. Although easier to process than the indigenous sorghum and millet, it is less resistant to drought. Peasant farmers also produce a wide variety of other crops, both for domestic consumption as well as for sale. Depending on the area, these might include beans, cassava, sweet potatoes, gourds, pumpkins, groundnuts, rice, cotton, cowpeas, tobacco, sugar cane, and various fruits and vegetables. The mix of crops depends on the availability of land and labor, the nature of the local environment, the prices to be paid for individual crops, as well as the ease of obtaining credit, advice and inputs. In many places inadequate transportation and marketing facilities, and sometimes a lack of goods to buy, discourage the sale of surplus crops and livestock, and the setting of low fixed government prices for produce has frequently had the same effect.

Commercial farming

Most African peasants hold their land under communal tenure arrangements: land is allocated by the local headmen (who are frequently now replaced with some form of village committee). Land cannot be owned under this system. Largescale commercial farmers usually own or lease their land and hire labor. The production methods of large farms are similar to those practiced in the developed world – mechanization, heavy application of fertilizers and pesticides, and the use of high-yielding varieties of both crops and animals, which require "scientific" management.

For a long time government help and loans for agriculture in many southern African countries were heavily biased toward white commercial farmers, greatly assisting their development. As nations in the region gained independence, more African-owned commercial farms began to emerge. State farms were also developed, particularly in socialist Angola and Mozambique, where most European

farmers left at independence in 1975.

The major commercial crops grown by largescale farmers vary from country to country in the region. Maize, wheat, sugar, fruit, vegetables and wine are produced in South Africa; maize, tobacco, wheat, sugar, cotton, fruit and vegetables in Zimbabwe. Tobacco, tea and sugar are grown in Malawi; sugar, fruit and timber in Swaziland; and maize and tobacco in Zambia. Coffee and cotton have been important in Angola, and cotton, cashewnuts, tea, sugar and copra in Mozambique. However, agriculture in these two countries has been disrupted by military conflict since independence. Commercial livestock production for meat, hides and wool is important in South Africa, Namibia, Botswana and Zimbabwe.

TRADITIONAL GARDENS

An important element of southern African peasant farming systems is the use of small valley wetlands for crop irrigation. Small gardens are supported on these wetlands, known as *dambos* in Malawi and Zambia, *bane* in Zimbabwe, and *vleis* in South Africa. During the rainy season the ground becomes waterlogged; and because the water table is close to the surface in the dry season, crops grown in these gardens can be irrigated, usually with the use of a watering can only, when other fields are too dry to be used. This water resource – which has probably been used for hundreds of years – is of great significance in a region where the rainfall is often scarce and unreliable. It contributes greatly to the farmers' security; the yields of maize from the *dambos* sometimes rival those of the best commercial farms, and vegetables and other cash crops grown here are important sources of income.

In Zimbabwe the use of the wetlands for cultivation is restricted by a law that came into force in the period before independence. Although such an action is usually justified on environmental grounds, it also served to protect the white commercial farmers from peasant competition. Approximately 20,000 ha (50,000 acres) of *bane* land is presently under cultivation in Zimbabwe. This area could be increased fourfold if the restrictions were to be relaxed.

THE STRUGGLE TO EXPORT

In parts of southern Africa where land is plentiful traditional methods of managing crops and livestock are still appropriate. In many places, however, peasant farmers face problems as a result of land shortages: the white-owned commercial farms have appropriated large tracts of land, and population increase has added to the strain. When soil fertility is damaged by overuse and overgrazing it becomes increasingly difficult to produce subsistence and cash crops.

Governments want cash crops and livestock for urban consumers and to export as a source of foreign exchange; farmers want cash to improve their living standards and educate their children. There is consequently strong pressure to increase yields by using both fertilizers and high-yielding seeds such as hybrid maize. While largescale commercial farmers can usually afford to apply such methods, they are often too expensive for smallscale farmers to employ.

Consequently an ever-present dilemma is whether to opt for costly hybrid seeds, which offer the prospect of better yields but are liable to fail totally in poor rains, leaving farmers with little to eat and with crippling debts. Irrigation would minimize this risk, but the river systems of southern Africa are generally far less suited to the massive lowland irrigation networks that are so productive in Asia, for example.

Export markets

High technology irrigation is very successful on the Orange and Vaal rivers in South Africa; it enables large commercial farmers to grow a variety of crops from winter wheat to fruit. Commercial irrigation is also important in Zimbabwe. The rising European demand for fruits and vegetables in winter has generated significant exports from both these countries. Much of South Africa's fruit production is geared to this lucrative market, and Zimbabwe now produces high-value perishable items such as sugar peas (mange touts or snow peas), green beans and flowers for airfreighting to Western Europe. These products require very exacting standards of quality control as well as packaging and transportion arrangements that are usually well beyond the capacity of small farmers.

Governments are keen to diversify their export bases and so encourage peasant farmers to produce a wider range of crops. In Malawi, for example, sugar cane has become important since independence, and various new schemes have been developed to encourage small farmers to produce both air-cured (burley) tobacco, as well as the more traditional fire-cured variety. Malawi, whose domestic and export revenue is derived entirely from agriculture, has the advantage of a good network of marketing depots.

The production of peasant cash crops, however, fluctuates according to the prices set by Malawi's government-run central marketing board. These are often very low, and the profits have not usually been reinvested in the peasant sector. There has been considerable international pressure to free the market from central control by making privatization a condition for being granted loans. It is hoped that this will lead to greater incentives to increase productivity levels.

Within the region as a whole marketing boards have frequently mismanaged the

A tea estate in Malawi (*right*) Tea, which was introduced as a plantation crop during colonial times, is one of Malawi's principal agricultural exports. White commercial farming has been detrimental to Africa's indigenous peasant farmers, but considerable efforts are now being made to encourage cash crop production outside the plantation sector.

Farming on the edge of the desert (*below*) Water from the Orange river in Namibia is diverted to fields growing cash crops. Not all river systems in the region can be used for largescale irrigation schemes.

delivery of essential resources, or have delayed transporting crops or paying for them. As a result, peasant farmers may make the rational decision to stop producing the desired crops, or to enter into "illegal" marketing, including smuggling. In Zambia, for example, where minerals dominate the national economy, agricultural development has been neglected or poorly implemented, and consequently there has been widespread migration from the country to the towns.

Successful peasant farming

Zimbabwe provides a good example of peasant commercial success. Here cash crop production has boomed since independence. This expansion has partly been due to improved access to credit and advice, which has gone some way to restoring the imbalance between peasants and largescale farmers. In recent years, roughly half the marketed maize and cotton has come from the peasant sector, and Zimbabwe is able to export maize to neighboring countries.

Land shortages still pose a serious problem; resettlement on land purchased from the commercial farming sector is progressing very slowly. Some white farmers argue against land redistribution, claiming that peasant farms are less productive than the largescale operations. Peasants often use the land more intensively, however, and some commercial land is at present underutilized. The experience of Kenya in eastern Africa shows that productivity can remain high on redistributed land. The long-term political stability in Zimbabwe, Namibia and a future post-apartheid South Africa will depend to some extent on land inequalities being redressed.

Only South Africa and Zimbabwe have regular major food surpluses to export. Angola, Mozambique, Lesotho, Botswana, Namibia and Zambia all import food, though Zambia is sometimes self-sufficient in maize. The need for regular imports should not always be construed as a problem or failure: Namibia and Botswana, for example, are extremely arid countries, with small populations and very successful mineral-exporting sectors. Yet Angola, Mozambique and Zambia certainly have the potential to produce food and other agricultural surpluses, though in Angola and Mozambique any improvements depend on bringing an end to civil war.

FAMINE AND DESTABILIZATION

Farmers in parts of Mozambique and Angola find it very difficult, or even impossible, to farm their land because of the fighting between antigovernment forces and the army. As a consequence of the conflicts, which have continued almost without break since independence in 1975, millions of people have become refugees (*"deslocados"*), and famine remains an ever-present threat.

Guerrilla fighting causes widespread rural destruction and loss of life. In Mozambique young men and boys are abducted from their villages and moved to areas hundreds of miles away. People in the countryside are tortured and murdered, buildings such as schools and clinics demolished, and villagers' crops and possessions stolen or destroyed. Fields are often mined, and many farmers and their children have lost limbs while tending their crops.

The impact on agriculture has been enormous, bringing an end to food production in many places and making necessary massive imports, often (in the case of Mozambique) in the form of food aid. Some peasant farmers, nevertheless, try hard to maintain some form of production. Refugees in Dedza district of Malawi, for example, work on their fields in Mozambique during the day, retreating back over the border at night. Commercial and peasant farmers have even been attracted to some relatively secure areas, such as the well-guarded Beira corridor, which maintains a vital overland link between land-locked Zimbabwe and the port of Beira on the Indian Ocean.

Game ranching

Archaeological evidence suggests that domesticated cattle, sheep and goats have been present in large numbers on the grasslands of Africa south of the Sahara for only about 5,000 years. While these animals have undergone some evolutionary changes to cope with the local conditions of seasonal drought, heat, poor grazing and disease – N'dama cattle, for example, are resistant to tsetse-borne trypanosomiasis, and zebu cattle require little water – the indigenous herbivores are far better adapted to this environment. In recent decades, some commercial farmers with large ranches in southern Africa have begun to exploit indigenous species of game animals. This form of range management involves economic and ecological considerations. As well as being a profitable way to use the land, managed game ranching is a means of ensuring that local species are conserved.

Adaptation to the savanna

A major advantage of game animals is that they exploit all the natural vegetation of the savanna grasslands – each species fills a niche by feeding on plant life not favored by others. All strata of vegetation are used: elephants and giraffes browse on trees and bushes, eland and impala graze on the lower and middle levels, and steenbuck and the tiny duiker feed close to the ground. Cattle, on the other hand, exploit only a narrow range of vegetation, while goats and sheep will strip all vegetation unless they are kept on the move. Game animals enjoy other advantages: they are resistant to many of the diseases that affect domestic breeds and are better able to withstand drought. Some species are able to derive most of their water requirement from vegetation.

These ecological advantages mean that the sustainable biomass of game – the weight of animals that can be sustained without damaging the range – often exceeds that of the domesticated animals. Evidence from Zimbabwe suggests that in low rainfall areas, the land can support about twice the weight of game compared to cattle on a well-managed ranch. Game ranching has expanded: land devoted to the commercial use of wildlife increased by about 70 percent between 1974 and 1984. In the southeastern savanna lowveld 24 percent of land was used solely for game by 1984. Game ranches have also been developed in Zambia, South Africa and Botswana.

Game auction (*above*) A profitable specialist trade has developed in the sale of live game to ranches, game reserves or zoos. Game ranching is a rapidly expanding sector of agriculture that satisfies both commercial and ecological criteria.

Eland ranching (*right*) Indigenous herbivores offer the rancher considerable advantages over domesticated livestock such as cattle or sheep. They are better adapted to survive drought and are also resistant to many of the diseases that domestic breeds suffer from.

Hunting for profit

Only a part of the profit from game animals comes from their meat – the trade has to overcome long-standing consumer preferences and the established meat distributors often refuse to deal in it. Hunting is a far more lucrative activity. A trophy animal is very valuable, and provides a source of hard currency as many of the hunters are foreigners.

Live game is sold to other ranches or to game reserves. Sometimes a mixed system operates that integrates the farming of semidomesticated species such as springbok, impala and kudu with other kinds of farming. Experiments are underway to extend the commercial utilization of game to peasant farming areas in Zimbabwe, and it looks set to expand throughout southern Africa.

LAND OF LONG TRADITIONS

ENVIRONMENTAL INFLUENCES · A FARMING WAY OF LIFE · FOOD FOR THE MILLIONS

Agriculture has been practiced in the Indian subcontinent from very early in human history, and many of its peasant farmers still maintain a farming technology and way of life that has changed little over the millennia. Its earliest origins are unclear, but in the hot dry valley of the lower Indus the people of the Mohenjo Daro civilization (2500–1600 BC) are known to have grown barley and wheat, irrigating it with river water. In the more humid east, shifting cultivation was practiced, and still survives in some remote hill districts. Elsewhere it was replaced by rice farming. In the coastal areas of the Bay of Bengal coconuts have long been important, providing food, fibers, and wood for building. In the dry interior coarse millets are grown, and cattle, sheep and goats are grazed by both settled and seminomadic groups.

COUNTRIES IN THE REGION

Bangladesh, Bhutan, India, Nepal, Pakistan, Sri Lanka

Land (million hectares)

Total	Agricultural	Arable	Forest/woodland
413 (100%)	223 (54%)	198 (48%)	80 (19%)

Farmers

258.7 million employed in agriculture (66% of work force)
0.9 hectares of arable land per person employed in agriculture

Major crops

Numbers in brackets are percentages of world average yield and total world production

	Area mill ha	Yield 100kg/ha	Production mill tonnes	Change since 1963
Paddy rice	52.7	22.3 (68)	117.7 (25)	+61%
Wheat	32.0	18.2 (78)	58.2 (11)	+274%
Millet/sorghum	30.3	6.3 (55)	19.1 (21)	+10%
Pulses	24.9	5.3 (66)	13.2 (24)	+3%
Cotton lint	9.1	2.8 (52)	2.6 (16)	+81%
Maize	7.2	10.9 (30)	7.8 (2)	+30%
Groundnuts	6.8	8.4 (73)	5.8 (27)	+11%
Sugar cane	4.1	539.5 (90)	222.4 (23)	+73%
Vegetables	—	—	53.0 (13)	+107%
Fruit	—	—	30.3 (9)	+94%

Major livestock

	Number mill	Production mill tonnes	Change since 1963
Cattle	247.4 (19)	—	+15%
Sheep/goats	235.8 (14)	—	+74%
Buffaloes	93.7 (68)	—	+48%
Milk	—	25.2 (5)	+158%
Fish catch	—	4.4 (5)	—

Food security (cereal exports minus imports)

mill tonnes	% domestic production	% world trade
−2.1	1	1

Exploiting the land Throughout the subcontinent farmers exploit almost every available piece of land. High in the Karakorum mountains on the border of India and Pakistan they store water from the summer snowmelt to irrigate their terraced fields.

ENVIRONMENTAL INFLUENCES

Water storage and irrigation have a long association with farming throughout the Indian subcontinent, where the rainfall is highly seasonal and in places unreliable; in addition some staple crops, notably rice, make high demands on water. The region embraces an extraordinary diversity of environments, including some of the world's highest mountains, tropical rainforests, hot sandy deserts and extensive river plains. It can, however, be divided into three main areas for agriculture: the triangular Deccan peninsula in the south and center; farther north, the vast plains formed by three large rivers, the Indus, the Ganges and the Brahmaputra, and their tributaries; and, in the extreme north, the Himalayas, which form an important dividing line for civilizations, climates and agricultural practices.

In the fertile, irrigated valleys of the west and central Himalayas rice is grown; on higher slopes, as temperatures fall, hardier crops are planted, predominantly maize and barley. Commercial orchards for almonds and apricots are common in the lower valleys, particularly in Kashmir in the northwest; at higher altitudes apples are grown. Above the treeline animal husbandry is important, with yaks providing transport, milk and butter.

The great plains of the northern subcontinent, where the greatest densities of

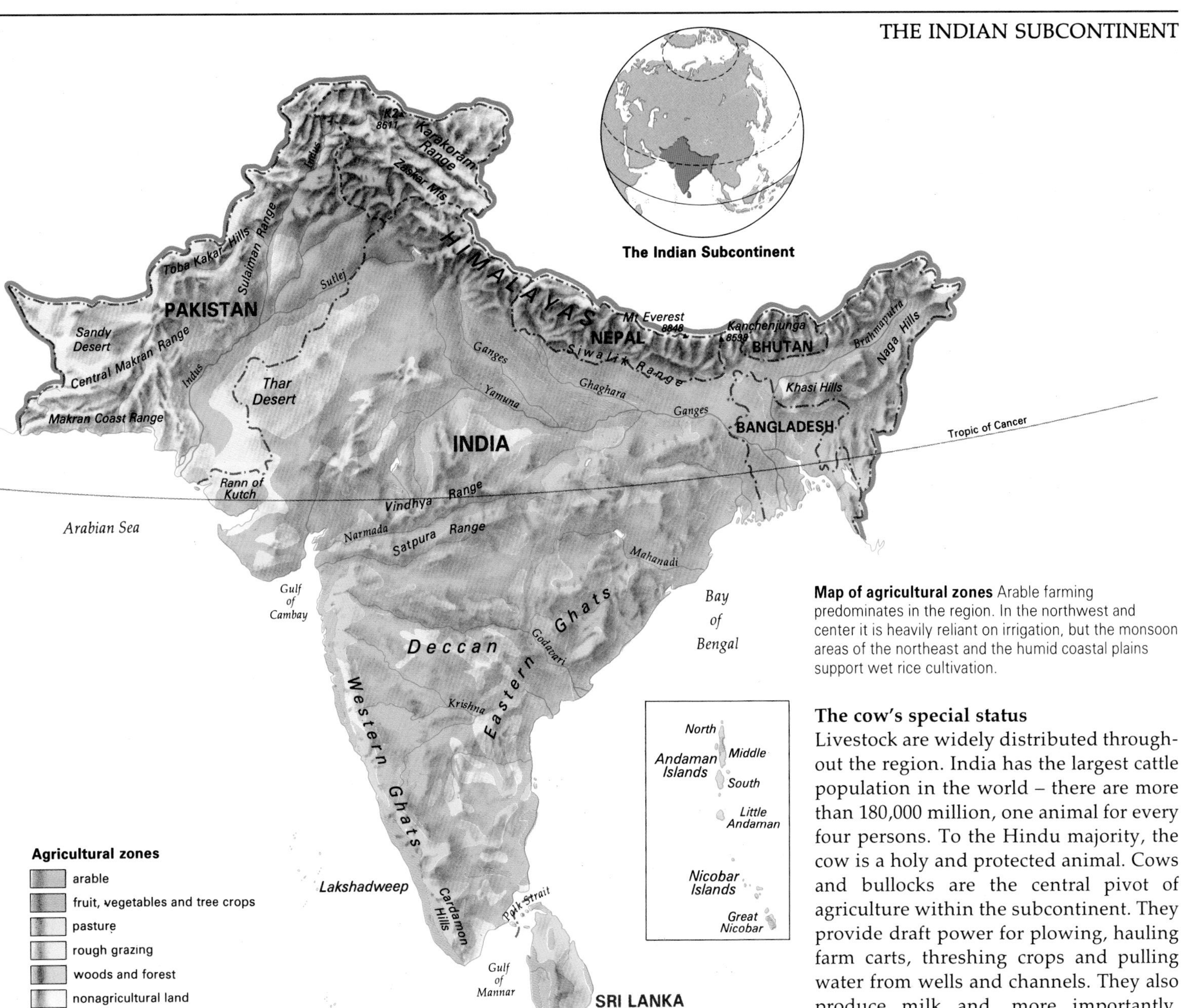

Map of agricultural zones Arable farming predominates in the region. In the northwest and center it is heavily reliant on irrigation, but the monsoon areas of the northeast and the humid coastal plains support wet rice cultivation.

population are found, contain the best agricultural land in the region, though aridity in the northwest and flooding in the Ganges delta in Bangladesh and northeast India present difficulties for farming. The cultivation of crops in the fertile Indus valley of Pakistan and in the upper Ganges is heavily dependent on irrigation, through an extensive network of canals. Traditionally, wheat and cotton have been grown on these lands, but in recent years there has been an upsurge in the cultivation of new, high-yielding rice varieties, introduced in the 1960s as part of the Green Revolution to boost grain productivity. Sugar cane is also important in this area.

Irrigation is less essential all year round in the center and east of the plains, where much heavier and more reliable monsoon rain falls. In the lower Ganges monsoon rice is chiefly grown in flooded paddy-fields. Where rainfall is unreliable, the rice is transplanted from small, irrigated seedbeds, but broadcast cultivation is more common in areas where plentiful rainfall is guaranteed.

Farther south, rice is the main crop of the rainy coastal plains of the Deccan peninsula. It is also cultivated in the interior, wherever irrigation tanks have been constructed. Where conditions are more arid, cotton and coarse grains such as millet are grown. Goats and sheep are numerous, scavenging widely over poor wasteland.

Agriculture in the island of Sri Lanka includes huge plantations of tea, rubber and coconuts, interspersed with smallholdings where rice, sugar cane, cassava, sweet potatoes, cashew nuts and other subsistence crops are grown.

Coconuts are the principal crop of the Maldives, together with millet, breadfruit and tropical fruits and vegetables, some of which are exported. Fishing is the major industry; locally caught tuna and other fish are canned or frozen for export.

The cow's special status

Livestock are widely distributed throughout the region. India has the largest cattle population in the world – there are more than 180,000 million, one animal for every four persons. To the Hindu majority, the cow is a holy and protected animal. Cows and bullocks are the central pivot of agriculture within the subcontinent. They provide draft power for plowing, hauling farm carts, threshing crops and pulling water from wells and channels. They also produce milk and, more importantly, dung, which is used as fertilizer and, dried, as fuel for cooking. Mixed with mud, it makes a slurry that dries as hard as concrete but does not crack – a necessity in making a threshing floor.

As well as cattle there are millions of buffaloes (a major producer of milk), goats and sheep. Fodder is scarce, mostly consisting of the residue from harvested crops, so animal health is generally poor. This is one reason why plows are simple and do not turn the soil deeply, and also why milk yields are low: 150 kg (330 lb) per cow per year in India compared with 4,000 kg (8,800 lb) in Western Europe. It has been estimated that present fodder resources are sufficient to feed adequately only two-thirds of India's livestock. Most people here are vegetarians and, since cows are venerated, they are usually allowed to die a natural death. Although India is a major exporter of cattle hides, it in fact exploits only a small part of this market because touching carcasses is regarded as unacceptable, except for those of the lowest status of caste.

A FARMING WAY OF LIFE

Most people in the Indian subcontinent are involved in farming. It provides 93 percent of employment in Nepal and Bhutan, 75 percent in Bangladesh, 70 percent in India, 55 percent in Pakistan and 53 percent in Sri Lanka.

Nearly all types of farming are involved, much in subsistence cultivation. Most farmers are tenants who rent land for cash; some are short-term sharecroppers who rarely take home half the crop they grow. Those who are landless may receive payment in food, but this is now declining in favor of cash wages.

Peasant cultivators are sometimes freeholders, though they generally possess little land. Most holdings are fragmented into tiny plots. In the Indian Punjab, for example, in the northern plains, an average holding is about 4 ha (10 acres), and will be divided into four or five plots, whereas in Bangladesh, with its high density of population, an average holding of perhaps 1 ha (2.5 acres) may be split into as many as 12 or 15 plots. In just a few places on the subcontinent there are major plantations or estates, nearly all producing crops for export.

The task of feeding the subcontinent's rising populations thus falls on an overwhelmingly peasant workforce, farming the land with mainly traditional methods. The land under cultivation is unevenly distributed; a very large percentage of farmers are concentrated on a very small amount of the total area. Small farmers are hampered by a lack of credit, and the land tenure system provides little incentive for improvement.

Staple grains

Rice is the most important food crop throughout the region. It occupies about one-third of all the land under food grains. In traditional agriculture the same variety is never planted twice running on the same plot. This inhibits the buildup of pests and diseases, as each variety has differing resistance to pests, virus diseases, fungi and drought. A farmer in Bangladesh may consequently have 15 or 20 varieties in store for planting on his 10 or so plots, each of them suited to the local microenvironmental conditions. In the state of Bihar, in northeast India, there are thought to be more than 50,000 local rice varieties in cultivation.

In the 1960s high-yielding varieties of rice, bred to be responsive to fertilizers, were introduced as part of the Green Revolution. Greater use of fertilizers, pesticides, insecticides and irrigation also formed part of this program to modernize agricultural practices and to make each country in the region self-sufficient in food. Wheat production – the region's second grain crop, occupying one-fifth of the area under grain in India – increased rapidly in the early years of the Green Revolution, helped by the expansion of irrigation in the wheat-growing areas. This was made possible by new canals, and by using electric or diesel-powered tube wells to pump groundwater from much deeper beneath the surface than traditional wells could do.

Tea picking India is the world's largest producer of tea. The best tea grows in warm climates at altitudes of 1,000–2,000 m (3,000–7,000 ft). The tea plant takes 3–5 years to mature, after which the flushes – new shoots – are picked.

Pulses, mainly gram (chickpeas) and lentils, are the most common source of protein in India, though fish is eaten in coastal and river areas. Beef and mutton are also eaten in Pakistan and Bangladesh. Pulses are agriculturally advantageous, as they are able to fix nitrogen in their roots, and so enhance soil fertility. They are intercropped with millet in the Deccan peninsula. Other grain crops include sorghum, maize and barley.

Fruit and vegetables are grown for domestic consumption in gardens surrounding most villages. Since packaging and preservation is little developed in the subcontinent, supplies of fresh produce – including meat, milk, vegetables and fruit – have to be immediate. This means that towns take up the surplus of local farmers, who often undertake the marketing themselves. Goats and chickens are taken live to the market and slaughtered on the spot if there is a buyer. Milk from stall-fed cows kept in urban areas is supplied direct to the consumer from churns carried on bicycles. Fodder for the cattle is imported from the countryside – often in loads balanced on the head.

Crops for cash

India is a major producer of tea, pepper, groundnuts and sugar cane, grown for domestic use and for export. Other significant cash crops are coffee, coconuts, cotton, jute and tobacco. Coconuts provide both food and fiber (coir), most of the latter for export. Sri Lanka is a particularly important producer of this commodity; it also grows pepper and other spices for export. Coffee production, for the domestic market, is restricted to south India; rubber, produced on larger plantations in both south India and Sri Lanka, is sold locally and exported.

Tea is grown on large estates in the mountains of northeast India, northern Bangladesh and Sri Lanka. In the past these estates were notorious for their abuse of migrant laborers, and they still pay their workers extremely low wages.

Timeless agriculture (*above*) Farming technology in much of the region has altered little for thousands of years. Many operations, such as winnowing wheat, are still performed by hand.

The farming year (*below*) In the Ganges delta a pattern of farming has developed to exploit the high rainfall of the monsoon season, when fields can be flooded for rice planting and growing.

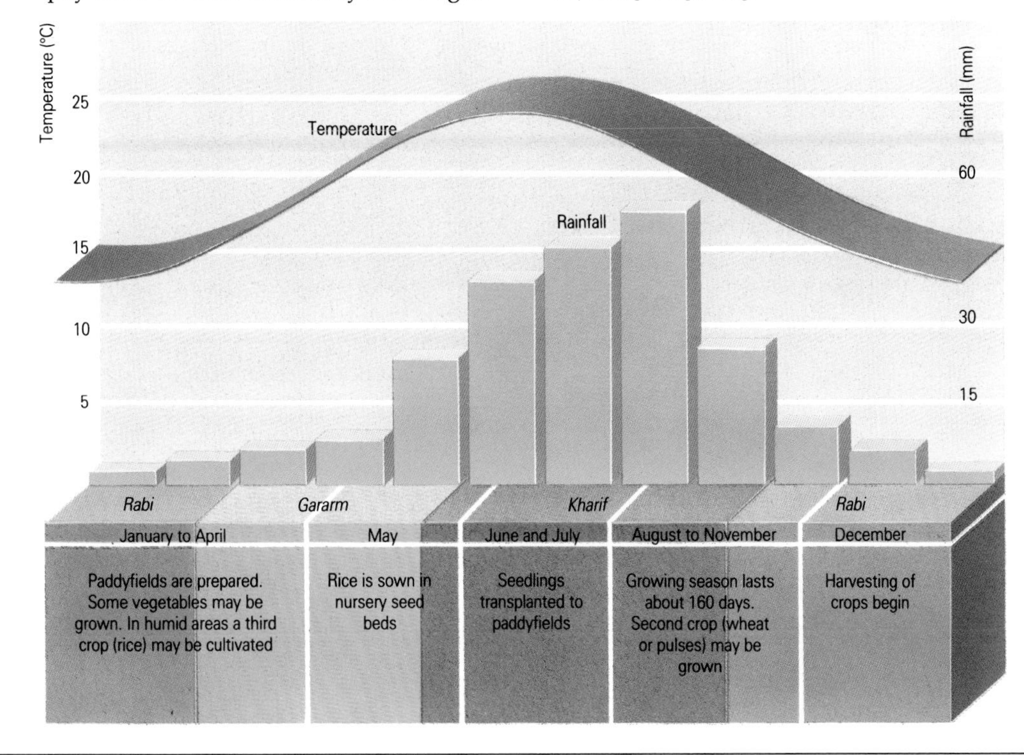

THE THREE-SEASON YEAR

The natural vegetation of much of the Indian subcontinent is tropical deciduous; the trees drop their leaves not (as in temperate regions) when it is cold, but in the hot season when conditions become arid. This characteristic cycle is reflected in the agricultural practices of the region, where there are three distinct seasons of the farming year: *kharif*, *rabi* and *garam*.

The season of rains, known as *kharif* – lasting from late May or early June until September – is the time for planting wet, or monsoon, rice: it is grown in flooded paddyfields, standing in depths of 5 to 10 cm (2 to 4 in) of water. The rains are then followed by the cooler, dry *rabi* – from October to about late February – when cereal crops such as wheat, requiring smaller amounts of water and plentiful sun, and pulses are grown. The third season of the year, *garam* – meaning "hot" or "heat" – lasts from the end of February until June. In the Ganges valley a searing wind known as the *loo* may start: if crops are not gathered in by then they will become parched and withered. In some humid coastal areas it is possible with the aid of irrigation to cultivate a third crop, usually rice, in this season.

FOOD FOR THE MILLIONS

Since 1947, when India and Pakistan gained their independence, the population in each country within the region has at least doubled, putting immense strain on food self-sufficiency. In India, for example, the population grew at an annual rate of 2.8 percent for the first 20 years after independence, unmatched by similar increases in food output. By the mid 1960s several areas had experienced near famine and the country was dependent on cereal imports, mostly of wheat from North America. Population growth rates have since declined, and the rate of increased food output has begun to rise; as a result, imports are no longer needed in most years.

Pakistan, too, started to face food deficits in the 1960s as government policy to keep food prices low in the cities depressed farm output. In the early 1980s, however, the trend began to reverse, and there was a dramatic rise in rice production. By the end of the decade it was accounting for 8 percent of all exports.

The increases in food production were largely due to the adoption of new high-yielding strains of cereal plants and new fertilizers, but the Green Revolution, for all its successes, has had mixed results. The new varieties of dwarf rice, with their shorter stems for greater strength so that the ear can grow fatter without the plant falling over, proved unsuitable for areas that were frequently deeply flooded. Many of the early new varieties were susceptible to pests and diseases. Not only was this an inherent characteristic, but because the varieties were not constantly rotated, as they are in traditional farming methods, conditions were created that allowed nonspecific diseases to flourish, necessitating the use of even more pesticides and fungicides.

In the initial stages of the Green Revolution it was thought that the new technology, together with the increased cost of pesticides and other inputs, would impoverish small landholders and benefit larger ones, who would be able to amass even larger holdings at the expense of the unsuccessful. To some extent this has been circumvented by a great drive to improve rural banking and give small farmers access to credit. Cooperative irrigation schemes have increased the supply of water. Although many small farmers did fail, the technology has been adopted by small and large alike. But progress in general education is needed too – an illiterate farmer cannot read the instructions on a packet of chemicals; neither can he understand the written conditions on a banking loan. Although slow, but real, advances have been made in these areas, anxieties must remain over the long-term ecological implications of imprecise chemical use.

Acting as a brake on the speed of progress is the overwhelming amount of rural poverty. While agricultural productivity remains low, rural incomes cannot rise. Many people in rural areas are therefore dependent on cash remittances from members of their family living in the cities. About half the rural population of India and Pakistan live below the poverty line (and it is a low one); in Bangladesh 90 percent do so.

A mixed pattern of success

The increases in food output since the Green Revolution have been greater in some areas than in others. Particular crops have benefited more conspicuously than others, too. In India the production of protein-rich pulses such as gram has

FARMERS' TRADITIONAL KNOWLEDGE

When modern scientific methods of farming fail to achieve the results that are expected of them, the peasant farmer is often blamed for lacking the necessary skills to put them into effect. Such a view overlooks the wealth of knowledge that all farmers, even illiterate ones, possess. Increasing efforts to study and understand traditional farming behavior is leading to a growing awareness that it may often be as soundly based as scientific research, if not superior to it.

Most agricultural research programs have been, and still are, carried out by scientists who specialize in one crop only, and devote their effort to improving just one part of that crop – for example, the grain head of a specific cereal. However, peasant farmers look for more than one benefit from a crop. The new short-stemmed rice varieties, for example, produce less straw (used as cattle feed) than the traditional ones, and it does not keep as well: these are strong reasons for farmers preferring the older varieties.

Furthermore, many new crops are grown as monocultures, which have to be treated with specific insecticides and herbicides. However, traditional intercropping, in which two or more crops are grown simultaneously on the same plot, inhibits the spread of disease without the farmer needing to resort to chemicals. Different plants growing within a field or plot may form a small ecosystem that has many side benefits for the farmer. The ragged and irregular crops growing in mixed fields throughout the subcontinent may actually be doing much better than the scientist, who tends to prefer uniformity and order, supposes.

Rice: the staple food (*left*) Rice is grown all over the subcontinent wherever there is enough water to do so successfully. Here a farmer in the rainy coastal plain of Tamil Nadu in the south of the peninsula is preparing a field for planting in the time-honored way, using oxen for plowing.

Irrigation technology (*below*) Irrigation is one of the most ancient agricultural arts and has long been used in the region: many irrigation systems have hardly changed over the centuries. The increasing use of wells lowers the water table, so that diminishing water reserves become more difficult to exploit.

stagnated, causing the quality of many people's diet to decline. The Green Revolution rice varieties proved much better adapted to the hot, dry areas of Pakistan and northwest India, where wheat production also showed dramatic improvement, than to the traditional rice-growing areas in the wetter east. Within India, the states that have the greatest areas of poverty and need do not have the purchasing power to buy surpluses from the more fortunate states. Consequently there is considerable government intervention in grain trading between states.

Bangladesh has always traded jute in exchange for imported rice. As its population rises its position becomes increasingly precarious, but significantly it did manage to achieve self-sufficiency for the first time in 1989 during a very good year. Here, more than anywhere in the region, the benefits of the Green Revolution have been most thinly spread, and increases in rice yields have proved hard to achieve. Wheat, very little grown in the past, is becoming increasingly important as a second (winter) crop, following the monsoon rice. Even so, wheat – the most widely available traded grain – accounts for more than 7 percent of its imports, and vegetable oils a further 5.7 percent.

Greater cropping intensities

India and Pakistan now achieve food self-sufficiency in most years, and their food supply in the near future would seem assured. But if increases in yields should taper off, as rice yields in Pakistan are doing at present, and population growth does not slow down still more, the longterm situation is by no means secure. There is little remaining land to bring into cultivation. However, in many parts of the region some land is cropped more than once a year, and efforts are being made, by increasing the extent of irrigation schemes, to make the practice more widespread.

Irrigation can bring with it many problems, however. In many areas large canal schemes have proved very difficult to run efficiently, with water never reaching farms at the tail end, and supplying other farmers in unpredictable quantities at irregular times. Irrigation can also cause damage to the land through waterlogging and salinization: some 22 percent of land irrigated by Pakistan's canal system is now unusable because of salinity, and 13 percent in India.

Growing jute for the world

Preparing the fiber The long thin stems of the jute plants are soaked in any available piece of water to soften the bark. This is then laboriously stripped away by hand.

Jute is the common name of two species of a tall, woody plant (*Corchorus capsularis* and *C. olitorius*). It is similar to a tall reed or rush that, when cultivated and cropped each year, reaches a height of between 3 and 4 m (10 and 12 ft). It grows throughout the subcontinent, but is best suited to the warm, moist conditions of Bengal in northeast India and Bangladesh.

Its bark produces an excellent fiber for sacking, rope making, and carpet and linoleum backing. It is separated from the woody jute stem by soaking (or retting) it in water in a pond or specially built tank for about twenty days. The fiber is then stripped by hand before being sold to a bulking merchant, who presses it into 180 kg (400 lb) bales for export.

The plant has other uses as well. The woody stem is suitable for making fences, for the base layer of roof thatch, and for a fuel to cook with. It also makes perfect charcoal for gunpowder, which gave it great value in the past. The fresh green leaves of one species can be eaten, and are somewhat similar to spinach.

Jute has always been part of the rural economy of the Indian subcontinent. It rose to commercial importance when India was under British rule in the mid 19th century; the jute grown in Bengal became the world's leading source of fiber for sacks, required for carrying the burgeoning international trade in grains, cotton and wool. Most of it was shipped to Scotland, in northern Britain, for manufacture; the town of Dundee is still a center of the jute trade. Manufacturers there are in regular touch with the jute auction houses in Calcutta, where a manufacturing industry also grew up under the British.

Trade rivals

The partition of Bengal in 1947 between India and East Pakistan (later to become Bangladesh) abruptly severed the relationship between Calcutta and the jute suppliers of East Bengal, and the situation was exacerbated by the subsequent trade war between India and Pakistan. Production increased in both countries; jute grown in East Pakistan was Pakistan's most important export in the years before the province gained its independence as Bangladesh in 1971. It also commanded 60 percent of the world jute market.

However, world demand has consistently declined in recent decades because of the growing use of synthetics in carpeting and the replacement of sacks by bulk containers in the international carriage of goods. However, jute and jute products still account for about 30 percent

A jute mill in Bangladesh (*above*) Bundles of jute fiber are prepared for spinning into coarse rope. A large proportion of the raw fiber is shipped to processing plants in Western Europe and Japan, where it is used to provide backing for carpets and linoleum flooring. The growing use of synthetics has seen a fall in demand, but jute still remains an important export.

Exhausting work (*left*) An umbrella has been tied to a jute stem to provide shelter from the sun for a farmer weeding his plot of young jute plants. Jute is an annual herbaceous plant; it requires humidity and an average monthly rainfall of 7.5–10 cm (3–4 in) to reach its full height of 3–4 m (10–12 ft), and is therefore ideally suited to the climatic conditions of West Bengal and Bangladesh.

of Bangladesh's total export revenue; this is matched in importance only by the finished garment industry. However, its share of the world market has fallen to less than 40 percent, and it has been overtaken by India and China.

Although jute cultivation remains very large in terms of the volume produced and the value of export revenue earned, it is not run by big businesses or carried out on extensive plantations. It still remains the main support of a very large number of small peasant producers, who cultivate it for cash on their smallholdings and ret the bark in the nearest pond. Its cultivation is in direct competition with rice for the farmer's land and labor. If improved Green Revolution varieties begin to increase rice profitability in east India and Bangladesh, jute growing will probably decline still further in both countries.

The vegetable market

Vegetables are an extremely important element of people's diet throughout the Indian subcontinent: in India about half the population are vegetarian.

Most of the region's peasant farmers have only tiny surpluses of produce for sale at any one time. As a result the marketing network, though appearing unsophisticated, is quite complex and may involve several links. Typically, a farmer will deliver a small amount for sale to his local market. This is usually within walking distance of his farm, and he will often carry what he has for sale in a basket on his head.

Each local market will have specialized areas for produce within it, where local people can buy what they need for their own use. There will also be a large number of middlemen who visit the different producers, purchasing enough from each of them in turn to make up a truckload. This is then delivered to other wholesalers and retailers in nearby towns and cities. Each of India's and Pakistan's large cities is surrounded by a complex web of collection routes, which radiate out over wide areas of the countryside.

Mountains of cauliflowers in a section of a vegetable market in Rawalpindi, Pakistan.

FEEDING ONE-FIFTH OF THE WORLD

THE TRADITIONAL PATTERN · THE CHANGING ORGANIZATION OF AGRICULTURE · A DELICATE BALANCING ACT

China covers only one-fifteenth of the world's land area, yet it succeeds in feeding all its people, totaling one-fifth of the world's population. Its principal problem is in ensuring that agricultural production – particularly of the key grains, wheat and rice – at least matches the needs of this large and growing population, which in 1987 was 1.08 billion and by 2000 is expected to be 1.2 billion. Failure to increase production will result in hunger, malnutrition and famine for the Chinese, most of whom live on a subsistence diet. Some 80 percent of the people belong to peasant or rural households, and their lives are still lived in ancestral villages growing their traditional crops. However, since 1949 communism has significantly transformed Chinese agriculture, and rural life is now experiencing fundamental changes.

COUNTRIES IN THE REGION

The People's Republic of China, Taiwan

Land (million hectares)

Total	Agricultural	Arable	Forest/woodland
933 (100%)	414 (45%)	94 (10%)	117 (12%)

Farmers

451 million people employed in agriculture (69% of work force)
0.2 hectares of arable land per person employed in agriculture

Major crops
Numbers in brackets are percentages of world average yield and total world production

	Area mill ha	Yield 100kg/ha	Production mill tonnes	Change since 1963
Paddy rice	32.7	54.1 (165)	177.0 (38)	+106%
Wheat	28.8	30.5 (131)	87.8 (17)	+295%
Maize	20.3	39.5 (109)	80.1 (17)	+254%
Roots/tubers	9.2	158.7 (126)	146.1 (25)	+32%
Soybeans	8.5	14.4 (76)	12.2 (12)	+14%
Rapeseed	5.3	12.5 (88)	6.6 (29)	+538%
Cotton lint	4.8	8.8 (160)	4.2 (26)	+246%
Millet/sorghum	4.6	22.0 (192)	10.1 (11)	−41%
Pulses	4.4	12.1 (150)	5.3 (10)	−27%
Vegetables	—	—	110.0 (26)	+134%

Major livestock

	Number mill	Production mill tonnes	Change since 1963
Pigs	344.6 (41)	—	+75%
Sheep/goats	166.5 (10)	—	+41%
Cattle	71.3 (6)	—	+16%
Milk	—	3.4 (1)	+25%
Fish catch	—	9.6 (10)	—

Food security (cereal exports minus imports)

mill tonnes	% domestic production	% world trade
−8.6	2	4

Irrigated agriculture Water from the foothills of the Himalayas is diverted into narrow channels to irrigate crops in this terraced upland valley in Yunnan province in southern China. The majority of Chinese still live in villages like this one, engaged in the centuries-old cycle of crop cultivation.

THE TRADITIONAL PATTERN

Much of China, particularly its northern and western plateau, supports at best poor scrub pasture; only about 10 percent of the country's vast land area – the third largest in the world – is suitable for arable cultivation. Such areas are mainly in the valleys and plains in the east and south, in what is historically termed Inner China. They are farmed by the Chinese, or Han, people, while nomads, including the Mongols, Uighurs and Tibetans, tend their herds in the north and west, in Outer China.

Outer China is a land of plateaus, high mountains, and a number of basins in which irrigated cultivation is possible. Rainfall is limited, and much of the land is desert or has poor sandy soils. It comprises about half the country's land area, but supports only 6 percent of its total population.

Since the communists succeeded to government in 1949, the nomads have been under pressure to form permanent settlements, and the Han migrants have created a market, stimulating increased irrigated crop production. The pressure to increase regional self-sufficiency in grain production caused many areas unsuitable for tilled agriculture to be plowed, and soil erosion has become a major problem. More recently, however, the traditional role of these specialist meat-producing areas has been recognized.

Land of wheat and rice

Inner China has a very different agricultural landscape. Han culture emerged in the lands around the upper Huang (Yellow) river, and was based on wheat and rice production. Over 4,000 years this

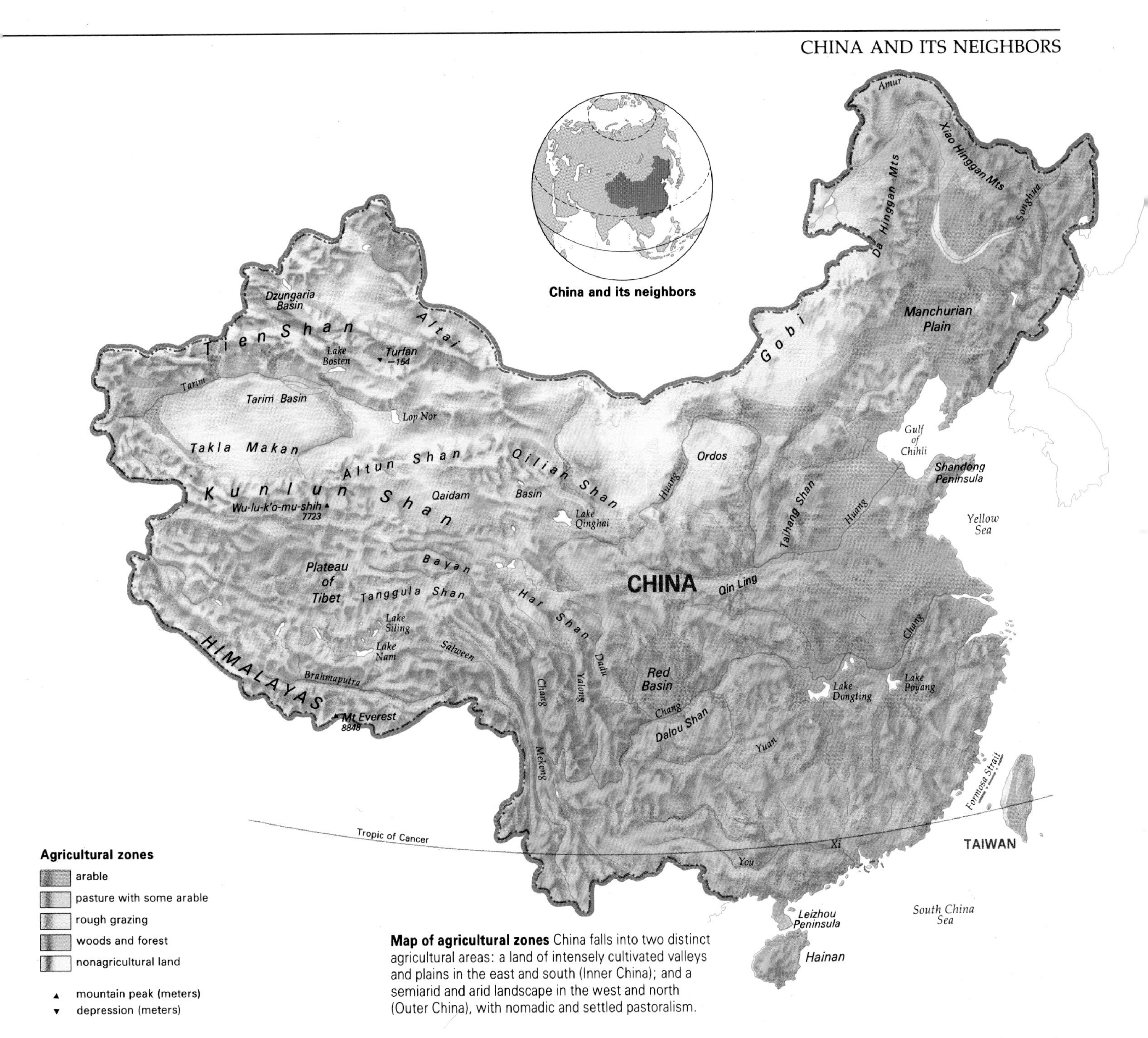

Map of agricultural zones China falls into two distinct agricultural areas: a land of intensely cultivated valleys and plains in the east and south (Inner China); and a semiarid and arid landscape in the west and north (Outer China), with nomadic and settled pastoralism.

culture has spread east and south to colonize and control Inner China. Intensive agriculture was practiced by family farmers on small plots of land; the main crops were grains and vegetables.

Water control, to prevent flooding and provide irrigation, has become an important feature in this area. The valleys and hills were once cloaked with forests, but these were steadily cut back as Han Chinese colonized the land, and now only 10 percent of China is forested, compounding the problems of severe soil erosion.

The north–south divide

Within Inner China there is an important agricultural contrast that is marked by the Chang (Yangtze) river and the Qin Ling mountains. The northern part is dominated by level plains and elevated plateaus and has low, unreliable rainfall. This is predominantly a wheat-producing area, with maize, millet and potatoes; the loose, limy soils are generally not suitable for paddy rice. However, the introduction of quick-growing rice varieties has enabled some rice production here.

South of the Chang the growing season increases, rainfall is more abundant and reliable, and the generally acidic soils are well adapted to paddy rice. Here are found the landscapes of plains and terraced hillsides that make up most outsiders' image of China. Life revolves around the seasonal demands of rice cultivation, which is very labor intensive.

Animals have always played a limited role in the agriculture of Inner China. These were mainly restricted to draft animals such as water buffaloes, and to scavengers such as pigs, chickens and ducks. These would eat remnants of family food and provided – together with fish – an occasional but important addition to a grain- and vegetable-based diet.

This agricultural system was in many ways very efficient. Although an ancient agriculture, it was not an unchanging one. Improved techniques for irrigation stimulated specialist crop production of tea, tobacco, mulberry trees (for silk production) and fruits such as lychees. These became important specialist crops, particularly in parts of south China. From the 16th century maize, potatoes, groundnuts and tobacco were introduced from the Americas, enabling more sandy, less productive soils to be brought into cultivation. However, life for many people remained very precarious; the traditional greeting "Have you eaten?" provides some indication of just how close life was to the margins.

THE CHANGING ORGANIZATION OF AGRICULTURE

China's agriculture has traditionally emphasized grain and vegetable production. Today it is one of the world's largest grain producers. In 1986, for example, China's leading crops included rice (177 million tonnes), wheat (89 million tonnes) and maize (65.5 million tonnes).

Rice is the largest crop by extent of sown area, by production and by importance to the diet. Most strains of this tropical grass require large supplies of moisture; the paddy fields are flooded to a depth of some 10 cm (4 in) during much of the growing season. It also requires very heavy labor inputs – to create the paddy fields and dams, for weeding, planting and transplanting the young plants, and for harvesting the crop. Water buffalo are extensively used to prepare the rice fields for cultivation. Dry upland rice strains are important in moist areas that cannot readily be flooded. In mountainous parts of south and central China terracing has created extra land for both types of rice.

Before 1949 most rice was produced by family farmers for their own consumption; even during the period of collective agriculture most rice was eaten locally. However, historically and since 1949 all Chinese governments have tried to ensure that enough rice and other food grains are sold through state and private markets to feed the urban population.

Wheat is China's second major crop, and has traditionally been important in the drier conditions to the north of the Chang river. Its range has been extended to south China, where wheat is now grown as a winter crop in annual rotation with rice. Millets, sorghum and maize are also grown.

Much of China's vegetable crops are grown for the farming families' own consumption, though in areas close to large towns they are significant cash crops. More specialized crops, such as teas, fruits and cotton, are locally important; and more of this production is sold through state and private markets, both domestically and for export.

Livestock production still plays only a small part in Chinese agriculture. Pigs, however, which are efficient scavengers and converters of vegetable matter into proteins, are widespread. Even at the height of collectivization in the 1960s most families had one or two pigs; these and the family's chickens provided the limited amount of meat in the diet.

The need for change

The success of traditional Han Chinese agriculture lay in the high productivity it achieved per hectare and in its ability to increase productivity, at the same time accommodating the growing work force. This enabled it to feed a rising population, which had reached 400 million by the mid-19th century. The reasons for its success were perhaps also its fundamental weakness: productivity per person was low, and only limited mechanization had been achieved. There was a narrow margin between a harvest that gave the household bare sufficiency and one that led to indebtedness and hunger.

In the 1950s the communist government carried out a major program of land reform, redistributing land, animals and machinery from richer to poorer households – the beginnings of collectivization. In 1958, as part of the Great Leap Forward, rural China was organized into large-scale collective units (communes). These were subdivided into brigades, and then into production teams. Generally the team – often the same as the traditional village community – was the basic unit that organized household labor. Team leaders assigned household members to various tasks, each being worth so many work points.

A vivid green carpet *(above)* Growing rice in the paddyfields requires a great amount of labor. Most of it is still carried out by hand, using draft animals – water buffaloes – for motive power.

Weighing radishes *(right)* After the needs of the farming family have been met, surplus vegetables are sold in local markets. During collectivization, most households were allowed a private plot where they could grow vegetables, which are integral to the Chinese diet.

AGRICULTURE IN TAIWAN

Taiwan, though part of the Chinese cultural area and similarly oriented to rice production, now has a markedly different agricultural system. Between the years 1948 and 1953 the government of Taiwan undertook a radical program of land reform that distributed land equitably to family farmers. Economic aid from the United States enabled government agencies to ensure that these farmers were supplied with capital to invest in fertilizers, irrigation equipment, machinery, and improved strains of rice and other crops. Agricultural productivity increased, and profits helped to finance the industrialization on which Taiwan's economy has increasingly depended.

The agricultural economy has become correspondingly less important. In 1952 32 percent of all goods and services originated in agriculture; by 1986 the figure had fallen to 5.6 percent. Many people left the land for urban-based industrial and service sector employment. Most agricultural production is now for the domestic market. In 1987 the principal crops were sugar cane, vegetables and rice; fruit, beef and pork are also key products. Government policy aims to protect agriculture from competitive world market prices, while encouraging a second phase of land reform in which small family farms are being converted into large, mechanized commercial units.

Devolved decision-making

Once or twice a year households received cash income and grain supplies from the production team – the amount depending on the harvest from the collective fields and the number of work points the household had achieved. In addition, households were generally allowed a very small private plot on which they could grow vegetables and probably keep a pig. The decisions about when to plant the crops were made by team leaders; what to plant was decided by commune leaders and local officials in accordance with production targets set by central government and the Chinese Communist Party (CCP) officials.

The targets set each year for the various crops, particularly for wheat and rice, had essentially to be met. The state determined the price at which it would buy the produce; any surplus might be bought back by team members at a higher price or be distributed among them.

During "slack" periods – when there was little or no work to be done in the fields planting or harvesting crops – teams, brigades and commune labor could be mobilized to transform the land and increase agricultural production. The construction of irrigation and flood control works both large and small, for example, were undertaken by collective labor. Such tasks were ordinarily far beyond the labor or capital capacities of an individual household. Similarly, the collective could afford to purchase the machinery, such as small tractors, needed to modernize China's agriculture.

After 1978 the dominant group within the CCP totally changed agricultural and rural organization yet again. The collective fields were broken up and the land distributed on long-term leases to peasant households. Although the state still guided agricultural production – through setting production targets and quotas and defining the price at which it would buy the various crops – the "free market" was given room to shape what peasants grew and how they marketed the produce.

At first this resulted in significant national increases in production – though good weather certainly helped. However, the long-term problem of increasing productivity has remained; the destruction of the collective system means that there is no effective organization to mobilize the labor and capital to achieve technical modernization. The private market may

The collective system at work A team of laborers prepares the paddyfields for later planting. Despite the changes of direction in agricultural organization and policy, production has risen since 1949. The decision to encourage peasants to abandon the countryside for the towns may alter this, however.

tap the initiative of individual families to produce more, but those who gain most will be the richer peasants with larger landholdings and those living in areas close to larger towns, who can concentrate on vegetables and other cash crops. This will increase the divergences in wealth between different areas of the country, and between rich and poor peasants. The market system may also stimulate such cash crops as vegetables, tobacco and cotton, but it is less certain that it will result in increased production of the key grains needed to feed China's ever-increasing population.

A DELICATE BALANCING ACT

Balancing the needs and interests of urban and rural dwellers has been a constant problem since the communist government came to power in 1949. The CCP has often stated that agriculture should be a priority, and indeed it was singled out as one of the Four Modernizations in the economic reform program in the early 1970s.

The rhetoric, however, has not been matched by action. Only limited state investment has gone into agriculture, and this has been mainly in the form of fertilizers and machinery, the costs of which the state has recouped by charging the peasants high prices for these inputs while keeping down the price that the state pays the peasants for their produce. If the state were to increase grain and other crop prices it would stimulate increased production, but this would also result in higher prices for urban consumers and less capital being available for industrialization programs.

China's attempts to modernize its agriculture continually come up against a problem familiar in so many other Third World states: capital for investment is limited, many demands are made on it, and industry generally offers both higher and quicker returns on capital investment than agriculture can provide.

Significant modernization

Despite these problems there has been significant modernization in Chinese agriculture since 1949, and production, particularly of grain, has increased. This expansion has been assisted by the use of new higher-yielding seed varieties – especially of rice, wheat and maize – that have been developed both locally and abroad. These new plant hybrids, together with the greater use of fertilizers and irrigation, have enabled more land to be devoted to intercropping (growing two or more different crops together) and to the production of several quick-ripening crops during the year.

Traditional agriculture had maintained

Holding up the traffic Nomadic herdsmen with their flocks on a remote road in Xinjiang in Outer China. Cultivation is mostly impossible on these wind-swept mountains and plateaus, where the summer heat is intense, the winters cold and there is year-round aridity.

high productivity with the help of heavy inputs of organic fertilizers, including human and animal manure. However, since 1949, when many small fertilizer plants were built, and especially since the 1970s, the use of chemical fertilizers has steadily increased.

Productivity has also been improved by the introduction of many water control schemes to alleviate the effects of uncertain monsoon rains, which can bring floods or drought. Electricity has been extended to many rural areas; more than half the total arable land in China is now irrigated, using electric or diesel pumps – the world average for irrigated land being just over 10 percent.

Despite efforts to reclaim new land for cultivation and the use of improved techniques, the amount of new land that has profitably been brought into production is very limited. Indeed, it is likely that overall there may have been a small net loss of arable land because of industrial and urban expansion. Lack of attention to environmental protection has led to many lakes being silted up as a result of soil erosion, and uncontrolled chemical emissions have imperiled the fertility of many lakes and ponds.

Improved transportation (particularly the use of small tractors) has enabled the better marketing of peasant crops and livestock to develop. However, the mechanization of field operations has been minimal; long, hard and intensive labor remains the norm. Even though many aspects of rice production can be mechanized while maintaining small family farms – as in Taiwan, Japan and Korea – labor in China is not in short supply, and here mechanization is unlikely to make significant improvements to overall productivity.

The move to the towns

The government's current policy, however, is to reduce the agricultural workforce, and its target is for a third of all peasants to abandon farming and seek productive work in industry in nearby towns. This will put further pressure on agriculture to feed not just a growing population, but an increasing number of urban dwellers in what promises to be one of the largest-scale movements of people from rural to urban areas in the history of the world.

Agricultural production will need to rise sharply to feed this large and growing urban population. In the 1980s farmers were encouraged to produce more for sale in urban markets by the removal of price controls and by allowing demand to set the prices for goods such as pork and vegetables. Although this succeeded in raising production, it also resulted in inflationary prices. Between 1988 and 1989 vegetable prices in the cities increased by a third; this was one factor causing the urban unrest that resulted in demonstrations in Beijing and other major cities in June 1989.

KEEPING FAMINE AT BAY

In 1980 Deng Xiaoping, China's leader, stated that "the basic problem facing the Chinese people is having enough to eat". This uneasy balance between levels of food production and rising population numbers has been a constant factor throughout China's history. Between 1949 and 1990 the population rose by more than 500 million – an increase that exceeds the combined total populations of the United States and the Soviet Union.

Wheat and rice provide 90 percent of the calories and 85 percent of the protein in the Chinese diet. Production of these grains rose between 1955 and 1977 – the years of collectivization – but not sufficiently to increase consumption per head. This remained low, on a par with many Third World countries.

China has always been vulnerable to famine caused by natural disasters such as flood or drought. In 1949, with the central planning and modernization programs of the communist regime, many hoped that this would no longer be the case. But between 1959 and 1961 China suffered one of the worst famines known in human history, which caused between 25 and 30 million deaths. Flood and drought undoubtedly played their part, but most experts now blame the scale of the disaster on the government's attempts to push through its collectivization program.

Since the early 1970s strict birth control policies have limited the growth of population. But the early successes in increasing food production through the freeing of the market have not been sustained, and China's vulnerability to the vagaries of the climate remains. In 1981 a major famine after the failure of the monsoon rains was averted by buying grain on the world market and invoking international aid.

The success of aquaculture

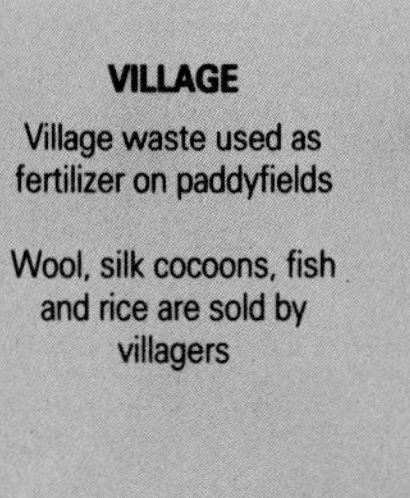

Fish provides an important supplement to the grain-based Chinese diet, and China is also one of the world's leading deep-sea fishing countries. Each year it lands many tonnes of fish caught in the seas off east and south China. In addition, as much as a third of total fish consumption is met by aquaculture – the raising of fish and other marine and freshwater animals in controlled coastal or inland environments.

The first known treatise on aquaculture, or fish farming, was written by Fan Li in 473 BC. Thereafter the cultivation of fish – particularly carp – was developed as an important source of protein throughout China. It was and is particularly important in the lower Chang river (Wuxi is a major center) and along the lower Xi (Pearl) river. Fish yields in the south of China are higher than in the colder north; in the extreme south, in the New Territories of Hong Kong, 1 ha (2.5 acres) of fish ponds makes twice as much money as a paddyfield of similar size.

Freshwater fish are cultivated in many of China's inland lakes, rivers and ponds, where their environment can be readily controlled and supplementary food given. Fish are sometimes produced in close combination with other agricultural products. In certain areas, for example, young fish are placed in paddyfields just after the rice seedlings have been transplanted into the fields. The fish eat some of the insects that threaten the young rice plants, and they stir up the mud, releasing nutrients that are taken up by the plants. Fish excrement also benefits them. However, to combine rice and fish production conditions have to be ideal: the water supply, for example, must be assured and the drainage effective.

The drive for efficiency

The general drive for increased productivity since 1949 has encouraged aquaculture to develop as a specialized activity in some areas. In others it has been closely integrated with complementary rural activities. Near Suzhou in the lower Chang delta, for example, there are important silk-producing areas where sheep and fish are also raised as sidelines.

In this effective combination of activities, silkworms feed on the mulberry trees, whose leaves in winter are fed to the sheep. The sheep droppings provide mineral-rich fertilizer for the trees, and the organic waste products from the silkworms and mulberries provide extra nutrients for the fish ponds. Other organic fertilizers such as grass, pig, chicken and duck manure (and in controlled conditions human excrement or "night soil") are also recycled as fish food. When the fish ponds are drained in order to harvest the fish the fertile mud can then be spread onto the paddyfields.

Many new practices, closely geared to the needs of peasant farmers, have been encouraged by the government. Many are quite small in scale – such as artificial fish spawning in specialized pools and the use of small hatching areas – and can readily be adopted by peasant farmers using simple techniques they can operate themselves. However, the expansion of aquaculture into new areas is hindered by the lowered fertility of many lakes as a result of industrial pollution.

Aquaculture is an excellent example of a long-established form of agriculture that has benefited from scientific research and modern organization. Because it can be closely geared to the needs of peasant farmers, it could be easily integrated into the agricultural organization of other countries as well, offering an efficient way of increasing world food production.

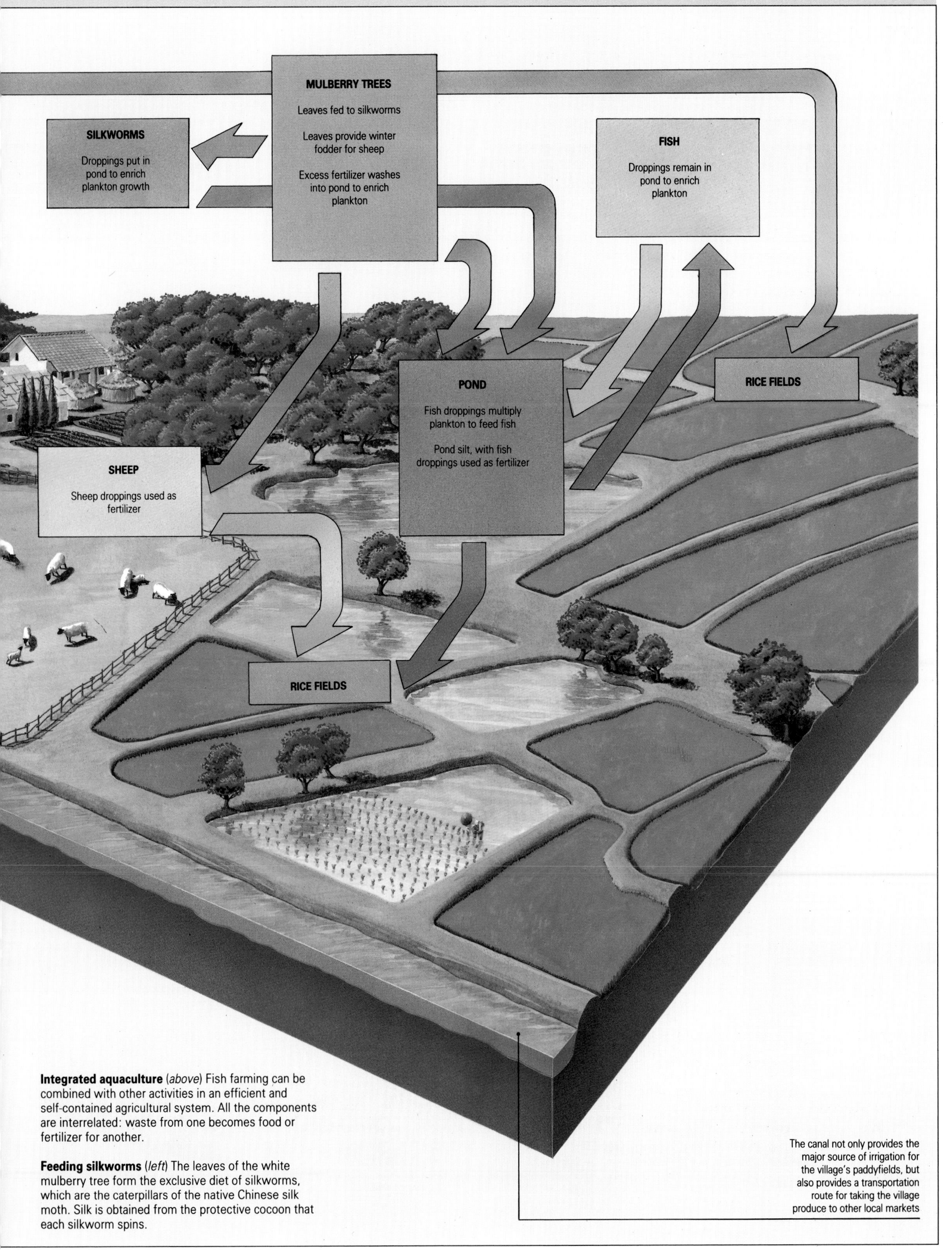

The canal not only provides the major source of irrigation for the village's paddyfields, but also provides a transportation route for taking the village produce to other local markets

Integrated aquaculture (*above*) Fish farming can be combined with other activities in an efficient and self-contained agricultural system. All the components are interrelated: waste from one becomes food or fertilizer for another.

Feeding silkworms (*left*) The leaves of the white mulberry tree form the exclusive diet of silkworms, which are the caterpillars of the native Chinese silk moth. Silk is obtained from the protective cocoon that each silkworm spins.

Farming on the roof of the world

The village threshing floor at Samyai, south of Lhasa in Tibet, is a center of activity as the cereal harvest is gathered in. First of all the grain is separated from the stalks of the crop by beating them on the ground with simple handtools (sometimes oxen or water buffaloes are used to trample the plants). The grain is then winnowed to remove the seed husks, or chaff. One method is to throw the grain up into the air so that the wind carries off the chaff, but here the grain is being tossed in a simple drum-like machine; most winnowers of this kind are turned by hand, but this one has instead been hitched up to a rotavator to power it.

The cleaned grain is then spread out to dry in the sun. The threshing floor needs to be clean and free of cracks. Animal dung mixed with mud and baked in the sun makes a very good surface for a threshing floor, though nowadays concrete is often used.

Even with the improvised mechanization being used here, a great number of people are needed to complete all the work of the harvest. Women play an equal part in all stages of agricultural production in China and in Tibet, to which Chinese methods of organization and management were introduced following the invasion in 1951.

High mountain peaks make a dramatic backdrop to this scene of harvest activity in Tibet.

DIVERSITY IN THE TROPICS

FROM EARLY BEGINNINGS · SMALLHOLDINGS, PLANTATIONS AND SWIDDENS · PRESSURE FOR CHANGE

The complex interaction of environment, economics, politics and society gives Southeast Asia its variety of agricultural systems. These range from the capital-intensive estates of the Malay Peninsula, where plantation crops such as rubber – introduced from South America by British colonists – and palm oil are grown for export, to the socialist agriculture of Vietnam, where the state maintains a firm hold on production and marketing, and on all other aspects of rural life. Extensive upland areas and unproductive forest soils mean that much of the region is unfavorable for farming. The mountains of Borneo, for example, remain largely forested; farmers are shifting cultivators, maintaining a traditional farming system in which they clear a forest plot and grow crops for a few years before abandoning it.

COUNTRIES IN THE REGION

Brunei, Burma, Cambodia, Indonesia, Laos, Malaysia, Philippines, Singapore, Thailand, Vietnam

Land (million hectares)

Total	Agricultural	Arable	Forest/woodland
436 (100%)	90 (21%)	59 (13%)	240 (55%)

Farmers

97 million employed in agriculture (54% of work force)
0.6 hectares of arable land per person employed in agriculture

Major crops
Numbers in brackets are percentages of world average yield and total world production

	Area mill ha	Yield 100kg/ha	Production mill tonnes	Change since 1963
Paddy rice	35.2	28.5 (87)	100.2 (22)	+103%
Maize	8.3	15.8 (44)	13.2 (3)	+140%
Cassava	3.4	116.3 (125)	39.3 (29)	+153%
Pulses	2.0	7.8 (97)	1.5 (3)	+126%
Soybeans	1.6	10.3 (54)	1.6 (2)	+255%
Sugar cane	1.3	547.9 (92)	71.2 (7)	+143%
Bananas	—	—	9.7 (15)	+128%
Other fruit	—	—	14.3 (6)	+196%

Major livestock

	Number mill	Production mill tonnes	Change since 1963
Cattle	28.9 (2)	—	+36%
Fish catch	—	9.0 (10)	—

Food security (cereal exports minus imports)

mill tonnes	% domestic production	% world trade
+1.5	1	1

FROM EARLY BEGINNINGS

Some of the world's earliest agriculture appears to have developed in Southeast Asia. Evidence found at two important archaeological sites, at Baan Chiang in northeast Thailand and around the Red river in northern Vietnam, suggests that the change from hunting and gathering to agriculture on the mainland took place between 5500 BC and 3500 BC. The practice and technology of farming appeared later on the islands, probably having spread from the mainland; there are signs of food production in Sulawesi dating back to 2500 BC.

Rice, the key to life in most areas of modern Southeast Asia, was probably not domesticated until 3000–2000 BC. Since then its cultivation has spread throughout the region, and it has become the dominant agricultural crop. It is only in the more arid locations such as the Indonesian island of Timor that the inhabitants rely on other crops like cassava to meet their needs. These alternative staples are regarded locally as "poor man's food".

Examples of all three stages through which farming has developed – hunting and gathering, shifting cultivation and settled agriculture – are still to be found in the region. No type is more advanced than another; they are merely geared to the diverse environments and to different levels of population pressure. The commercialization of farming dominates much of the region's agriculture (though less so in socialist countries in mainland Southeast Asia); it is a recent development in most areas. Until shortly after World War II subsistence was the keynote of agriculture. Farmers used traditional technology, family labor and locally procured seeds and manures. Subsistence farming has now almost entirely disappeared. Farmers grow crops for the market: they buy and use chemicals, hire labor, use machinery, and plant newly developed, high-yielding crop varieties.

Some forms of agriculture have become industrialized following this commercialization. The forests of the region are extensively logged, and crops such as sugar cane, rubber, oil palm and pineapples are grown under industrial conditions on large estates that are often owned

Irrigated rice terraces in Bali, Indonesia reflect the evening light. These fertile volcanic lowlands are among the most intensely cultivated in the world, often bearing three crops of rice a year. Rice remains the staple crop throughout the region, as it has been for thousands of years.

Map of agricultural zones Arable land is concentrated in the major river valleys of Indochina and the lowland plains of Indonesia and the Philippines. Commercial farming of tropical crops has developed on cleared forest land; shifting cultivation is still practiced in some remote upland forest areas.

by multinational companies. The destruction of Southeast Asia's forests is now viewed internationally as an environmental disaster. The tropical forest soils are shallow; once the vegetation cover is removed nutrients are leached downward through the soil, which is then easily eroded by the wind and rain.

Rivers and rains

Southeast Asia has few favorable niches for agriculture – often a result of too much or too little water. Even on the limited lowlands farming is plagued by swampy conditions. However, suitable places do exist, particularly along the valleys of major rivers such as the Chao Phraya in Thailand and the Red in North Vietnam. The volcanic soils of Luzon in the Philippines and central Java in Indonesia are also fertile; here farmers grow up to three crops of rice a year, and agricultural population densities often exceed 1,000 people per sq km (0.39 sq mi).

Southeast Asia lies entirely within the humid tropics. Close to the Equator rain falls evenly throughout the year, and in some places exceeds 4,500 mm (177 in). However, traveling north and south from the Equator, it decreases in amount and becomes increasingly seasonal. Over much of the mainland between 80 and 98 percent of the year's rain falls within the seven months from April to October – the period of the southwest monsoon. It is this seasonality of rainfall, rather than the total quantity, which presents farmers with their greatest problems. In northern Thailand various systems of irrigation have been developed over many hundreds of years to overcome the vagaries of the rainfall. In other places farmers have developed sophisticated strategies of rainfed cultivation. They carefully select and grow local varieties that are resistant to floods or drought, matching the plants to the conditions of the microenvironment in which they will be cultivated.

SMALLHOLDINGS, PLANTATIONS AND SWIDDENS

Farming in Southeast Asia is predominantly based on wet rice production, and is carried out by family farmers on small areas of land. In Java nearly two-thirds of farm households have less than 0.5 ha (1.24 acres) of land, and yet they still often manage to produce enough for all their subsistence needs. The way in which farmers have continually managed to squeeze that little bit extra out of their land is called "agricultural involution".

Where the environment allows, they not only keep a diverse and well-stocked home garden and raise livestock, but also cultivate cash crops such as maize, cotton and cassava. Children help to herd buffaloes and cattle, and both husband and wife are involved in the day-to-day running of the enterprise. Land tends to be owner-occupied, though tenancy and landlessness are both on the increase throughout the region. In central Thailand 25 percent of land is rented, while in the Philippines some 30 to 35 percent of the rural population is landless.

Although agriculture is becoming increasingly commercial, farmers nevertheless still largely operate a strategy that will minimize risk, rather than maximize profit. Their one essential guiding principle is that a certain critical level of production must be maintained from year to year, through conditions of flood and drought. Such farmers have sometimes been criticized for falling behind technologically, but this is unfair – they are conservative. To experiment with new methods and techniques may not only be risky, but could also make all the difference between survival and starvation.

Farming for profit

Plantation systems differ radically from smallholdings. They are large, nontraditional, highly commercialized and geared to export. They also use new scientific techniques, employ large numbers of wage laborers and are organized and run as commercial enterprises, not as family concerns. What determines decision making is the profit margin – and because the crops are sold all around the world, profit is related to prices in the fluctuating international economy.

Plantation agriculture accounts for over two-thirds of all cultivated land in Malaysia; rubber and oil palm alone contribute 18 percent of total exports. But the men and women who work on these largescale plantations do not own the land. To counteract the social and political disadvantages arising from this, governments have been promoting smallholder production of the traditional plantation crops. In Malaysia, nearly 80 percent of land planted to rubber is now operated by smallholders. They often do not have the capital to invest in modern cultivation techniques and processing facilities, but building central processing plants has married the social and political advantages of smallholder production with the agricultural and economic advantages of estate management. Unfortunately, smallholder production is of very variable quality, which makes processing considerably more complex.

At the other end of the political spectrum is the state-controlled agriculture of Vietnam. In the early years of the communist regime, huge state-owned communes were set up, each supporting some 5,500 farming families. These stifled incentive: available figures suggest that commune members produce over 50 percent less than private farmers on the same amount of land. In 1988 a disastrous rice harvest threatened the northern provinces with famine, and the government was forced to reform the system, allowing families tenures of 15 years or more on the land they farmed. They now pay tax based on the fertility of the land, decide what to grow, and sell their surpluses on the free market.

Productive in its own way

Shifting cultivation is another common farming system in Southeast Asia. Land is partially cleared through burning, and crops such as hill rice are grown in the fertile ash. After the temporary plot (a swidden) has become exhausted it is abandoned for between 15 and 30 years before being cultivated once again. Some shifting cultivators live in settled communities. They practice an extensive form of rotation in which the fields shift but the settlement does not. Other groups are migratory, shifting fields and settlement.

Clearing land and then moving on may seem to be primitive and environmentally ruthless, but it has proved sustainable for thousands of years and can be carried out on steep slopes. Shallow cultivation, partial clearance, multiple cropping and the small size of the plots all help to reduce erosion because they imitate the natural forest ecosystem. The Kantu people of Kalimantan in Indonesia plant an average of 17 different rice varieties and over 20 crops on their small cleared plots each year. They know the land intimately, and what it can support.

Although shifting cultivation is an extensive system of farming, the yields are low. It has therefore been characterized by some as inefficient; however, land was not formerly a scarce resource in

Traditional agriculture (*above*) Shifting cultivation is still practiced by people in upland areas. Pigs and poultry are often raised by these subsistence farmers, and a wide range of crops are grown, typically including rice, maize and manioc or yam.

Fish ponds in Vietnam (*right*) Fish plays an important part in local diets, and the region provides a wide range of habitats for marine and freshwater fish. Inland fish breeding in specially adapted ponds and lakes can provide a profitable alternative to farming.

MALAYSIA'S RUBBER CROP

Rubber is among Malaysia's most important exports – annual earnings from the crop are about $2 billion, representing about 8.5 percent of total exports. The rubber tree *Hevea brasiliensis*, from which most commercial rubber is obtained, was introduced into colonial Malaya from Brazil in the late 19th century, and rapidly spread as international demand for the commodity grew. Initially it was grown on estates that were under European control, but Chinese and Malay smallholders quickly appreciated its profit-making capabilities, and also began to plant the crop. By 1940 1.6 million ha (4 million acres) were planted to rubber, about 40 percent on smallholdings. Today the figure is some 2 million ha (5 million acres), of which only 22 percent is grown on large estates.

The new, quick-maturing and high-yielding rubber trees take some five years to come into production, and last about 25 years before yields begin to decline. The latex, the milky-white fluid from which the rubber is processed, is drawn from the tree through cuts made in the bark, and is collected in bowls strapped to the trunk. After collection the rubber is processed into sheets or blocks and sold.

Over half the workers on rubber estates are Indians, the descendants of indentured laborers brought in to meet a shortage of workers at the beginning of the 20th century. Wages are low, conditions poor, and the rubber estates remain sinks of poverty in a fast-developing country. Fluctuations in the international price of rubber and competition from synthetic materials, as well as the high costs of modernization and replanting, are responsible for the fall in estate rubber.

the upland forested areas of Southeast Asia. It was people who were scarce. Now the pressures of population growth, commercialization and deforestation are changing the way that upland areas are being used. Shifting cultivators have been forced to settle and to intensify production; fallow periods are being reduced as land becomes scarce. Traditionally the land was communally owned through customary law or *adat*, but now individual landownership is spreading. The whole system of shifting cultivation is being radically reworked and, in some cases, destroyed.

PRESSURES FOR CHANGE

Farming is still the most important sector in Southeast Asia's economy, despite rapid urban and industrial growth; it involves 54 percent of the population. At a superficial glance it may appear to cling to its traditional roots and follow the age-old rhythms and patterns, but beneath the surface the changes have been dramatic. Population growth has put pressure on farmers to raise yields. At the same time, technological advances have provided farmers with new crops and methods. Increasingly, land and labor are owned, operated, hired and paid for in new ways and the aspirations of farmers have changed dramatically.

Meeting new needs

In 1800 Southeast Asia's population was 32 million. By 1990 it had risen to some 450 million. In addition to this rapid population growth, farmers are now demanding far more goods and services if they are to live, by their own standards, a creditable existence. These twin pressures have altered the very nature of farming, while urban growth means that there is less land available for farming.

Farmers have responded to the pressing need to raise yields by planting new crops and using increasing amounts of agricultural chemicals such as fertilizers and pesticides. In 1968 yields of foodcrops in Indonesia averaged 1.4 tonnes per hectare (0.5 tonnes per acre), while chemical fertilizer application amounted to a mere 8 kg per hectare (7 lb per acre), and was very unevenly used. Twenty years later yields had risen to 2.5 tonnes per hectare (1 tonne per acre) and fertilizer use to 107 kg per hectare (95 lb per acre). As a result countries such as Indonesia and the Philippines have been able to reduce their dependency on imported cereals, and also to export large

Sun-dried cassava in Java *(above)* Cassava, a traditional subsistence crop in many tropical regions, is now also grown as a cash crop. It is favored because it is very easy to grow, and gives high yields. The root is soaked in water, then dried in the sun to leave the white starch, which has many commercial uses.

Destruction of the forests *(left)* These dipterocarp trees in Sabah, Malaysia are being stripped of their bark. Forests are being felled more quickly than they can grow, with serious environmental consequences.

quantities of high value crops such as cocoa, coffee, fruit, spices and tea.

While technical advances have helped to offset the decline in available land, the fragmentation of farms, as plots are subdivided between family members, and the rising number of tenancies, are creating serious problems. Inequalities have widened, and some farms have become so small that subsistence needs can no longer be met. Farmers either sink into poverty, or search for alternative employment outside agriculture. The pressure to increase yields sometimes leads to a decline in fertility as more is taken from the soil than is put into it. Increased mechanization has deprived farm workers of their livelihoods, forcing many to migrate to the rapidly expanding urban areas in search of work. The population of Bangkok, for example, rose from 1 million in the early 1950s, to 7 million by 1990. Rural decline is a source of concern to policymakers.

Farming as big business

The move toward more commercialized farming means that more and more nontraditional crops are being grown for cash. Rubber, oil palm, jute, coffee and tobacco are all widely cultivated. The ratio between the area planted to food and to nonfood crops is 70 to 30, and even traditional subsistence crops such as rice are now partially grown for sale.

Farming, then, is making the transition from being a way of life to becoming a business. Money is borrowed (in some cases at rates of interest of 100 percent a year), profit margins have to be carefully calculated and the relative prices of different crops assessed as part of the decision-making process. Farmers can now earn incomes to buy consumer goods and to pay for services. One effect is that they have been drawn more tightly into the international cash economy, making them more vulnerable to fluctuations in the international market place. Bankruptcy is all too common.

Within this context of change, government policies provide a framework through which the transition process can be helped or hindered. For example, government agricultural research stations and extension programs have aided the promotion and diffusion of new technologies and techniques. In Malaysia the government-funded Rubber Industry Smallholders' Development Association has been instrumental in modernizing and revitalizing smallholdings by providing advice, technology and interest-free loans. Nearly all smallholdings that grow rubber have been replanted with new high-yielding varieties.

Throughout Southeast Asia governments have supported prices and helped to establish cooperatives and marketing agencies for both crops and livestock. However, some government actions have not always been to the benefit of farmers. The taxes on certain crops, such as the rice premium that has operated in Thailand for many years, for example, significantly lowered farmers' returns.

Maintaining the pace

The pace of change is not slackening. New technologies and crops will be still more widely adopted: they include new chemical pesticides and herbicides, the adoption of genetically engineered crop varieties, and improved labor-saving machinery. Commercialization is likely to proceed in tandem. These changes will bring about fresh problems, two of which stand out: the need to raise yields and absorb a surplus rural population as land becomes an increasingly scarce resource; and the need to promote sustainable agricultural growth so that production can continue to rise to feed the populations of the future. Careful and balanced development of Southeast Asia's farm sector is required if the high yields grown today are to continue in years to come.

NEW LAND FOR FARMING

Indonesia is the world's fifth most populated country. Nearly 60 percent of its 185 million inhabitants live on the island of Java, only 7 percent of the country's land area. Elsewhere, in Irian Jaya and Borneo, huge areas of land are forested and seemingly underpopulated and underutilized. Since 1950 the Indonesian government has tried to redress this population imbalance by promoting settlement on the "outer islands".

This exceptionally large transmigration scheme has resettled over 5 million people. Huge areas of forest have been cleared to accommodate and feed them – but there are problems. The shallow, heavily leached, infertile forest soils deteriorate under cultivation, so yields quickly decline. The wet rice system that dominates the scene is thought by some experts to be environmentally inappropriate: soils are more suited to upland crops and shifting cultivation, while irrigation facilities are either nonexistent or perform poorly. Settlers find it hard to market their produce or buy agricultural inputs in these remote areas. To add to the difficulties many of the areas, far from being unpopulated, are inhabited by shifting cultivators.

Inevitably there has been a certain amount of conflict between the old and new inhabitants, and some settlers have returned to Java. However, the Indonesian government has recently made much greater efforts to adapt its schemes so that they do not encroach on the traditional way of life of the indigenous inhabitants. It is also considering ways of basing the agriculture of these new areas on more suitable tree crops, such as rubber or coconuts.

The "rice bowl" of Asia

Preparing the rice fields The rice cycle begins as soon as enough rain has fallen to soften the soil. The banks (bunds) are repaired, and the weeds removed and left to rot. When the fields are covered with water plowing begins, ready for transplantation of the seedlings from the nursery bed.

Transplanting seedlings (*above*) The young rice plants are carefully pulled up and bundled to be taken for replanting in the prepared rice field.

Threshing with water buffalo (*right*) Trampling by people or animals is the usual way of separating the grain. This method retains the protein- and vitamin-rich part of the seed (endosperm).

Thailand is a kingdom of rice farmers. Nearly two-thirds of crop land – some 1.2 million ha (3 million acres) – is planted to rice. It is both the staple crop of the country's 5.2 million farm households and the most important cash crop. In 1989 Thailand exported 6 million of its 21 million tonnes, making it the world's largest rice exporter. Most of this is produced on small, family-owned farms: over half the average holding of 4.5 ha (11 acres) will usually be given to rice.

The cultivation of rice by the wet or paddy system begins between May and July during the southwest monsoon. The land is carefully leveled, producing small plots where water collects. Farmers first broadcast pregerminated rice seed into a meticulously prepared nursery bed. The main field is plowed and puddled using either buffaloes or, increasingly, hand-held rotavators. After about a month the seedlings are transplanted. The harvest of the main rice crop takes place from November through January. Once the rice has been stored safely in the rice barns the farmers turn out their livestock to graze the stubble. They then begin to prepare for the work and festivals of the dry season. "Having rice in the barn," say the villagers, "is like having money in the bank."

Treasures of the paddyfield

Although rice production is at the core of the system, the paddyfield is also used to raise fish; the frogs and crabs that live in the bunded fields provide the people with an important source of animal protein as well. At the end of the rice season the farmer sometimes uses the field to grow a second crop such as beans or sesame, taking advantage of the moisture that has been retained in the soil.

More than any other staple crop, wet rice is heavily dependent on an ample and stable supply of water. In the central plains and the north of Thailand, where irrigation helps to control the water, double-cropping is sometimes possible. In comparison to systems in Java and the Philippines, however, irrigation methods are crude. There is only limited control over water supplies, and deep inundation remains a problem in the wet season, while insufficient water handicaps farmers in the dry season. Consequently, yields in the central plains average only 2.5 tonnes per hectare (6 tonnes per acre), poor in comparison to Java's 5 tonnes per hectare (12 tonnes per acre).

In most areas, though, farmers still grow rice in rainfed conditions where water availability is less certain. This means that yields tend to be both lower and more variable. Nevertheless, using sophisticated cultivation methods, and one of the thousands of traditional varieties of rice, some of which can grow in several meters of water, farmers can usually grow enough to meet their own needs and to contribute to Thailand's substantial rice exports. In the delta areas of the central plains a system particularly characteristic of Thailand has evolved – deepwater rice cultivation. Seed is broadcast directly onto the dry field; it grows in dry conditions until the rains arrive in earnest perhaps 50 days after germination. Then the plants must compete with the rising water, which can eventually exceed 1 m (3 ft) in depth. Yields are low, perhaps only 1.5 tonnes per hectare (3.7 tonnes per acre), and the system is being replaced in many areas by transplanted rice as water control improves.

Like all farming in Southeast Asia, wet rice cultivation has undergone many drastic changes. Production is becoming increasingly mechanized, the new high-

The rice plant Rice is one of the world's most important food crops. It grows best in shallow water; the hollow stem allows oxygen to reach the roots submerged in the flooded soil. The plants grow to a height of 80–180 cm (31–70 in), and have several stems, each one bearing a head (panicle) of rice kernels. The grains mature 110 to 180 days after planting. Rice is a good source of carbohydrate, though it is slightly lower in protein than other cereals.

yielding varieties are being more widely planted, and farmers may use large quantities of chemicals.

These modern techniques are not without their problems. Social inequalities have widened as some households within a village have benefited at the expense of others, causing friction between neighbors. At the same time, some scientists now believe that farmers are courting ecological disaster by planting the same varieties of rice over large areas, instead of rotating a large number of varieties as traditional farmers did. Genetic variability is reduced, the chance of catastrophic pest attack increased, and the land becomes prone to deterioration as excessive demands are placed upon it. All the same, modern techniques and irrigation have raised yields in environmentally favored areas by up to 6 tonnes per hectare (2.5 tonnes per acre), four to five times higher than in the less advantaged provinces. Whatever methods are used, rice remains central to traditional rural life.

A REGION OF RICE

THE CHANGING FACE OF FARMING · AGRICULTURE WITHOUT ANIMALS · INTERVENTION BY GOVERNMENTS

Agriculture in Japan and Korea has always been dominated by the struggle to maximize limited opportunities to grow a single crop, rice – the staple food of the region. Despite advances in farming technology, changing markets and diversification into growing a wider range of crops, this emphasis in the region's agriculture has hardly changed. Because of the mountainous terrain, less than a fifth of the region is under cultivation, half of which is given over to rice production. Farmland – a scarce commodity – is fast becoming even scarcer, consumed by urban sprawl. People are now, therefore, having to look to other sources of food, particularly to fish farms. The mountain forests, which still cover almost two-thirds of the region, are a traditional source of foods such as berries, nuts and fungi.

COUNTRIES IN THE REGION

Japan, North Korea, South Korea

Land (million hectares)

Total	Agricultural	Arable	Forest/woodland
60 (100%)	10 (17%)	8 (14%)	41 (68%)

Farmers

12.8 million employed in agriculture (14% of work force)
0.7 hectares of arable land per person employed in agriculture

Major crops

Numbers in brackets are percentages of world average yield and total world production

	Area mill ha	Yield 100kg/ha	Production mill tonnes	Change since 1963
Paddy rice	4.3	63.2 (193)	27.1 (6)	+14%
Soybeans	0.6	14.7 (77)	0.9 (1)	+43%
Barley	0.6	25.9 (111)	1.5 (1)	−46%
Pulses	0.5	10.2 (126)	0.5 (1)	−2%
Wheat	0.5	34.6 (148)	1.7 (—)	+5%
Maize	0.5	63.6 (175)	3.0 (1)	+105%
Millet	0.4	12.6 (165)	0.6 (2)	+39%
Vegetables	—	—	26.8 (6)	+71%
Fruit	—	—	8.8 (3)	+117%

Major livestock

	Number mill	Production mill tonnes	Change since 1963
Pigs	17.8 (2)	—	+182%
Cattle	8.7 (1)	—	+66%
Milk	—	8.8 (2)	+222%
Fish catch	—	16.4 (18)	—

Food security (cereal exports minus imports)

mill tonnes	% domestic production	% world trade
−34.7	99	16

Planting out young rice plants in flooded paddyfields in South Korea. This sort of production is very labor-intensive. In the past the feudal societies of Japan and Korea provided a pool of peasant labor to work the land. Today the agricultural work force in both countries is declining.

THE CHANGING FACE OF FARMING

A key factor in Japan and Korea's specialization in rice cultivation has been the climate of warm summers with two rainy seasons between June and October. Over much of the region the climate allows two harvests a year. Unfortunately, level, low-lying ground and good soils are in very short supply.

Feudalism and subsistence farming by peasants largely dominated early farming in the region. From the 15th century onward, however, the slow growth of towns necessitated a more commercial approach to farming. At the same time society became even more divided into an elitist landowning class and the laboring majority. While the masses in Korea were further classified as "the low-born" (those engaged in manufacturing and trade) and "the good people" (those engaged in agriculture), farm workers in Japan had to endure a great deal of feudal oppression. This was exemplified by the old Japanese saying that "peasants are like rape seed; the more you squeeze them the more you get out of them," particularly in the form of taxes.

Expanding urban appetites

The circumstances of farming in Japan changed significantly after 1870 when the country's centuries of isolation from the rest of the world ended. The subsequent modernization of Japan hinged on the development of industry, with agriculture acting as its springboard. The task now was to feed an increasing number of people, many of whom lived in towns and purchased their food out of wages earned in factories and shops; more food had to be produced by a proportionately shrinking farm population.

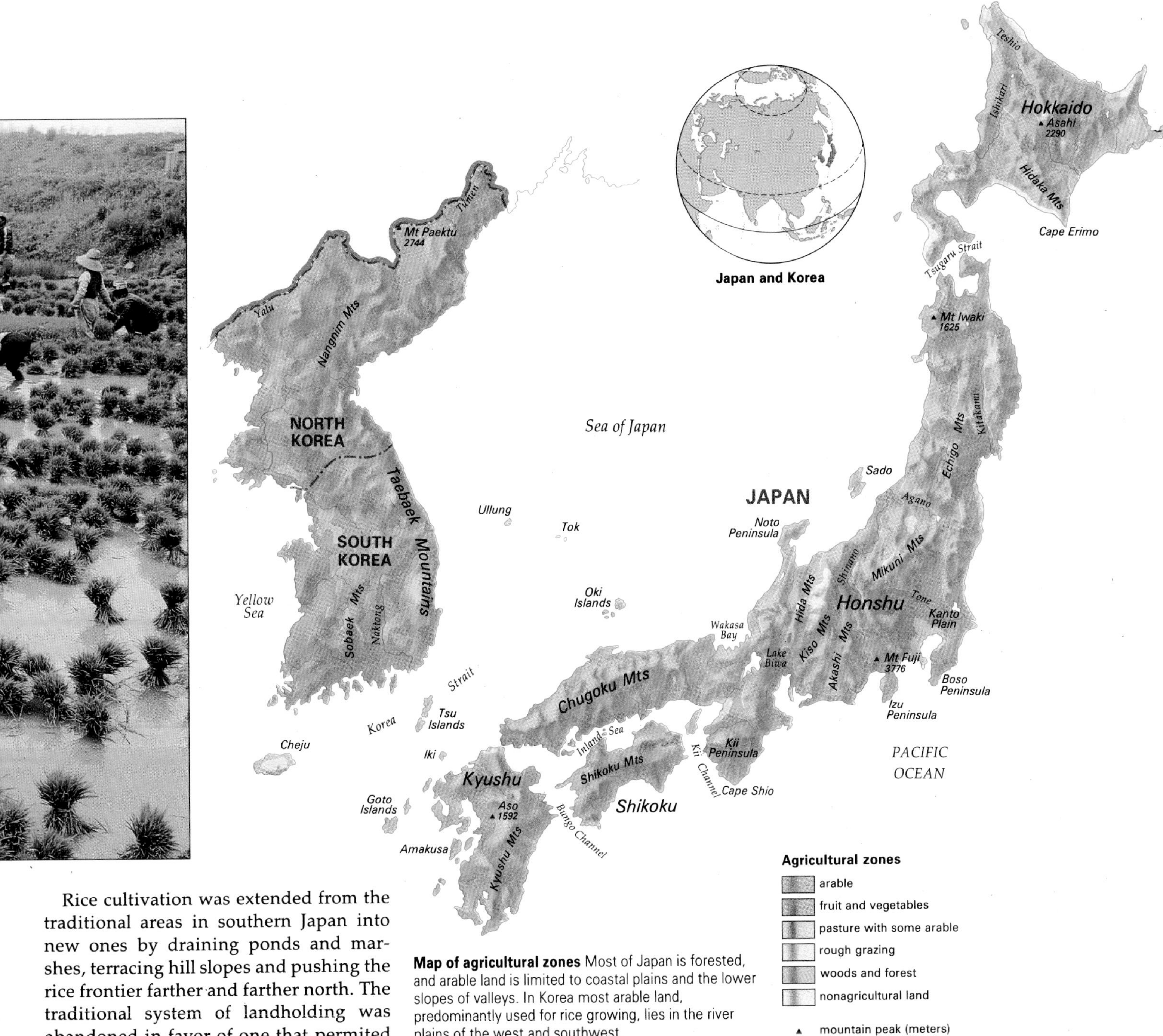

Map of agricultural zones Most of Japan is forested, and arable land is limited to coastal plains and the lower slopes of valleys. In Korea most arable land, predominantly used for rice growing, lies in the river plains of the west and southwest.

Rice cultivation was extended from the traditional areas in southern Japan into new ones by draining ponds and marshes, terracing hill slopes and pushing the rice frontier farther and farther north. The traditional system of landholding was abandoned in favor of one that permited the sale of farmland. This significant change quickly relegated most farmers to the status of tenants exploited by absentee landlords, and the imposition of even heavier taxes spearheaded a mass exodus from the countryside.

As a result, Japan became unable to feed its growing population, and it took steps to acquire an overseas empire in order to provide a supply of both food and industrial raw materials. Korea, annexed at the turn of the century, was soon being exploited by the Japanese: traditional Korean systems were abandoned and its farming, like that of Japan, became subservient to the expansion of modern industry and commerce.

Deprived of its overseas empire at the end of World War II, Japan has since had to make every effort to raise domestic food production. As a consequence agriculture has undergone much change, and this has resulted in a paradox. The oppressive landlord–tenant system has now been replaced by one of small owner-occupiers; mechanization and diversification have taken place; and food productivity has improved. As a result of these important changes the nation is much more self-sufficient than before. Despite all this, agriculture's contribution to the national economy has declined. Only 3 percent of domestic wealth now comes from farming (in 1930 it was 22 percent) and only 7 percent of the work force is employed on the land (a reduction from 48 percent in 1930).

Postwar partition of Korea

The partition of Korea in 1945 has led to a divergence in agricultural development. South Korea's path has been broadly similar to that of Japan, but farming still employs 27 percent of the workforce. In North Korea the adoption of socialist practices means that all farming is now conducted either by cooperative (collective) ventures or by state farms; here agriculture still employs 35 percent of the labor force. The government is pursuing an inward-looking development policy, which sets self-sufficiency as a national priority. Its agriculture continues to be based on subsistence farming, unlike that of Japan and South Korea, where it has been modernized and commercialized.

AGRICULTURE WITHOUT ANIMALS

About a quarter of Japan's land area is suitable for agriculture, and only half of that can be cultivated; the figures for the two Koreas are slightly higher. Small fragmented lowlands, mostly scattered round the coasts, are the main areas under cultivation. In Japan there is only 0.04 ha (0.10 acre) of cultivated land per head of population; the figures for South and North Korea are 0.05 ha (0.12 acre) and 0.11 ha (0.27 acre) respectively.

The inherent shortage of farmland seems to have been mainly responsible for two outstanding and long-established agricultural features in the region: the small size of individual farm holdings, and the intensive methods of cultivation. In the 1950s almost three-quarters of Japanese farms were less than 1 ha (2.47 acres) in extent. By 1990 amalgamation of farm holdings had reduced that figure to two-thirds. This consolidation has, in turn, further contributed to the overall reduction in the size of the farming population: between 1950 and 1990 it had declined by about a third.

The agricultural reforms introduced in North Korea have gone some way toward altering this regional tradition of small-scale, intensive, family-based farming. The new collectives average 500 ha (1,250 acres) in extent, and the state farms 120 ha (300 acres). Much effort has also gone into extending irrigation to land not used for paddyfields and into reclaiming tidal land for agricultural use.

The bias of farming in the region has always been toward the growing of low-land crops, especially rice. In North Korea rice now accounts for more than 40 per-cent of crop production; in South Korea the figure is nearly 60 percent, and in Japan 33 percent. Since the mid 1970s annual rice production in Japan has fallen from a peak of 13 million tonnes, since Japan now produces more rice than it needs. This resulted from remarkable advances that were made in developing higher-yielding strains, as well as ones capable of growing well in the harsher climate of northern Japan. In the late 19th century the chief rice-growing area was the southern island of Kyushu; today it has shifted to northern Honshu.

Now that rice is in surplus farmers

Mechanical rice harvesting At the end of the 19th century Japan could not produce enough food to feed its growing population. Today, thanks to improved, higher-yielding varieties and the introduction of mechanization, rice is in surplus.

Tending the rows of tea bushes Tea has been grown in Japan since the 6th century. The ritual of the formal tea ceremony occupies a central place in Japanese culture and is still widely practiced. Japanese tea is usually green as the leaves are not fermented in the manufacturing process, unlike other teas.

SILK AND SILKWORMS

One of the first cash crops to become important in Japanese agriculture was raw silk. The production of raw silk by raising silkworms – a process known as sericulture – had long been established within the Japanese peasant economy. The farmers and their families grew mulberry trees (the leaves being the silkworms' staple food) and reared the silkworms (the caterpillars of the silk moth) in sheds. The caterpillars produce a single filament of silk (which may be 600–900 m/2,000–3,000 ft long) to make a cocoon. These were then unwound, or reeled, using very simple equipment, to produce a yarn, which could be woven into a fine cloth.

When, in the later 19th century, Japan began to trade more extensively with the rest of the world, it soon found that there was a huge foreign demand for silk; it was increased by the spread of silkworm disease in Europe. The boom in silk production that ensued saw a rapid increase in the amount of land planted with mulberry trees. Output peaked in about 1930, when sericulture accounted for 12 percent of the value of agricultural production, and mulberry trees covered some 0.7 million ha (1.7 million acres).

The collapse of the silk industry, which came about mainly as a result of fierce competition from other textiles and changing fashions, was as spectacular as its rise. Annual cocoon production plummeted from 225,000 tonnes in 1930 to 40,000 tonnes in 1990, when only 90,000 ha (200,000 acres) of land remained planted with mulberries, and sericulture accounted for less than 1 percent of agricultural output. Mulberry cultivation and silk production are today largely confined to central Honshu.

have diversified in order to meet a rising demand from the towns and cities for fresh fruit and vegetables. As a consequence, paddyfields have increasingly been given over to the growing of a whole range of vegetables such as tomatoes, eggplant, carrots, sweet potatoes and onions, as well as fruits such as watermelons, melons and strawberries.

Terraced mountain orchards

Above the paddyfields, on the lower mountain slopes, the character of farming is somewhat different. Although the slopes have been terraced, there is little irrigation. Instead of cereals there are fields of bushes and trees. Traditionally these were tea and mulberry, but since the decline of the silk industry in the 1930s the mulberry trees have been largely replaced by a range of other trees. In the north apples and pears are important; vineyards are also to be found on the sunnier slopes. To the south various species of citrus fruit – such as clementines, mandarins and tangerines – are now widely grown, along with peaches, nectarines and persimmons. There is a growing export business associated with these lucrative tree crops.

Changing attitudes to livestock

Traditional farming in Japan and Korea has often been described as agriculture without animals. It was widely believed in the countryside that milk and eggs were food suitable only for invalids and babies. Livestock was not kept for food, apart from poultry and occasionally pigs. Most farms would own one or two horses or cows for transporting heavy goods, plowing and as a source of manure, but red meat did not form part of the normal diet. In North Korea today this tradition remains largely unaltered.

In Japan and South Korea, however, the postwar years have seen quite radical changes. Poultry farming – that is, the production of dressed chicken and eggs – is now widespread, but is particularly concentrated close to the main urban markets. Beef production, too, is important in the southern half of Japan and in parts of South Korea; indeed, one of Japan's most popular dishes – *sukiyaki* – requires a particular kind of beef that has been fattened on maize. Dairy farming is also increasing in importance, especially in Hokkaido with its cool, moist climate and rich pastures.

INTERVENTION BY GOVERNMENTS

A salient feature of Japanese and Korean agriculture since 1945 has been government intervention – but for different reasons. In socialist North Korea all farmland was initially requisitioned by the government and then redistributed to the peasants, thereby abolishing tenancy. Later the ideological drive for collectivization became the priority, and by 1958 all private farming had completely disappeared. In 1990 some 3,800 cooperatives, or collectives were in existence, working more than 90 percent of the cultivated area. The remainder is worked by the employees of about 180 state farms.

In capitalist South Korea the government also abolished farm tenancy, but instead of cooperatives has encouraged the development of small owner-operated farms. The South Korean government also intervenes in other ways, such as subsidizing the high costs of fertilizers and pesticides, encouraging the farmers to form groups to purchase farm machinery, and giving grants to raise rice production to ensure that the country becomes self-sufficient in its staple food.

In postwar Japan a government-led drive toward a self-sufficient food supply has also been a major target. Defeat at the end of World War II deprived Japan of overseas territories such as Korea, Taiwan and Manchuria in southeast China, which had become important sources not just of rice but also of other foodstuffs such as sugar and soybeans. The drive for self-sufficiency has been helped by a whole range of government activities.

Grants, subsidies and tax incentives have been offered directly to farmers in order to persuade them to improve their efficiency and to raise their levels of productivity. At the same time agriculture has been protected from outside competition. By imposing high import taxes and stringent import quotas, the Japanese government has been able to ensure that home-produced food has priority in the domestic market place; it has not allowed the market to be swamped by imports of

Intensive farming over a river plain in Kyushu. In Japan's increasingly urban and affluent society, there is mounting pressure on the country's limited area of agricultural land. People living in the towns seek to use the countryside for a growing number of leisure pursuits, and more farmland is lost.

Massaging his cattle Changes in Japan's dietary habits means that beef production has grown enormously in recent decades. Particularly lean meat is needed for certain beef dishes, and this farmer is ensuring that his animals will provide it.

PART-TIME FARMING

One recent feature of Japanese agriculture has been its relegation from a full-time to a part-time activity, despite the rising levels of agricultural output. In 1935 only about 25 percent of farming was undertaken on a part-time basis, whereas by 1990 this figure had climbed to 90 percent. This change has been encouraged by the small size of the average farm holding. Given the high cost of living, farms are no longer large enough to produce sufficient income to support a whole family. The profits made from producing rice and other foodstuffs are not enough to enable the farmer to buy those material items that most people now regard as an essential part of modern life – a car, washing machine or holiday abroad.

At the same time the buoyant economy has meant that plenty of part-time jobs are available in nearby towns and cities. The heads of many farming households, therefore, have either had to take on some form of part-time work to supplement their income or have virtually ceased to be farmers at all. In the latter case the male head of the household will perhaps help on the family farm after work and during the weekends. The actual running of the farm will be left largely to his wife and any elderly relatives living at home. Older children may also assist after school and during the school holidays.

cheap food from countries such as Australia and the United States. By 1990 Japan produced over 70 percent of its food requirements; more significant, however, is the fact that it is now more than self-sufficient in rice production.

Pressures from urbanization

Although agriculture may have been carefully protected from competitive forces outside Japan, it has nonetheless been increasingly threatened from within. Rapid urbanization in the postwar period – to the extent that two-thirds of the Japanese population now lives in cities containing more than 100,000 people – has created a series of pressures. In Japan and South Korea alike, the spread of builtup areas in the lowlands inevitably reduces the scarce supply of good farmland: buildings replace crops. As the growing urban population becomes more affluent, so, beyond the city, the pressure builds up to convert fields into recreational amenities – golf courses, picnic areas and country clubs.

Through the mass media the young people of rural areas have become aware of the urban way of life; for many of them life appears to be better in the cities and they are persuaded to move. Although this rural-to-urban migration provides the chance to consolidate farm holdings into larger and more viable units, there is mounting concern that the countryside will all too easily become deprived of farm labor.

Dietary fashions

Another important influence on postwar agriculture in Japan and South Korea has been the Westernization of the diet. This change was probably initiated by the presence of American troops in both countries during the late 1940s and 1950s, and was later encouraged by the growth of the Western media, particularly in television, advertising fast foods such as hamburgers, french fries and Coca-cola.

The traditional diet of the region was originally based on three basic foods – fish, miso (produced from soybeans) and rice. Since 1945, however, the Japanese in particular have developed a taste for beef, beer, dairy products and bread. This, in turn, has had a big impact on farming: it has, for example, encouraged some Japanese farmers to turn to the rearing of livestock and the cultivation of cereals other than rice, most notably wheat and barley for bread and livestock feed. Supplies of good quality barley are also needed to satisfy the nation's growing thirst for beer and lager, which has given rise to a large brewing industry.

The parts of Japan most suited to these new lines of agricultural production have been those with a cool temperate climate, particularly the northern island of Hokkaido. Traditionally Hokkaido, with its long winters and short summers, had been the backwater of agricultural Japan. Thanks to these changes in diet, it is now an important agricultural district. It is able to support livestock to give meat and dairy products, and to grow a range of cereals and root crops.

Harvesting the seas

Japan and Korea have for hundreds of years relied on the sea as a source of food, possibly because of the shortage of viable farmland. Fish has always had an important place in the diet of the region; indeed it still provides half of the animal protein in the Japanese diet, despite the recent fashion for meat eating. The raw fish dishes of *sushi* and *sashimi* remain as popular as ever. Fish is eaten at most meals and in many different forms: not just raw, but steamed, dried or salted; as a paste, sausage or as soup stock.

The convergence off the coast of central Honshu of the warm, northeast-flowing Kuroshio (Japan) Current with the cold, southern-flowing Oyashio Current produces ideal conditions for both bottom-living fish such as cod, haddock and plaice and for surface species such as tuna, bonito, mackerel, squid and sardine. Crustaceans of all kinds, such as shrimps, lobsters, oysters and whelks, are also harvested, as are many different varieties of seaweed – an important source of salt and vitamins.

It is only since World War II that Japan has become the world's leading fishing nation, stimulated by the desperate need to find alternative food sources once all its territories overseas had disappeared. New fishing grounds were therefore exploited, fishing technology improved and the fishing fleet enlarged. The increased annual catches that resulted have been achieved by a dramatically contracting labor force. Since 1945 the number of full-time fishermen in Japan has declined by 80 percent; in 1990 less than 1 percent of the national work force was involved in the fishing industry.

Enormous fishing vessels, which put to sea for weeks on end, fish in distant international waters – not only in the Pacific, but also in the Indian and the Atlantic Oceans. Despite an international moratorium that was introduced to conserve fast-dwindling whale stocks, some of these ships still take part in whaling; the Japanese have an enduring passion for whale meat. Offshore fishing takes place from small- to medium-sized vessels working within a day or two's sailing from home ports. Much smaller craft are involved in coastal or inshore fishing, working within sight of land.

Fish farming

The rearing of freshwater species, such as eels, trout and carp, in tanks and ponds has been carried on for centuries in both Japan and Korea. It was, and still is, often engaged in as a sideline to farming. What is new is the huge scale of fish farming today and the fact that it is increasingly conducted in coastal waters. Here the commercial farming of fish has risen in importance as the natural stocks have become depleted by overfishing. The change has gained in pace since the mid 1970s when the maritime nations extended their offshore territorial limits to 320 km (200 mi). This international ruling denied Japanese fishermen access to what they had come to regard as their traditional fishing grounds.

The Japanese and Koreans also recognize that the hunting of fish over great expanses of ocean can be very wasteful of time, effort and money when similar fish – such as tuna and bream – can easily be reared and harvested in the numerous sheltered bays that characterize much of their native coastline. Aquaculture has also been extended to include other forms of marine life such as shellfish: oysters, clams and shrimps are being commercially reared, as are several species of seaweed eaten by the Japanese, notably the red algae known as nori.

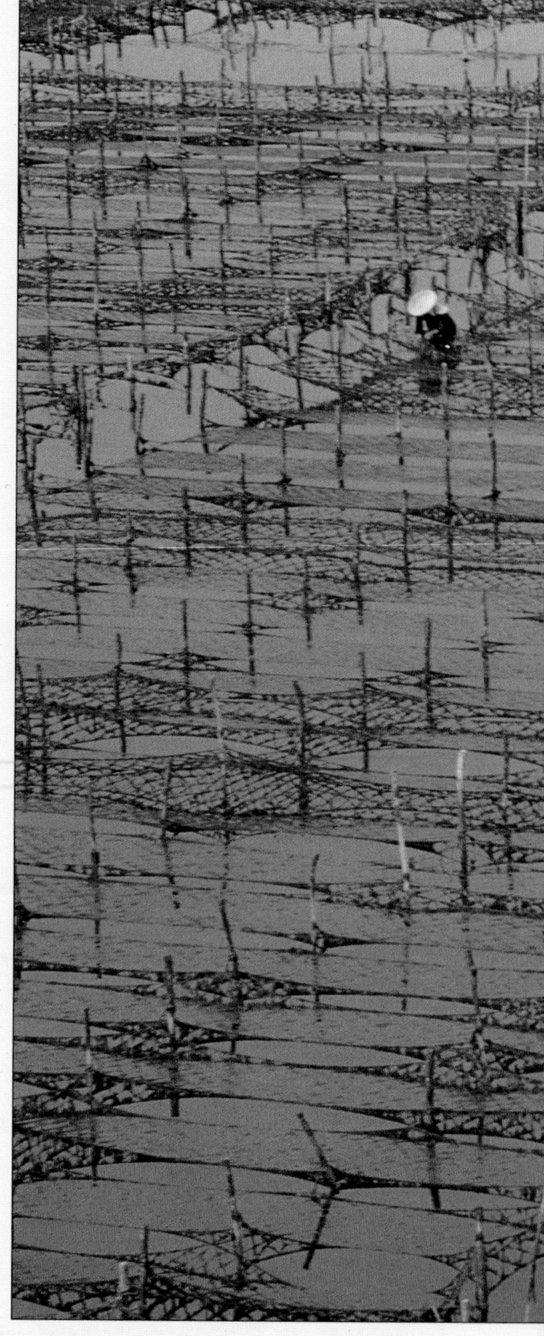

Seaweed harvest (*above*) Rich in vitamins and minerals, seaweed is much used in Japanese cuisine. It also has a number of industrial uses. Here seaweed is being commercially grown on a structure of nets in the shallow waters of a coastal inlet.

Stocking the fish farm (*left*) Thousands of young fish pour out into the waters of a commercial fish farm. Japan's indented coastline is very suitable for this kind of farming, which has grown in importance as overfishing has depleted natural fish stocks.

Ocean fishing (*right*) The efficiency of modern fishing methods causes concern about the environmental damage inflicted by the world's ocean-going fishing fleets. Trawl nets, which are dragged along the sea bed to ensnare bottom-dwelling fish, sweep up many other species and cause widespread damage. Seine and gill nets are pulled through the water to catch shoals of ocean fish. Known as "walls of death", gill nets can stretch for several kilometers, trapping dolphins and sea birds as well as fish. International pressure to regulate net size and impose catch quotas to prevent overfishing is mounting.

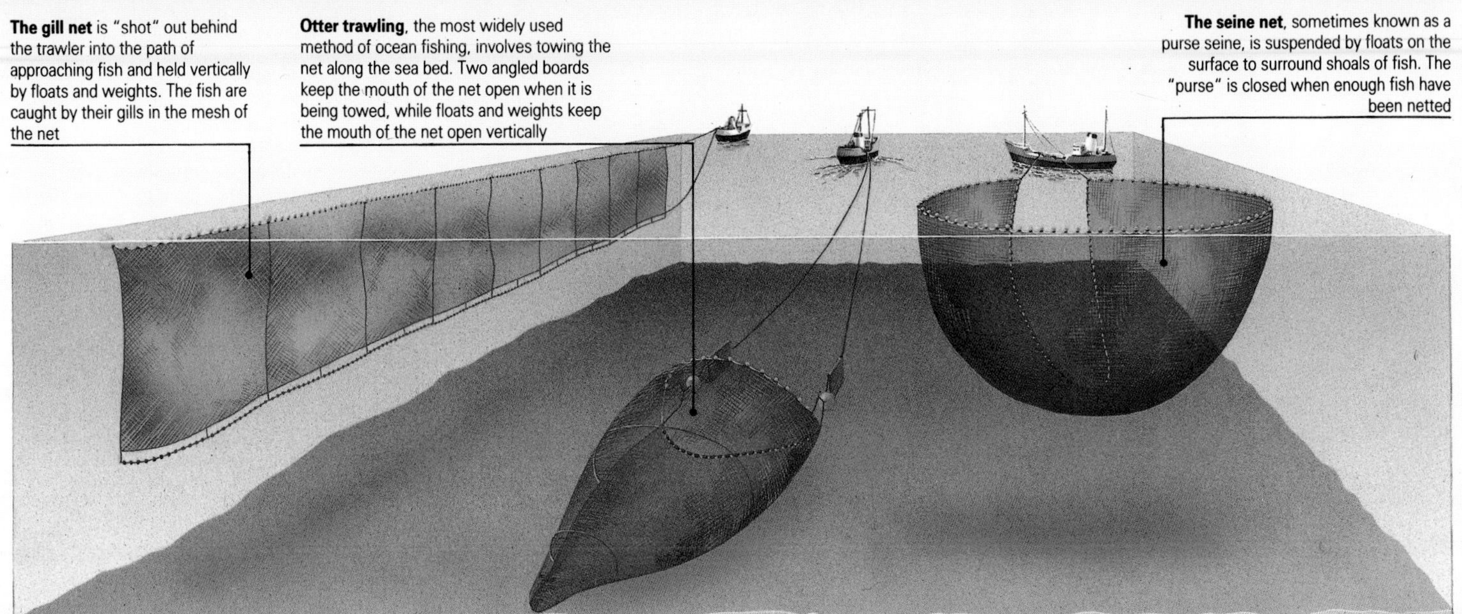
The gill net is "shot" out behind the trawler into the path of approaching fish and held vertically by floats and weights. The fish are caught by their gills in the mesh of the net
Otter trawling, the most widely used method of ocean fishing, involves towing the net along the sea bed. Two angled boards keep the mouth of the net open when it is being towed, while floats and weights keep the mouth of the net open vertically
The seine net, sometimes known as a purse seine, is suspended by floats on the surface to surround shoals of fish. The "purse" is closed when enough fish have been netted

SHEDDING A COLONIAL LEGACY

EARLY SETTLERS · GIGANTIC RANCHES TO MINUTE FARMS · PROBLEMS AND SOLUTIONS

As a result of their colonial past, Australia and New Zealand, and many of the Pacific islands, have been economically dominated by Britain and France over the past two centuries. Their agriculture was consequently geared almost entirely toward supplying the needs of the former colonizing powers, with whom they enjoyed a secure market and special terms for their exports of food, timber and wool. Major changes have occurred since 1973, when Britain joined the European Community (EC). This ended Commonwealth trade preferences for Australia and only marginally protected New Zealand exports. Japan and other countries in the Pacific instead became their main trading partners. France's territories in the Pacific, on the other hand, are still given favorable trading terms within the EC.

COUNTRIES IN THE REGION

Australia, Fiji, Kiribati, Nauru, New Zealand, Papua New Guinea, Solomon Islands, Tonga, Tuvalu, Vanuatu, Western Samoa

Land (million hectares)

Total	Agricultural	Arable	Forest/woodland
843 (100%)	499 (59%)	48 (6%)	156 (19%)

Farmers

2 million employed in agriculture (17% of work force)
24 hectares of arable land per person employed in agriculture

Major crops
Numbers in brackets are percentages of world average yield and total world production

	Area mill ha	Yield 100kg/ha	Production mill tonnes	Change since 1963
Wheat	9.1	13.9 (60)	12.6 (2)	+49%
Barley	2.4	15.6 (67)	3.8 (2)	+255%
Oats	1.3	13.7 (75)	1.8 (4)	+47%
Pulses	1.3	12.4 (154)	1.6 (3)	+2,314%
Sorghum	0.8	17.4 (118)	1.4 (2)	+540%
Sugar cane	0.4	743.7 (125)	28.6 (3)	+87%
Fruit	—	—	4.2 (1)	+55%

Major livestock

	Number mill	Production mill tonnes	Change since 1963
Sheep	213.4 (19)	—	+1%
Cattle	30.5 (2)	—	+20%
Milk	—	13.7 (3)	+10%
Fish catch	—	0.8(1)	—

Food security (cereal exports minus imports)

mill tonnes	% domestic production	% world trade
+21.1	87	10

EARLY SETTLERS

Until European settlement started in 1788 the South Pacific could, agriculturally, be divided in two: Australia, where there was no settled agriculture and the only domesticated animal was the dog; and elsewhere in the region, where various forms of intensive manual gardening were practiced using a small range of tropical and subtropical trees and subsistence crops such as coconut and taro. This was supplemented by hunting, gathering and fishing. Pigs were the main domesticated animal in Melanesia and much of Polynesia, but not in New Zealand, where there were no mammals until they were introduced by the British.

Whalers and sealers, who followed the European explorers out to Australia and New Zealand, founded shore settlements to process their catch. In the Pacific islands, which they reached in the late 18th century, these traders bought fruit, vegetables, pork and later beef from the indigenous farmers, who were also encouraged to dry coconut meat to produce copra – an oil extract used in soap and cosmetic preparations. The islands gradually came under the protection of the colonial powers, who encouraged the growth of other crops for cash.

The most sought-after product to come from the Pacific islands since has been the natural phosphate deposits on Ocean Island and Nauru. Converted to superphosphate, this has provided the main fertilizer for the grain- and grasslands of Australian and New Zealand farms for over a century.

The earliest commercial farming in Australasia was sheep and cattle rearing, started in the early 19th century in the temperate south of Australia. The pastoral industry spread rapidly into the interior to meet the increasing demand from Britain for leather and wool. Squatters, or

Distance no object Good transportation is vital in Australia, where farms are huge and often remote from the main markets. This road train is thundering across the outback to deliver livestock. Cattle are often moved great distances, either to new pastures for fattening, or to be slaughtered.

Map of agricultural zones Less than 10 percent of Australia's land area sustains arable farming. Much of the arid interior remains unutilized for agriculture, the rest provides grazing for sheep and cattle. Dairying is restricted to areas of high rainfall in the southeast.

herdsmen, claimed their huge tracts of arid land, where their finer-wooled merino sheep thrived. Cattle were later introduced into all but the driest parts of the continent. In most of New Zealand sheep and cattle were farmed extensively, and the export of agricultural products formed the basis of its economy. Fats and oils were extracted from the carcasses to be used for dripping and candle-making, until refrigerated ships were introduced in 1882, and thereby allowed the profitable export of frozen and chilled meat and dairy products.

Success with new crops

The original European settlers introduced wheat and barley into Australasia, where they have prospered, particularly in New Zealand's South Island and in southeast Australia. Sugar cane (which was first cultivated in the Pacific area, probably in New Guinea or the adjacent islands) has also thrived in southeast Queensland. Coffee and cocoa were introduced to New Guinea by German and Australian planters in the late 19th century.

In the wetter areas of Australia, huge farms were gradually broken up earlier this century and allocated to other settlers, especially to servicemen from both world wars, for more intensive use. Railroads were built to serve these new farms, and irrigation schemes were introduced, especially in the Murray valley in southeast Australia. Commercial ranching of kangeroos, horses and rabbits was started, and of exotica such as buffaloes and crocodiles in the Northern Territory and camels in the central desert.

Adapting to changing markets

Similar innovations, such as deer and Angora goat farming, have appeared in New Zealand since Britain joined the EC. Forestry and fishing have also become major export earners as well as being useful for the domestic market. Enormous plantations of Monterey pine in the volcanic areas of central North Island supply nearby pulp mills; this fast-growing Californian species thrives on the young volcanic soils hostile to other land use.

Both New Zealand and the Pacific islands have benefited from the insatiable appetite of the Japanese for seafood, and their local fishing industries have been given a new lease of life since the 1970s.

GIGANTIC RANCHES TO MINUTE FARMS

Farm systems in Australasia and Oceania are dominated by the widely different physical characteristics of each country. They range from Australia's huge farms, or ranches, of 40,000 ha (100,000 acres) or more, on which cattle and sheep are reared, to Oceania's small intensive gardens of a few hectares where cash and subsistence crops are grown.

Australia's vast central desert is surrounded by rough grazing land. In the drier parts, an area of about 400 ha (1,000 acres) is needed to support a single animal, and helicopters and scrambler motorcycles are used to round up the vast, widely dispersed herds. The hardier beef cattle are grazed in most of the climatic conditions found in Australia; tropical breeds such as the Brahmin are reared in the north, while European breeds are more common in the temperate south. Massive truck and trailer combinations, known as road trains, transfer cattle rapidly to fattening areas on the humid or irrigated fringes of the deserts, and to coastal slaughterhouses.

Initially these vast Australian ranches were run with the use of convict labor, but Aboriginal stockmen now form an important part of the work force, much of which is contract labor. Most of these ranches are owned by large companies and are managed by tenants.

In the less arid areas between the central desert and the coastal arable lands the large farms, also 30,000–40,000 ha (75,000–100,000 acres), are best suited to sheep. Australia is the world's major producer and exporter of wool, mainly from merino sheep. Gangs of shearers travel from sheep station to sheep station to remove the heavy fleeces at the end of winter. More intensive sheep rearing, with lamb production as an important by-product, is carried out within the mainland arable areas and in Tasmania.

In the arable areas, mainly in the east and southeast of Australia, farms are slightly smaller. Even so, some wheat farms still comprise hundreds and often thousands of hectares, which are plowed, seeded and harvested by giant machines. Aerial spraying and fertilizing by contractors is commonplace, and the labor force is minimal, often living in townships a long way from the grain fields.

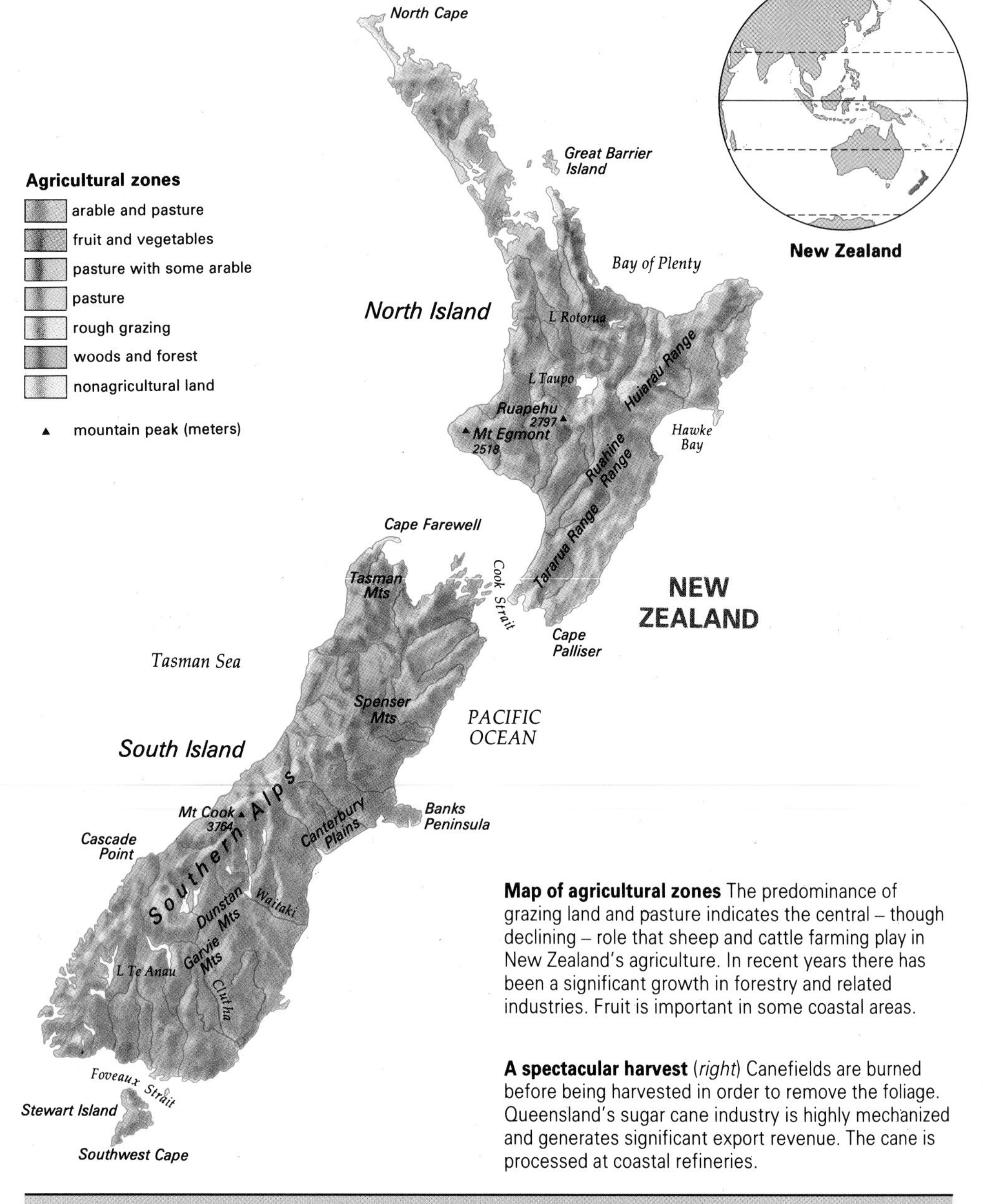

Map of agricultural zones The predominance of grazing land and pasture indicates the central – though declining – role that sheep and cattle farming play in New Zealand's agriculture. In recent years there has been a significant growth in forestry and related industries. Fruit is important in some coastal areas.

A spectacular harvest (*right*) Canefields are burned before being harvested in order to remove the foliage. Queensland's sugar cane industry is highly mechanized and generates significant export revenue. The cane is processed at coastal refineries.

CREATING A MARKET FOR EXOTIC FRUIT

When fruit farmers in New Zealand were faced with a cutback in the demand for apples and peaches in the 1970s, they were forced to look for new markets. Recent developments in refrigeration techniques meant that containers of perishable produce could be rapidly airfreighted anywhere in the world. Fruits such as Chinese gooseberries and tree tomatoes had long been grown for local consumption by market gardeners – many of them of Chinese, Indian, Lebanese and Yugoslav origin – on the coast around Auckland, Bay of Plenty and Hawke Bay on North Island and Nelson on South Island. The renaming of these fruits as kiwi fruit and tamarillos or tamangoes was an inspired piece of marketing; skillful advertising and the devising of special recipes, backed by regular supply of the produce in first-class condition, helped to create a large international market within just a few years. The market has continued to expand, and now extends into Japan and Southeast Asia, the Middle East, North America and Western Europe.

The frost-free, humid coastlands of northern North Island have proved most suitable for these crops, which are grown on holdings of less than 50 ha (125 acres). These are surrounded by quick-growing windbreaks to protect the fruit from weather damage and preserve their external appearance. Mechanization means that family labor is sufficient to work the holdings, except at harvest time when local housewives and students help with picking and packing. Further exotics such as nashi pears from Japan and passion fruit have diversified the output. Most customers are in the northern hemisphere, and although many of these countries have followed New Zealand's lead in growing these fruits, the reversal of the seasons means that New Zealand can supply the market at times when other growers cannot.

Intensive farming
More intensive cultivation with higher yields, on yet smaller farms, occurs in the irrigated areas of southeast Australia. The principal exports from these areas are orchard fruits (apples and pears) and grapes (both dried and as wine). In the coastal valleys of Queensland, where rainfall is high, there is specialized cultivation of sugar cane. In the past, contract labor (predominantly Italian) was used, but the sugar industry is now intensively mechanized, with railroads connecting the fields to the mills.

Small farms are also characteristic of New Zealand, which has an equable moist climate with warm summers and mild winters; irrigation is common only in the South Island on the Canterbury Plains and central Otago area. Sheep and dairy farming are by far the most important types of agriculture. Although New Zealand's land area is only 4 percent that of Australia its sheep flock is nearly half as large as its neighbor's and its cattle herd almost a third the size. In the driest areas merino sheep are most common, but elsewhere the heavy-wooled Romney cross is favored, producing carpet and knitting wool as well as tender meat.

Butter was traditionally an important by-product of the New Zealand dairy industry, but cheese, the milk protein casein and milk powder are becoming increasingly important. Most dairy farms are family holdings of 100 ha (250 acres) or less, sometimes combined with sheep rearing in more hilly areas. Fruit farming has also become an important export industry in New Zealand since Britain joined the EC – apples, pears and exotic fruits such as kiwi, tamangoes and nashi pears being prominent.

Small subsistence plots
In the Pacific islands agriculture is dominated by large companies that control plantations of, for example, sugar cane, coconuts or cotton, using contract labor. The volcanic islands generally tend to be patchily fertile, and the indigenous farmers mainly cultivate their own small plots of land, in which they grow subsistence crops such as taro, yams and sweet potatoes, as well as a cash crop such as coffee, cocoa, pineapples or bananas. Fertile areas of land on the lowlying coral atolls are rare, and agriculture here is limited to subsistence farming, supplemented with keeping pigs as well as fishing.

PROBLEMS AND SOLUTIONS

Shortage of labor has always been characteristic of agriculture throughout the region, and has therefore influenced the farming techniques adopted since the European settlers first arrived. Australia and New Zealand are essentially urban cultures, from which contract labor is sought to shear sheep, harvest wheat, cut cane or work in the slaughterhouses.

The farms have been mechanized as much as possible and labor-saving devices introduced. Inventions such as the stump-jump plow (for use on recently cleared forest land), mechanical harvesters, herringbone milking sheds and mechanical grape pickers have all helped to overcome the shortage of labor.

Early pastoralism in Australasia used convict labor from Britain, and people were also taken from the Pacific islands to Queensland in the 1860s to work as laborers, in ships known as "blackbirders". Whole islands in Melanesia and Polynesia were depopulated in a system that resembled slavery. In Fiji, to redress the resulting labor shortage, indentured laborers were brought from India to work the cane fields. This has led in recent years to some tensions between the local people and the more commercially successful Indians.

Coping with extreme weather

A major problem for Australian farmers is recurrent droughts and floods. Rainfall is not sufficient for the normal growth of crops in most parts of the continent. The irregularity of rainfall, even in better-watered areas, makes the problem worse. Today stock in the desert can be rapidly moved to an area where rain has recently fallen so that the animals can graze. In the interior, water for stock to drink is drawn from artesian wells.

Dry farming techniques are adopted in marginal arable land, where the annual rainfall is 300–600 mm (12–24 in) and wheat can be grown. Here frequent light plowing of the topsoil allows the soil moisture to build up over two to three years and then a crop is taken.

Irrigated land in the Murray river basin has been extended since the 1950s by diverting water from the eastward-flowing Snowy river into hydroelectric dams; from there it flows through tunnels, into the westward-flowing Murray/Murrumbidgee rivers. Even so only some 150,000 ha (370,000 acres) of land is intensively irrigated. Elsewhere in Australia another irrigation scheme, the Ord River Scheme, was developed in the northwest in 1959.

Too much water is rare in Australia, but monsoons and typhoons in northern Queensland and the Northern Territory produce heavy flooding and can cause serious damage to crops.

New Zealand suffers less from climatic

The irrigated southeast (*above*) The small, intensive farms of Mildura, Victoria, specialize in orchard fruit and grape production. The Murray river supplies the water needed for irrigation.

New markets (*right*) The profitable trade in meat exports to Britain began in the 19th century with the development of refrigeration ships, but was badly hit when Britain joined the European Community. Alternative Pacific markets have now been found.

extremes than Australia, but in exposed places the steep slopes and unstable, often deforested soils are prone to heavy erosion from very high rainfall. Reforestation has become necessary in the worst areas. The eroded material silts up lakes damned for hydroelectricity and, where it has been left unchecked, spreads across the river plains.

Reforestation has also taken place on central North Island over largely infertile areas that have been created as a result of volcanic activity. These areas have been enriched with superphosphate and nutrients and then planted with exotic pines. Reclamation has also produced extensive grazing land in the upper Walikato valley on central North Island.

The oceanic tropical climate of the Pacific islands has distinctive wet and dry seasons, exaggerated in the case of the sugar-growing areas by being sheltered from the southeast trade winds. From November to March rainfall in excess of 200 mm (8 in) a month and temperatures of over 25°C (77°F) can be expected; typhoons, too, sometimes occur. Thereafter the weather is usually slightly cooler and drier.

The need for exports

A further problem in the region as a whole is that most farming takes place in areas that are remote from their export markets, and often from domestic ones too. The small local market for food and raw materials has never absorbed more than a fraction of the total farm output, so the building of railroads and roads has been vital in bringing the surplus to coastal ports for local sale or export.

The problem has been exacerbated in the Pacific islands because of the Polynesian and Melanesian people's aversion to commerce. This has made crop production for market even more difficult, even though economically it is of prime importance for the islands.

The need to export has made the region very reliant on its agricultural trade and has left it vulnerable to changes in export markets, especially in islands that depend on a single agricultural export crop, as Tonga does on copra. Even economies that are as well established as those of Australia and New Zealand had to make dramatic changes both to their traditional export markets and to the crops they produce after Britain joined the EC.

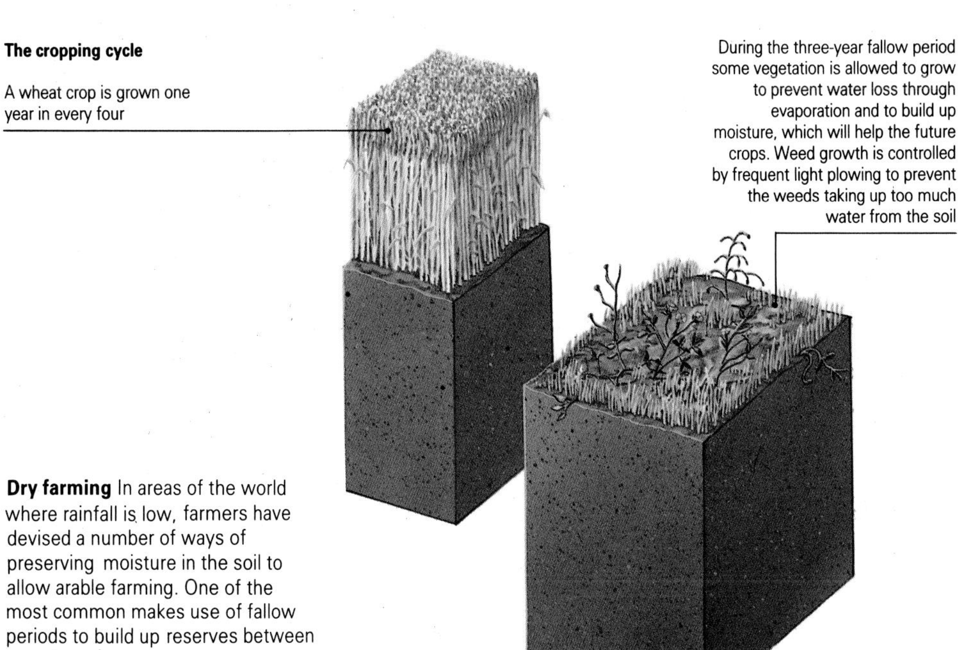

Dry farming In areas of the world where rainfall is low, farmers have devised a number of ways of preserving moisture in the soil to allow arable farming. One of the most common makes use of fallow periods to build up reserves between cropping.

AGRICULTURE AND THE EUROPEAN COMMUNITY

Much of the agricultural trade from Australasia and Oceania was traditionally directed toward Britain, thanks to preferential tariffs and quotas for goods from Commonwealth countries. However, since joining the EC in 1973, Britain has imported most of its remaining needs from within the EC.

This has had a devastating effect on, for example, New Zealand's trade in dairy products. In 1968 two-thirds of these were exported to Britain. By 1977 this had fallen to a third and, in 1988, to below a fifth. The type of cow kept has also been affected: whereas Jersey cows, which yield a very creamy milk, were dominant until the 1970s, there has since been a major change to Friesians in response to declining butter exports, and more recently to increased public concern about high cholesterol levels in dairy products.

New export markets have since been found in Japan, Southeast Asia and the west coast of the Americas. There has also been a great increase in the sale of sheep slaughtered by traditional Halal methods to Islamic countries.

Another development has been the processing of conventional agricultural goods and their sale for a higher price than unprocessed goods. Delicatessen butter packs, ready mixed with herbs and spices, command four times the price of the raw product. Farms have also entered the tourist industry, offering outback experiences and Polynesian island village stopovers.

Island agricultural economies

Coconuts are able to grow on volcanic coastal soils and on low sandy coral atolls: they are therefore grown throughout the Pacific islands. Since copra – the dried meat of the coconut – is nonperishable, it is an ideal export crop for scattered tropical islands. It can be carried by small trading vessels from outlying islands to larger centers, and has consequently become a principal cash crop of the region. Copra and coconut oil products form the basis of Tonga's agricultural economy, for example: 53,000 tonnes of coconuts are produced every year. The crop is extremely vulnerable to the unpredictable weather conditions that occur in the Pacific: hurricanes and tsunami (giant waves caused by volcanic eruptions) are liable to strike any of the archipelago's many scattered volcanic or coral islands at any time.

Tonga's economic dependence on its agricultural exports is made even more critical by a shortage of land for farming: its 36 inhabited islands, which are scattered over 1 million km (386,000 sq mi) of ocean, consist of a total land area of 750 sq km (290 sq mi). Every male of 16 years or more is entitled to an allotment of about 3 ha (7.5 acres) for subsistence farming of crops such as yams, taro, cassava, maize, pineapples and watermelon, but many are deprived of this privilege. Unemployment is high among the population of 97,000, and this has resulted in heavy emigration to New Zealand, Australia and the United States.

Copra: an island specialty (*above*) Coconuts contain a high percentage of oil. The coconut meat is dried in the sun to make copra, which is pressed to produce the valuable oil used in making goods such as candles, soap, cosmetics and margarine.

Shifting cultivation (*below*) In some of the islands, where there is land available, a form of shifting cultivation is practiced. Small garden plots, often some distance from the village, are cultivated for two or three years then abandoned to allow soil fertility to recover.

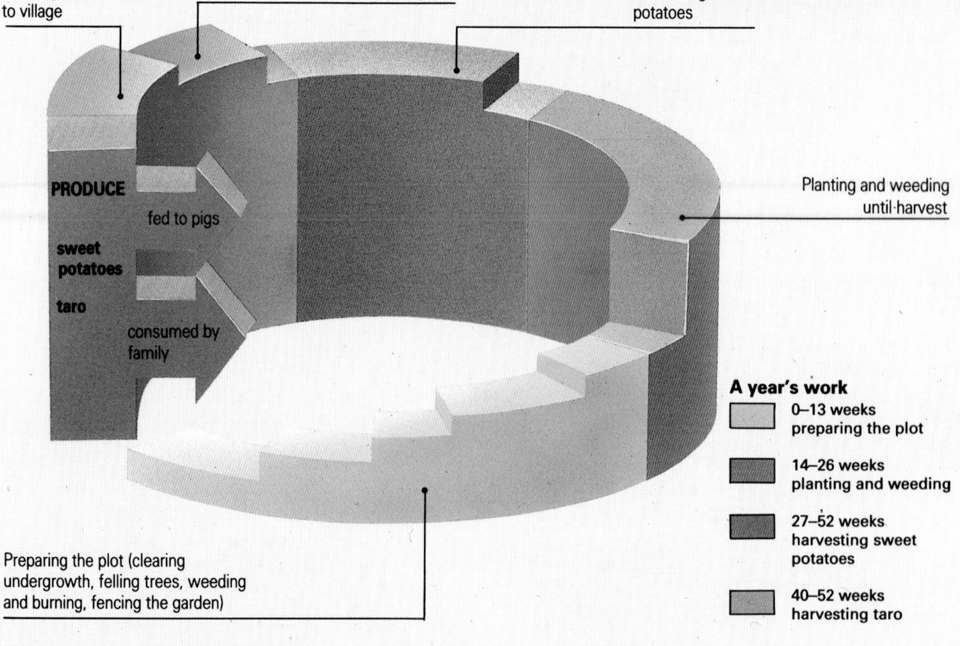

Two farming systems on Fiji

Some 800 km (500 mi) to the west of Tonga lie the Fijian islands. Here sugar cane, ideally suited to the extensive coastal plains of volcanic silts and the hot, wet climate, has become the principal crop. It was developed by the Australian Colonial Sugar Refining Company soon after Fiji became a British possession in 1879. Under the auspices of the CSR, indentured laborers from India were brought to the islands to work the cane fields, and today Indian tenant farmers and their urban counterparts form slightly more than half the population.

Under the CSR, sugar cane began to dominate the landscape, with coconut windbreaks and groves diversifying the scene. The north and west coastal plains of Viti Levu and those areas surrounding Labasa on the north coast of Vanua Levu were devoted to sugar plantations, and

A taro garden Vegeculture – the earliest form of agriculture – is widely practiced on Pacific islands for growing tropical root crops such as taro, sweet potato or yam. Most villagers have their own subsistence plots on the outskirts of the village, shaded by small tropical fruit trees.

Fertile coastal plains The best agricultural land on many of the Pacific islands is formed from volcanic silt soils. Subsistence crops predominate here, though some cash crops, such as bananas, coffee or pineapples, are also grown.

the CSR also controlled large refineries on both these islands. The company continued to bring in indentured Indian labor until 1916.

Once freed from their contracts, after ten years, the Indians leased land from the Fijians and grew sugar cane and coconuts. They were, however, forbidden to buy land. This restriction has continued to the present day, thus leaving the Indian majority largely landless but commercially rich. Sugar is now by far the most important export from the Fijian islands: in 1984 it was worth $106 million, coconut products $5 million and all other exports $29 million.

As well as producing exportable surpluses from the sugar and coconut harvests, the Indian tenant farmers tend market garden crops and cattle. Farming on these plots is often mechanized and linked to the sugar factories by narrow-gauge diesel railroads.

Meanwhile the Fijians still adhere to a modified traditional lifestyle. Their neat villages are surrounded by gardens and groves in which subsistence plots of taro, yam and sweet potato are shaded by coconut, pandanus, breadfruit and banana trees. Pepper root for the local traditional yanggona brew is also an important crop. Pigs are the main domestic animal. These Fijian village gardens are still often worked with hand tools. Villages frequently draw rent from the Indian community, enabling them to buy goods they cannot themselves produce.

Sheep country

Although sheep were introduced to New Zealand less than 200 years ago, today they greatly outnumber the country's people. They were brought to the islands from Australia where merino breeds, noted for their fine wool, had been introduced from Europe and had proved highly suited to the sparse grazing conditions there.

Sheep are bred for different environmental conditions and to satisfy human needs for clothing and food. Sheep with fine wool are generally raised for wool alone, while breeds with medium or long wool are raised primarily for their meat. The moist mountain pastures of New Zealand's South Island, improved with European grasses, proved ideal for fattening sheep: consequently breeds of merino were crossbred with longwool breeds to produce better meat-producing sheep. One successful crossbreed, the Corriedale, produces excellent lamb carcasses, and has been introduced from New Zealand to Argentina and Chile as well as the United States.

The development of refrigeration in the 1880s meant that sheep meat from the southern hemisphere could be transported to markets in Europe, and New Zealand subsequently rose to become the largest exporter of sheep meat in the world. Meat packing stations, where carcasses are prepared and chilled or frozen for onward shipment to overseas markets, are a major industry. The loss of its largest overseas market following Britain's accession to the European Community in 1973 dealt the meat industry a major blow, but alternative markets have been energetically developed. Farmers have diversified into other agricultural activities, and into forestry, but opportunities are fairly limited in the grasslands of the South Island, and sheep production remains important.

After the shearing Vast flocks of sheep on New Zealand's South Island are mustered once a year for shearing and for slaughter.

GLOSSARY

Agave (Agavaceae)
A group of plants, native to the south of the USA and tropical America, with long, thick, fleshy leaves. Some species, such as sisal, are grown commercially for the FIBER in the leaves.

Agribusiness
The various businesses that support farming, especially those that buy, process and distribute farm products.

Agriculture
The practice of cultivating land and farming DOMESTICATED animals.

Agrochemicals
The numerous human-made chemicals that are used in AGRICULTURE to control pests, diseases and weeds.

Agroforestry
Cultivation systems that integrate FORESTRY and farming. It allows a diversity of products to be produced without exhausting any particular component of the ENVIRONMENT.

Alfalfa (*Medicago sativa*)
Also known as lucerne; a plant that grows to about 1 m (3 ft), widely grown as FODDER for cattle. It fixes nitrogen in the soil, thus reducing the amount of fertilizer a farmer needs to use.

Alpaca (*Lama pacos*)
A hoofed, shaggy-coated animal native to the highlands of South America. It has been DOMESTICATED and is kept mainly for its long fleece.

Aquifer
A stratum of permeable rock, gravel or sand containing water, which can be brought to the surface from wells.

Arable
Land that is cultivated for crops. Arable farming is the activity of cultivating the land and growing crops.

Artificial insemination
Semen collected from the male is artificially placed into the vagina of the female at the time she is most likely to conceive. It is widely practiced in LIVESTOCK breeding.

Barley
A CEREAL grass belonging to the genus *Hordeum*, cultivated successfully from subtropical to subarctic regions. It is used in human and animal foods and in the brewing industry.

Battery farming
Originally used to describe the method of confining poultry in small cages within large sheds for EGG PRODUCTION, but now commonly used to describe other forms of INTENSIVE animal HUSBANDRY in which the movement of the animals is severely restricted.

Beef cattle
Hoofed, ruminating animals belonging to the family Bovidae that have been DOMESTICATED and are killed for eating. The word "cattle" is mainly used for the domestic ox (*Bos taurus*).

Biotechnology
Technology applied to biological processes, including genetic engineering, the manipulation of the genetic makeup of living organisms.

Boll
The fruit of plants such as COTTON. It consists of a spherical capsule containing seeds.

Boreal forest
The name given to the coniferous forests of the northern hemisphere, where there are long, cold winters and short summers.

Breadfruit
The edible fruit of the tropical tree *Artocarpus communis*. It can be as much as 30 cm (12 in) across, and beneath its tough skin is a starchy interior. It is baked or roasted, and has a texture similar to that of bread.

Broadcast
The scattering of seed or other agricultural INPUTS on to the land, especially by hand.

Buckwheat
Plants belonging to the genus *Fagopyrum*. The species most commonly cultivated for its seed or as animal FODDER is *Fagopyrum esculentum*.

Bush fallowing
The system of farming in tropical areas whereby the land is allowed to revert to bush to regain its fertility after a period of cultivation.

Carbohydrates
A large group of organic compounds, including sugars, that contain carbon, hydrogen and oxygen. Green plants manufacture carbohydrates during photosynthesis, and are an important source of food energy for animals.

Cash crop
A crop that a farmer grows for sale rather than for SUBSISTENCE.

Cassava (*Manihot*)
A genus of tropical plants that are cultivated for their edible, starchy, tuberous roots, which are processed and made into tapioca or cassava meal.

Center-pivot irrigation
A form of IRRIGATION in which water is fed into a horizontal pipe pivoted at one end. As it moves in a circular motion, water is drip-fed onto the land along the pipe's length.

Cereals
Cultivated grasses that have been selectively bred to produce high yields of edible GRAIN for consumption by humans and LIVESTOCK. The most important are WHEAT (*Triticum*), RICE (*Oryza saliva*) and MAIZE (*Zea mays*).

Chernozem
A type of soil, also known as black earth, found in the grassland areas of continental interiors. It is one of the richest soils for cultivation.

Chickpea (*Cicer arietinum*)
Annual plant belonging to the pea family (*Fabaceae*) widely grown for its nutritional yellow-brown seeds; known as gram in India. It may be used to make a kind of meal or flour.

Citrus fruit
Juicy fruit from trees belonging to the genus *Citrus* such as oranges, lemon, and limes. The trees are evergreen, have glossy leaves and grow naturally in tropical and subtropical regions.

Cocoa
One of the ingredients of chocolate, cocoa is derived from cocoa beans, which are the seeds of the cacao tree (*Theobroma cacao*). They are found in yellowish pods that grow directly from the trunk. The tree is native to tropical America, but is cultivated mainly in west Africa.

Coconut
The fruit of the coconut palm (*Cocos nucifera*), which grows in tropical areas. The nut is encased in a fibrous husk used to make rope and matting. Oil can be obtained from the white kernel. The residual material is fed to LIVESTOCK.

Coffee
The seeds, known as beans, of tropical evergreen trees belonging to the genus *Coffea*. Several species of *Coffea* are used, but the highest quality beans come from *Coffea arabica*.

Coir
The FIBER from the fibrous husk of the COCONUT used in making ropes and matting.

Collective
A form of COOPERATIVE farming introduced mostly in communist countries, in which the state nationalized the land, livestock and crop reserves of the PEASANTS and created large farming units. The state also determined what was produced.

Collectivization
The process of bringing small farms together to make COLLECTIVES in communist countries.

Combine harvester
An agricultural machine that harvests grain crops, combining the functions of reaping and THRESHING.

Commodity
An agricultural product that is traded on the commercial market.

Common Agricultural Policy (CAP)
The agricultural policy of the EUROPEAN COMMUNITY (EC) designed to ensure that farmers have a reasonable standard of living and the public has secure supplies of affordable food.

Contract labor
Workers hired temporarily to do a particular seasonal job on the farm, such as harvesting or shearing. They often use their own equipment.

Cooperative
A term used in agriculture to describe a variety of community schemes whereby farmers come together to share resources, expertise and profits. They aim to improve productivity, the quality of produce and opportunities to market it. They are often encouraged by government.

Copra
The dried white kernel of the COCONUT from which coconut oil is extracted.

Cotton
Various herbaceous plants belonging to the genus *Gossypium*. They grow naturally in tropical and subtropical regions. Some species are grown for the FIBER that surrounds the seeds in the BOLL. The longest fibers are obtained from *G. bardadense*.

Cultivate
To prepare and use the land for growing crops.

Dairying
The business of keeping cattle, goats and sheep for the PRODUCTION of milk. Some is used fresh and some processed into dairy products such as butter and cheese.

Deciduous
Trees or shrubs that shed their leaves each year during the winter or dry season.

Digging stick
A simple tool used to prepare the surface soil prior to planting.

Domesticate
Animal species that have been brought under human control.

Drill
The placing of seeds directly into the soil to aid germination, normally using a machine that makes a hole and drops the seed into it.

Dryland farming
A method of producing crops in dry areas so as to make maximum use of the limited moisture available. It can involve leaving a protective layer of crop residue to reduce evaporation, contour plowing, and alternating cultivation with a period of FALLOW.

Ecosystem
A community of plants and animals and the ENVIRONMENT in which they live and react with each other.

Eggplant (*Solanum melongena*)
An alternative name for aubergine. The plant is grown in warm areas for its large, egg-shaped fruit, which typically has as a dark purple skin and white interior. It is eaten as a VEGETABLE.
Environment
The surroundings in which all animals and plants live.
European Community (EC)
Formerly known as the European Economic Community (EEC) or the Common Market, the EC is an organization of Western European states set up to encourage economic cooperation, with the eventual aims of achieving economic unity and a high degree of political unity. It was created by the six countries that signed the Treaty of Rome in 1957. In 1990 it had 12 members.
Extensive farming
Farming practice that operates over a large area using minimal INPUTS of capital and labor. Yields per hectare are lower than in INTENSIVELY farmed areas, but this is compensated for by higher PRODUCTIVITY.

Fallow
Farmland that is tilled but left uncultivated for a period of time to allow it to regain its fertility.
Feedlot production
A system of farming LIVESTOCK in which the animals are confined to an enclosed area, fed and fattened.
Fertile Crescent
A fertile area of the Middle East, watered by the Euphrates and Tigris rivers, that was an early center of agriculture and human settlement.
Fertilizers
Natural and artificial substances added to the soil to maintain or improve its fertility. The most commonly used in conventional farming are ammonium nitrate, PHOSPHATE and potassium.
Fiber crop
A plant grown commercially for its fiber, which can be used in the manufacture of such products as linen, matting and rope.
Flax
A FIBER CROP belonging to the genus *Linum. L. usitatissimum* is cultivated for its stem fibers, used to make linen, writing paper and cigarette paper. The seeds contain linseed oil.
Fodder
Feed for LIVESTOCK.
Forage crop
A leafy crop grown for feeding to LIVESTOCK.
Forestry
The planting, management and harvesting of trees in an area of forest.
Free range farming
Non-intensive management of farm animals.

Grain
Small, hard seeds produced by CEREAL crops.
Grain elevator
A storage building for GRAIN; also the device for lifting grain into a building, usually using an endless belt fitted with scoops or crosspieces.
Gram
Alternative name for CHICKPEA.
Green Revolution
The introduction of HIGH-YIELDING VARIETIES of seeds and modern agricultural techniques to increase agricultural PRODUCTION in developing countries. It began in the early 1960s.
Greenhouse cultivation
Growing plants and crops in glass or transparent plastic structures where the growing conditions can be carefully controlled.
Greenhouse effect
The process by which radiation from the Sun passes through the atmosphere, heats the surface of the Earth, is radiated back into the atmosphere and is then absorbed by gases in the atmosphere. The buildup of carbon dioxide and other "greenhouse gasses" is increasing the effect. There are fears that the temperature of the planet may rise as a result.
Groundnuts (*Apios tuberosa*)
The edible underground tubers of the groundnut plant, native to North America; also known as peanuts.
Groundwater
Water that has percolated into the ground from the suface, filling pores, cracks and fissures. An impermeable layer of rock prevents it from moving deeper so that the lower levels become saturated. The upper limit of saturation is known as the water table.
Growing season
The period of the year when the average temperature is high enough for plants to grow. It is longest at low latitudes and altitudes. Most plants can grow when the temperature exceeds 5°C (42°F).

Hacienda
A ranch or large estate in Spain or Spanish-speaking countries.
Hardwoods
The TIMBER from broadleaf trees such as oak, ash and beech. It is a classification, not a description of the qualities of the wood, which may be hard or soft. See also SOFTWOOD.
Hay
Grass that has been cut and dried for use as FODDER for LIVESTOCK.
Hemp (*Cannabis sativa*)
An annual herb native to central Asia. It is cultivated for its FIBER, which is used for making ropes, sacking and sailcloth. It also contains a resin that is a source of marijuana.
Herbicides
Chemicals used by farmers to control weeds.
High-yielding variety (HYV)
Varieties of crop plants that have been specially bred to produce high yields. They are often HYBRIDS and require INPUTS of FERTILIZERS.
Hoe
A long-handled farming tool that has a small blade for TILLING surface soil.
Homestead Act
An act passed in the USA in 1862 that granted settlers up to 65 ha (160-acres) of public land.
Hops
Climbing herbs native to Europe and Asia belonging to the genus *Humulus*. The most commonly cultivated variety is *H. lupus*. The female flowers are used to flavor beer.
Hormones
Chemical substances that can have special effects on plants and animals, such as promoting growth.
Horticulture
The INTENSIVE commercial PRODUCTION of fruit and vegetables, and ornamental trees and shrubs.
Hunter–gatherers
People who obtain their food requirements by hunting wild animals and gathering the berries and fruits from wild plants.
Husbandry
Farming in a caring and skillful manner.
Hybrid
An animal or plant that is the offspring of two genetically different individuals. Hybrid crops are often grown because they give higher yields and are more resistant to disease.
Hydroponics
A method of cultivating plants without soil. The roots of the plant are generally held in an inert medium such as sand and fed a controlled solution of the nutrients that the plant needs.

Indentured labor
A form of CONTRACT LABOR in which farm workers make a contract for a specific period with a farmer or landowner. The conditions are often disadvantageous for the worker, involving surrendering freedom of movement and the right to quit work.
Inputs
All the resources required for agricultural PRODUCTION including, for example, seeds, FERTILIZER, fuel, labor and capital.
Intensive farming
Farming that uses high inputs of capital and labor to maximize PRODUCTION from a small area of land.
Intercropping
A method of farming in which two or more crops are grown on a piece of land side by side.
Intervention buying
The method used by the European Community's COMMON AGRICULTURAL POLICY to maintain the target price for agricultural COMMODITIES produced by EC farmers. If the price on the EC market falls below target price, to raise it commodities are bought and stored. If the price goes above the target, the stored commodities are sold, thus reducing the price.
Irrigation
Watering farmland artificially, for example by sprinklers or canals.

Jute (*Corchorus capsularis* or *C. olitorius*)
A FIBER CROP cultivated in Asia, used to make ropes, sacks, hessian, carpet backing and tarpaulin.

Kibbutz (im)
An agricultural settlement developed in Israel, in which the land and property are owned communally.
Kiwi fruit (*Actinidia chinensis*)
The edible fruit of a Chinese climbing herb, sometimes known as a "Chinese gooseberry".
Kulaks
The wealthy peasants of Russia who, after 1906, were proprietors of their own Russian farms.

Land tenure
Any of a number of arrangements by which farmers have the right to farm an area of land, for example as the owner or tenant.
Laterites
A group of deposits formed by the weathering of rocks in tropical regions and consisting mainly of aluminum and iron oxides.
Latifundia
Originally large estates formed when the Romans reallocated land confiscated from the conquered communities. Now, used to describe large estates, especially those in Mediterranean countries and their former overseas colonies.
Leaching
The process by which surface minerals are moved down through the soil by percolating water.

Legumes
Any plant of the family Leguminosae, characterized by bearing their seeds in a pod and their ability to fix nitrogen in their roots. They include peas, beans and ALFALFA.
Livestock
DOMESTICATED animals that are managed to produce milk, eggs, meat, wool, etc.
Llama (*Lama glama*)
South American animal kept for meat and wool, as well as for carrying goods.
Loess
A fine-grained sediment, usually buff or yellowish in color, deposited after being transported by the wind in dust storms. It makes fertile soil.

Maize (*Zea mays*)
A widely grown CEREAL native to the New World. It bears cobs, each with long, parallel rows of GRAINS encased in a leafy covering.
Mango
The edible fruit of the large tropical evergreen tree *Mangifera indica*. Weighing up to 2 kg (4.4 lb), it has tough skin and juicy, orange flesh.
Manioc
An alternative name for CASSAVA.
Manure
Most commonly used to describe animal excreta, often mixed with STRAW. It is spread on the land to FERTILIZE it.
Market gardening
An alternative name for HORTICULTURE.
Mechanization
The introduction of machines to replace human labor and animal power in farming.
Mediterranean climate
Areas that have a climate similar to that of the Mediterranean region, namely warm, wet winters and hot, dry summers.
Merino sheep
A breed of sheep originating from Spain, valued for its high-quality fleece.
Microclimate
The climate over a small area that is different from the general climate of the surrounding area because it is influenced by local conditions such as a hill or lake.
Millet
One of various varieties of grasses cultivated for their seed. It is an important food crop, used to make flour or porridge, in Africa and in the driest parts of India and Pakistan.
Minifundia
Small SUBSISTENCE farms in Latin America.
Mixed farming
A farming system that combines ARABLE and LIVESTOCK farming.
Moldboard
The curved blade of a PLOW that turns over the soil and makes a furrow.
Monoculture
When one type of crop is grown on the land year after year rather than ROTATING the crops grown, for example a PLANTATION.
Monsoon
Generally used to describe winds that change direction according to the season. They are most prevalent in southern Asia, where they blow from the southwest in summer bringing heavy rainfall, and from the northeast in winter.
Mulch
A protective covering, often of partially rotted vegetable matter, sometimes of plastic, that is spread on the land to enrich it or to prevent moisture loss and soil erosion.

Nomadic pastoralism
The way of life of pastoral nomads, who travel with their LIVESTOCK according to the availability of grazing land.

Oats (*Avena satina*)
A grass grown in temperate regions for its GRAIN. It grows to a height of about 1 m (3 ft0 and has a branching cluster of seeds. It is used as a human food and for feeding LIVESTOCK.
Oil palm
A palm tree native to tropical regions, with dark, fleshy fruits from which oil is extracted for use in soaps, margarine and lubricants.
Oilseed
The seeds of plants from which edible oils can be extracted, such as COTTON, GROUNDNUTS, RAPE and SUNFLOWER.
Olive (*Olea europaea*)
An evergreen tree up to 12 m (40 ft) in height that is widely grown in Mediterranean and subtropical regions for its fleshy oval fruits. These can be pickled and eaten or pressed to produce a high quality edible oil.

Paddy
The harvested RICE kernel before it has been milled to remove the hull or husk.
Paddyfield
A flooded field in which RICE is planted and grows. The fields are drained to allow the rice grains to ripen and be harvested.
Papaya
Also known as pawpaw. The fruit of a small tree (*Carica papaya*) that grows in tropical regions. The orange flesh is sweet and juicy.
Pastoralism
A way of life based on raising and managing LIVESTOCK. Most pastoralists are seminomadic or nomadic, moving their animals between areas of seasonal PASTURE.
Pasture
Land under grass used for grazing LIVESTOCK. Most pasture is permanent; in temperate regions it may be grown in ROTATION with other crops.
Peasant
A smallscale agricultural producer, often a tenant farmer, usually engaged in SUBSISTENCE farming.
Pesticides
Chemicals used by farmers to control pests that can damage crops, such as insects and rodents. The word is often used as a general term for HERBICIDES, insecticides and fungicides.
Phosphate
A general name for FERTILIZERS that contain phosphorous compounds.
Pigeon peas (*Cajanus cajan*)
A tropical shrub (see LEGUME) that is cultivated for its nutritious pea-like seeds.
Plantation
A large estate where the original vegetation has bee replaced by a single CASH CROP.
Plow
An agricultural tool that is pulled or pushed through the soil to prepare it for planting crops.
Podsol
A soil that is typically found in cool areas of coniferous forest. It is characterized by a gray-white color in its upper layer from which minerals have been leached.
Polder
An area of level, lowlying land that has been reclaimed from the sea or a lake. They are normally cultivated as the soil is very fertile.
Potato (*Solanum tuberosum*)
A plant native to the Andes in South America, but now widely grown around the world as a VEGETABLE food crop under the ground.
Poultry
DOMESTICATED birds that are reared for their meat, eggs or feathers.
Prairie
The flat grassland in the interior of North America, much of which has been PLOWED and is used to grow CEREAL crops.
Production
Plant and animal products produced in farming.
Productivity
A measure of the output of a farm system compared with the INPUTS required.
Protein
Organic compounds essential to living organisms. They consist of one or more chains of amino acids linked by peptide bonds.
Pulse
The seeds of any plant (see LEGUME) belonging to the family Papilionaceae, including peas, beans and lentils.

Qanat
Underground tunnel systems of ancient origin dug to transport water long distances for IRRIGATION in desert areas.
Quotas
The setting of the share of total PRODUCTION that an individual farmer can or should supply.

Rainfed agriculture
Farming that relies on methods of preserving moisture from natural rainfall and does not use IRRIGATION.
Rainforest
Usually known as tropical rainforest and found in the equatorial belt where there is heavy rain and no marked dry season. Growth is very lush and rapid. Rainforests probably contain half of all the world's plant and animal species.
Rainshadow
An area that has relatively light rainfall because it is sheltered from the prevailing rain-bearing winds by a range of mountains or hills.
Ranching
The management of LIVESTOCK on a large farm or estate over which the animals graze at low density. It is an example of EXTENSIVE FARMING.
Rape (*Brassica napus*)
A plant of the mustard family grown for its oil-rich seeds and as FODDER.
Rice (*Oryza sativa*)
A CEREAL grass widely cultivated in tropical and subtropical regions. Lowland rice is grown planted in PADDYFIELDS, upland rice in dry fields.
Rotation
The practice of growing different crops in succession on the same piece of land to help control weeds, pests and diseases and make economic use of the nutrients in the soil.
Rough grazing
Land in mountainous or arid areas with sparse vegetation that supports LIVESTOCK such as sheep and goats.
Rubber (*Hevea brasiliensis*)
A tree that produces latex, which is made into rubber. Native to the Amazon rainforest, it is mainly obtained from PLANTATIONS in Southeast Asia.
Rye (*Secale cereale*)
A CEREAL grass grown in cool temperate areas for its GRAIN. It is milled to make black bread and used as feed for LIVESTOCK.

Salinization
The accumulation of soluble salts near or at the surface of the soil. This can occur when water used for irrigation evaporates. Eventually the land becomes worthless for cultivation.
Savanna
Areas of open grassland with scattered trees, found in tropical regions between tropical RAINFORESTS and hot deserts. There is a marked dry season, and too little rain even in the rainy season to support large areas of forest.
Schist
A hard metamorphic rock with coarse grains, formed from igneous and sedimentary rocks.
Seed bed
A specially prepared area of land where seeds are sown. Once they have grown into seedlings, they are transplanted.
Self-sufficiency
The ability of a farming family or nation to grow all the food it needs to support its members.
Sericulture
Management of silkworms to produce raw silk.
Set aside
The practice in the EUROPEAN COMMUNITY of taking farmland out of cultivation to help reduce surpluses of some farm products.
Shadoof
A simple construction consisting of a pivoted pole with a bucket at one end and a counter-balancing weight at the other, used to raise water for IRRIGATING fields.
Sharecropping
The practice of a farmer paying a proportion of the harvest to the landowner as rent.
Shifting cultivation
A method of farming prevalent in tropical areas in which a piece of land is cleared and cultivated until its fertility is diminished. The farmer then abandons the land, which restores itself naturally.
Silage
FODDER for cattle produced by storing undried grass or CEREAL crops in a SILO. It ferments and organic acids pickle the crop, making a feed that is nutritious and easily digested.
Silo
The airtight structure in which SILAGE is made.
Slash-and-burn farming
A method of farming in tropical areas where the vegetation cover is cut and burned to FERTILIZE the land before crops are planted.
Smallholding
A very small area of land with a farmhouse, used for agriculture. It is usually capable of providing enough food for the farmer's family with a small surplus for sale. Today, many are farmed on a part-time basis.
Softwood
The wood from coniferous trees. See also HARDWOOD.
Sorghum
Grasses belonging to the genus *Sorghum*. The seeds or GRAIN are used for making bread and as a source of edible oil, starch, and sugar.
Soybean (*Glycine max*)
A plant widely grown for its protein-rich seeds. They can be made into a variety of foods, including meat and milk substitutes. The plants fix nitrogen in their roots and are also grown as cattle FODDER.
Squash
Marrow-like, fleshy, fruits of the genus *Cucurbita*, usually eaten as a VEGETABLE. The variety most commonly grown is *C. maxima*.
Stall feeding
An intensive method of rearing LIVESTOCK in which animals are kept in a small stall.
Staple
The principal food crop eaten by a group of people.
Steppe
Open grassy plains with few trees or shrubs, found in temperate regions characterized by low and sporadic rainfall and experiencing wide annual variations in temperature.
Straw
The stalk of a CEREAL crop once the GRAIN has been removed. It can be used as LIVESTOCK bedding, as FODDER, or as a material for roofing.
Strip field farming
A system of farming in which the land is cultivated in strips, now used in areas prone to soil erosion.
Stubble
The short lengths of STRAW left in the field when a CEREAL crop has been harvested. It is sometimes burnt as a means of controlling pests and diseases and FERTILIZING the soil.
Subsistence farming
A type of farming in which the farmer's family consumes most of the farm output and there is little left for sale.
Sugar beet (*Beta vulgaris*)
A crop widely cultivated in temperate regions for its large bulbous root, from which sugar can be extracted. The leaves are used as FODDER.
Sugar cane
A tropical grass belonging to the genus *Saccharum*. Sugar can be extracted from the stem, or cane.
Sugar pea (*Pisum*)
A type of pea with edible pods, also known as mangetouts or snow peas.
Sunflower
A plant belonging to the genus *Helianthus*, cultivated on a large scale for its seeds, from which oil can be extracted.
Sustainability
The concept that human activities should aim to meet present needs without compromising the ability of future generations to meet their needs.
Sweet potato (*Ipomoea batatas*)
A tropical plant widely cultivated for its large edible tubers, which grow under the ground. They are cooked and eaten like a POTATO.
Swidden
A temporary agricultural plot produced by cutting back and burning the vegetation.

Taiga
Name given to the coniferous forest belt of the northern hemisphere, bordering TUNDRA in the north and forests and grasslands in the south.
Taro (*Colocasia esculenta*)
A plant widely cultivated in tropical regions for its edible tubers; also known as dasheen.
Tea (*Camellia sinensis*)
A bush or tree that grows in tropical and subtropical regions. Its shoots and leaves are picked, dried and then infused with hot water to make a drink.
Terrace
A level area of land cut out of a slope allowing crops such as WET RICE to be grown.
Threshing
The action of separating GRAIN from STRAW and husks using hand implements or a machine.
Till
To break the soil, by plowing, harrowing or hoeing, to prepare it for planting with seed.
Timber
Wood used in construction work rather than as fuel or for paper. It may be HARDWOOD or SOFTWOOD. It is also known as lumber.
Tobacco
A plant belonging to the genus *Nicotiana*. The large leaves are dried, fermented and made into cigarettes, cigars, etc.
Transhumance
A type of PASTORALISM in which the LIVESTOCK is moved seasonally between one area of PASTURE and another. The farmers usually move as well to stay with their animals.
Tree crop
Any tree cultivated as crop, usually in PLANTATIONS, such as COCOA or RUBBER.
Triticale
A hybrid CEREAL that is a cross between WHEAT (*Triticum*) and RYE (*Secale*).
Tube well
A pipe driven down into the ground to reach the water table and bring water to the surface.
Tundra
The land lying in the very cold northern regions of Europe, Asia and Canada, where the winters are long and cold and the ground beneath the surface is permanently frozen.

Vegeculture
The simple cultivation of root crops such as CASSAVA or SWEET POTATO.
Vegetable
Any of various herbaceous plants having parts that can be eaten, such as peas, cauliflower, beans and cabbage.
Vines (*Vitis vinifera*)
The name refers to a group of climbing plants; most often used in relation to the grape vine, whose fruit can be eaten fresh, dried or crushed and made into grape juice or wine.
Viticulture
The cultivation of grape VINES.

Waterlogging
Land that has become completely saturated with water, often as a result of IRRIGATION.
Wet rice
The variety of RICE that is transplanted into flooded PADDYFIELDS as seedlings. Upland rice or dry rice is not grown in flooded fields.
Wheat
A CEREAL belonging to the genus *Triticum*, native to the Mediterranean and western Asia. Varieties have been bred that can grow in a wide range of climatic conditions. Most wheat is milled into flour.
Winnowing
The separation of the GRAIN from the chaff using natural or artificially created currents of air.

Yam
A plant belonging to the genus *Dioscorea*. It is cultivated in moist tropical conditions for its edible tubers, which are eaten like POTATOES. The tubers are long and can weigh as much as 45 kg (100 lb).
Yield
The amount of produce from crops and animals, usually expressed as a unit such as "x" kg per hectare or "y" liters per day.

Zebu (*Bos indicus*)
A type of DOMESTIC cattle found in Africa and Asia, characterized by a hump over the shoulders, a loose fold of skin under the throat and long horns.

Further reading

Bayliss-Smith, T.P. and Wanmali, S. (eds) *Understanding Green Revolutions* (Cambridge University Press, Cambridge, 1984)
Bennett, J. *The Hunger Machine: The Politics of Food* (Polity in association with Channel Four Television and Yorkshire Television, London, 1987)
Briggs, D. *Agriculture and the Environment* (Longman, London, 1985)
Comoy, C. and Litvinoff, M. *The Greening of Aid* (Earthscan, London, 1988)
Currey, B. and Hugo, G. (eds) *Famine as a Geographical Phenomenon* (Reidel, Dordrecht and Boston, 1984)
Curtis, D., Hubbard, M. and Shepherd, A. *Preventing Famine: Policies and Prospects for Africa* (Routledge, London, 1988)
Dando, W. *The Geography of Famine* (Winston, Washington DC and Edward Arnold, London, 1980)
George, S. *How the Other Half Dies: The Real Reasons for World Hunger* (Penguin, Harmondsworth, 1976)
Goodman, D., Sorj, B. and Wilkinson, J. *From Farming to Biotechnology: A Theory of Agro-industrial Development* (Blackwell, Oxford, 1987)
Grigg, D. *An Introduction to Agricultural Geography* (Hutchinson, London, 1984)
Grigg, D. *Agricultural Systems of the World: An Evolutionary Approach* (Cambridge University Press, London, 1974)
Ilbery, B.W. *Agricultural Geography: A Social and Economic Analysis* (Oxford University Press, Oxford, 1985)
Jackson, ?. *Against the Grain* (Oxfam, Oxford, 1982)
Lappe, F.M. and Collins, J. *World Hunger: Twelve Myths* (Earthscan, London, 1988)
Lipton, M. and Longhurst, R. *New Seeds and Poor People* (Unwin Hyman, London, 1989)
Morgan, D. *Merchants of Grain* (Weidenfield and Nicolson, London, 1979)
Nicholson, B.E., Harrison, S.G., Masefield, G.B. and Wallis, M. *The Illustrated Book of Food Plants* (Peerage Books, London, 1985)
Pacione, M. *Progress in Agricultural geography* (Croom Helm, London, 1986)
Parry, M. *Climate Change and World Agriculture* (Earthscan, London, 1990)
Sen, A. *Poverty and Famines: An Essay on Entitlement and Deprivation* (Clarendon Press, Oxford, 1981)
Singer, H., Wood, J. and Jennings, A. *Food Aid: The Challenge and the Strategy* (Clarendon Press, Oxford, 1987)
Tarrant, J.R. *Food Policies* (Wiley, Chichester, 1980)
Tudge, C. *Food Crops for the Future: The Development of Plant Resources* (Blackwell, Oxford, 1988)
Tudge, C. (1979) *The Famine Business* (Penguin, Harmondsworth)
Warnock, J.W. *The Politics of Hunger* (Methuen, London, 1987)

Sources of data
The data panels in this volume have been gathered from the Food and Agriculture Organisation of the United Nations *Production Yearbooks, Trade Yearbooks* and *Fisheries Yearbooks*, published in Rome. Data are for 1987.

Farmers include all economically active people in agriculture, fishing and forestry. This number has then been divided into the arable land area to give a measure of agricultural population density.

Major crops. The area is the area that was harvested in 1987. The yield is a measure of the production per unit area and production is the total production of that crop for the region. Percentage change in production volumes since 1963 have also been calculated. Where a crop is a recent introduction the percentage increases may be very large. For fruit and vegetable crops production data only are available.

Major livestock. Numbers in 1987, and the tonnage of milk and fish produced, are shown, with percentage changes in numbers and production since 1963.

Food security shows the trade balance in cereals for an average of the years 1985, 1986 and 1987. This is calculated as a percentage of domestic cereal production and as a percentage of all world trade.

Acknowledgments

Picture credits
Key to abbreviations: ANT Australasian Nature Transparencies, Heidelberg, Victoria, Australia; **BCL** Bruce Coleman Ltd, Uxbridge, UK; **E** Explorer, Paris; **FSP** Frank Spooner Pictures, London; **HDC** Hulton Deutsch Collection, London; **HL** Holt Studios Ltd, Hungerford, UK; **M** Magnum, London; **MIL** multi-Image Library, Kington St Michael, UK; **NF** Naturfotografernas Bildbyra, Osterbybruk, Sweden; **NHPA** Natural History Photographic Agency, Ardingly, UK; **OSF** Oxford Scientific Films, Long Hanborough, UK; **PEP** Planet Earth Pictures/Seaphot, London; **RHPL** Robert Harding Picture Library, London, UK. b=bottom, c=center, l=left, r=right, t=top.

1 ANT/Bill Bachman **2** ANT/C&S Pollitt **3** PEP/Ken Lucas **4** PEP/ Rod Salm **6–7** E/A. Sferlazzo **8–9** Zefa/Miller **12** HL/J. von Puttkamer **12–13** Ancient Art and Architecture Collection/R. Sheridan **14–15** E/P. Roy **15** HDC/Bettman **16–17** E/Hug **17l, 17r** HSL/Nigel Catlin **18–19** NHPA/N. Callow **19** ANT/Bill Bachman **20** Christine Osborne **20–21** ANT/P. Jeans **21** E/C. Delu **22** Bridgeman Art Library/Victoria and Albert Museum **23** Richard and Sally Greenhill **24** E/J.L. Gobert **25** PEP/D. Barrett **26–27** ANT/Kelvin Aitken **27** Zefa/T. Lancefield **28–29** E/Alain Thomas **29** PEP/Hans Christian Heap **30–31** Zefa/Starfoto **32** M/Philip J. Griffiths **32–33** FSP/Gamma/ Photonews/Marc Deville **34** Zefa/Heilman **34–35** Zefa/Armstrong **36–37** E/G. Boutin **36** ANT/Denis & Theresa O'Byrne **38** E/F. Gohier **38–39** E/P.Roy **39** Zefa/Heilman **40** E/F. Gohier **40–41** ANT/Jan Taylor **44** PEP/John Lythgoe **44–45** Zefa/Hunter **46–47** Michael Dent **47** M/Kryn Tacomis **48–49** Zefa/Damm **50,51** RHPL **52** Zefa/K. Goebel **53** Zefa/J. Pickerell **54** FSP/Gamma/Sander-Liaison **54–55** NHPA/John Shaw **56,57t** Zefa **57b** E/Photo Researchers/Gary D. McMichael **58** Art Directors/Craig Aurness **59** E/Arthus-Bertrand **60–61** E/F. Gohier **61** E/Photo Researchers/Eunnice Harris **62** Zefa/Damm **64** US Government Department of Agriculture **64–65, 65** E/Earl Roberge **66** Art Directors **66–67** E/F. Gohier **68–69** Art Directors/Chuck O'Rear **71l** E/J.P. Courau **71r** E/Basin **72** Zefa/Karl Kummels **73** OSF/Philip Sharpe **74** E/C. Beziau **74–75** OSF/A. Butler **75** E/J.P. Courau **76–77** OSF/Sean Morris **78** HL/B. Moser **79** E/D. Desjardins **80** Tropix Photo Library/D. Charlwood 80–81 E/Samuel Costa **82** E/Pierre Cheuva **83t** BCL/Nicholas DeVore III **83b** NHPA/G.I. Bernard **84–85** E/F. Gohier **86–87** E/Luc Girard **88** Tore Hagman **89t** E/L. Giraudou **89b** E/B. Gerard **90** Ulf Risberg **91** NHPA/Patrick Fagot **92** Lars Jarnemo **93** Tore Hagman **94** NHPA/E.A. Janes **96–97** PEP/John Lythgoe **97** NHPA/Picture Box **98** Christine Osborne **98–99** PEP/Ivor Edmonds **99** MIL/J. Roland **100–101** NHPA/E.A. Janes **102–103** RHPL **104** E/Berthoule **106–107** E/Jean Paul Nacivet **107** E/Pierre Tetrel **108–109** E/Francois Jourdan **109** E/P. Roy **110** E/Hug **110–111** OSF/Raymond Blythe **112–113** Zefa/ Rossenbach **115** E/A. Saucez **116, 117, 118, 118–119, 119, 120, 120–121, 122–123** NHPA/Picture Box **124** E/Berthoule **126** E/P. Tetrel **126–127** MIL/J. Roland**127** Zefa **128** Zefa/V. Wentzel **128–129** E/P. Roy **130** MIL/J. Roland **130–131** E/P. Tetrel **132–133** E/P. Roy **134–135** E/Ken Straiton **136** E/H. Veiller **137** Zefa/H. Wiesner **138–139** E/Louis-Yves Loirat **138** M/Ferdinando Scianna **139** Zefa/Konrad Helbig **140–141** Art Directors/David Barnes **142–143** E/P. Weisbecker **144t** Zefa/Konrad Helbig **144b** Zefa/Idem **145** E/P. Roy **146–147** Zefa/Luetticke **147** E/P. Delarbre **148** PEP/G. Lythgoe **148–149** NHPA/N.A. Callow **150** Zefa/UWS **152** Zefa/Strachil **153** M/Bruno Barbey **154** Zefa/Starfoto **155** E/E. de Malglaive **156t, 157** FSP/Gamma/Laurent Vanderstock **156b** E/Claude Nardin **158–159** M/Abbas **160–161** Jurgensfot **161** M/Peter Marlow **162** E/P. Roy **162–163** M/Abbas **164** E/P. Roy **165** Zefa/Dr. Hans Kramarz **166** Zefa/Maroon **168t** E/Nicholas Thibaut **168b** Richard and Sally Greenhill **169, 170** Christine Osborne **171** PEP/Hans Christian Heap **172** E/Jacques Perno **172–173** E/Daniel Riffet **174–175** Zefa/Havlicek **176–177** E/P. Roy **177** Victor Englebert **178** Christine Osborne **179** E/Fiore **180** HSL/Richard Anthony **180–181** M/Bruno Barbey **182** HL **182–183** E/Guy Philippait de Foy **184–185** Zefa/Victor Englebert **186** HSL/Richard Anthony **186–187** PEP/Jonathon Scott **188t** PEP/Jeannie Mackinnon **188b** HSL/Nigel Catlin **189** Zefa/Hans Schmied **190, 191** BCL/John Anthony **192–193** M/Bruno Barbey **194** NHPA/A.Bannister **196** Tropix Picture Library/John Schmid **196–197** E/Hervy **198** NHPA/Anthony Bannister **198–199** PEP/Sean Avery **200** Anthony Bannister **201** Nigel Dennis/Anthony Bannister Library **202** E/Luc Girard **204** Sally and Richard Greenhill **205, 206** E/Berthoule **207** Christine Osborne **208** Panos Pictures/Trygve Bolstad **208–209** Panos Pictures/Tom Learmonth **209, 210–211** Christine Osborne **212** E/J.L. Gobert **214–215** Sally and Richard Greenhill **215** ANT/B. Hodgkiss **216** HSL/Mary Cherry **216–217** E/Sophie Bacheuer **218** Sally and Richard Greenhill **220–221** Zefa/Damm **222–223** Tropix Picture Library/R. Kendal **224–225** ANT/Jürg Bär **225** FSP/Gamma/J.C. Labbe **226** PEP/Richard Matthews **227** E/Daniel Prest **228, 228–229** E/P. Roy **229** Zefa/Starfoto **230–231** E/Jacques Perno **232** Zefa/Orion Press **232–233** Zefa/W. Eastep **234** E/Kraft **235** RHPL **236** PEP/Robert Jureit **236–237** Zefa/B.Crader **238–239** ANT/Natfoto **241** ANT/Cyril Webster **242, 244** Christine Osborne **242–243** ANT/Otto Rogge **244–245** ANT/Bill Bachman **245** OSF/Waina Cheng **246–247** E/Pierre Gleizes

Editorial, research and administrative assistance
Nick Allen, Helen Burridge, Joanna Chisholm, Richard Cobb, Hilary McGlynn, Pamela Mayo

Artists
The Maltings Partnership, Nick Salmon

Cartography
Maps drafted by Euromap, Pangbourne

Index
Susan Kennedy

Production
Clive Sparling

Typesetting
Brian Blackmore, Catherine Boyd, Peter MacDonald Associates

Color origination
Scantrans pte Limited, Singapore

INDEX

Page numbers in **bold** refer to extended treatment of topic; in *italics* to captions, maps or tables

A

B

C

D

E

S

T

U

V

W

Y

Z